Quantitative Remote Sensing in Thermal Infrared

Springer Remote Sensing/Photogrammetry

For further volumes:
http://www.springer.com/series/10182

Huajun Tang • Zhao-Liang Li

Quantitative Remote Sensing in Thermal Infrared

Theory and Applications

Springer

Huajun Tang
Zhao-Liang Li
Institute of Agricultural Resources and Regional Planning
Chinese Academy of Agricultural Sciences
Beijing, People's Republic of China

ISSN 2198-0721 ISSN 2198-073X (electronic)
ISBN 978-3-642-42026-9 ISBN 978-3-642-42027-6 (eBook)
DOI 10.1007/978-3-642-42027-6
Springer Heidelberg New York Dordrecht London

Library of Congress Control Number: 2013957572

Printed on acid-free paper

Springer is part of Springer Science+Business Media (www.springer.com)

Contents

Symbols

Variables	Meanings	Units
a	Intercalibration coefficient or regression coefficient	–
a_i	The channel-specific constant for the power-law approximation of Planck's law	–
A_f	Area of the sub-pixel fire	km^2
A_{pixel}	Pixel area	km^2
b	Regression coefficient or intercalibration coefficient	–
B	Planck function	$W\ m^{-2}\ sr^{-1}\ \mu m^{-1}$
B^{-1}	Inverse of the Planck function	K
c	Specific heat capacity of the material	$J\ kg^{-1} \cdot K^{-1}$
c_1, c_2	First and second Planck constant, respectively	$W\ \mu m^4\ sr^{-1}\ m^{-2}$, $\mu m\ K$
c_p or C_p	Specific heat at constant pressure	$J\ g^{-1} \cdot K^{-1}$
d	Zero plane displacement height	m
dT	Surface-air temperature difference	K
dT_{dry}, dT_{wet}	Surface-air temperature difference at the dry and wet limits, respectively	K
D_1	The maximum canopy and air temperature difference for a stressed crop	K
D_2	The lower limit of the canopy and air temperature difference for a well-watered crop	K
e_a	Actual vapor pressure	kPa
e_h	Actual vapor pressure at water source height	kPa
e_s	Saturation vapor pressure	kPa
$e^*(T_c)$	Saturation vapor pressure at the leaf surface	kPa
E	Irradiance	$W\ m^{-2}$
E_s	Extraterrestrial solar radiation	$W\ m^{-2}$
ET	Evapotranspiration	$mm\ day^{-1}$ or $W\ m^{-2}$
ET_c	Canopy transpiration	$mm\ day^{-1}$
ET_d	Daily ET	$mm\ day^{-1}$
ET_i	Instantaneous ET	$mm\ h^{-1}$

(continued)

(continued)

Variables	Meanings	Units
ET_p or LE_p	Potential evapotranspiration	mm day^{-1} or W m^{-2}
ET_r	Reference evaporation	mm day^{-1}
ET_s	Actual soil evaporation	mm day^{-1}
ET_{pc}	Potential canopy transpiration	mm day^{-1}
ET_{ps}	Potential soil evaporation	mm day^{-1}
ET_rF	Reference EF	–
f_g	Fraction of the total LAI that is green	–
f_k	Fraction area of end-member *k*	–
f_M, f_T	Plant moisture and temperature constraint factors	–
f_{geo}, f_{vol}	LiSparse-R geometric and Roujean's volumetric kernels, respectively	–
f_{SM}	Soil moisture constraint factor	–
f_{wet}	Relative surface wetness	–
f_{APAR}	Fraction of PAR absorbed by green vegetation cover	–
f_{IPAR}	Fraction of PAR intercepted by total vegetation cover	–
$f_i(\lambda)$	Spectral response function of channel *i*	–
$f_i(\theta, \varphi, \theta_s, \varphi_s)$	Angular form factor	–
Fr or Fv	Fractional vegetation coverage	–
g	Acceleration of gravity	m s^{-2}
$g_i(\lambda)$	Spectral response function of channel *i*	–
G	Soil heat flux	W m^{-2}
h	Vegetation height	m
h_{pbl}	Height of the planetary boundary layer	m
H	Sensible heat flux	W m^{-2}
H_c, H_s	Sensible heat flux for canopy and soil, respectively	W m^{-2}
H_{dry}, H_{wet}	Sensible heat flux at the dry and wet limits, respectively	W m^{-2}
i	Channel *i*	–
I	Radiant intensity	W sr^{-1}
J_λ	Source function	W cm^{-2} sr^{-1} μm^{-1} or W cm^{-2} sr^{-1} cm
k	Von Karman's constant or band number or sub-element	–
kB^{-1}	$\ln(z_{om}/z_{oh})$	–
k_λ	Mass extinction cross section	m^2 kg^{-1}
k_{geo}	Coefficient of the LiSparse-R geometric kernel	–
k_{iso}	Isotropic scattering term	–
k_{vol}	Coefficient of the Roujean's volumetric kernel	–
K	Thermal conductivity	J $m^{-1}\cdot s^{-1}\cdot K^{-1}$
L	Latent heat of vaporization	MJ kg^{-1}
LE	Latent heat flux	W m^{-2}
LE_c	Latent heat flux for canopy	W m^{-2}
LE_e	Equilibrium evaporation	mm day^{-1} or W m^{-2}

(continued)

(continued)

Variables	Meanings	Units
LE_i	Interception evaporation	mm day^{-1}
LE_s	Latent heat flux for soil	W m^{-2}
LE_{RE}	LE corrected by the residual energy correction method	W m^{-2}
LE_{dry}, LE_{wet}	Latent heat flux at the dry and wet limits, respectively	W m^{-2}
L_{MO}	Monin–Obukhov length	m
M	Radiant exitance or number of channels	W m^{-2} or –
M_{BB}, M_{BCS}, and M_{SV}	Number of the onboard BB, BCS, and SV frame	–
$M_{\Delta Tb}$	Median of ΔT for the background pixels	K
n_i	The channel-specific constant for the power-law approximation of Planck's law	–
N	Number of end-members or Number of elements	–
N_f	Number of background fire pixels surrounding a potential fire pixel	–
N_s	Shear stress	kg m^{-1}·s^{-2}
N_v	Number of valid background pixels surrounding a potential fire pixel	–
N_w	Number of water pixels surrounding a potential fire pixel	–
N_{aw}	Number of water pixels within eight pixels surrounding a tentative fire pixel	–
N_{uw}	Number of unmasked water pixels surrounding a potential fire pixel	–
N_{SVS}	Number of the SVS frame	–
P	Current pressure of the air parcel	Pa
P_0	Standard reference pressure	Pa
p_i	Fractional area of the ith fire component	–
P_v	Fraction of vegetation	–
Q	Total radiant energy	J
r_a	Aerodynamic resistance	s m^{-1}
r_c	Canopy resistance	s m^{-1}
r_{cp}	Canopy resistance at potential evapotranspiration	s m^{-1}
r_{cx}	Canopy resistance associated with nearly complete stomatal closure	s m^{-1}
r_s	Bulk surface resistance	s m^{-1}
r_w	Resistance to water transfer	s m^{-1}
r_{xc}	Stomatal resistance	s m^{-1}
$r_{a,dry}$, $r_{a,wet}$	Aerodynamic resistance to sensible heat transfer at the dry and wet limits, respectively	s m^{-1}
r_h	Hemispherical–directional reflectivity	–
R	Gas constant of air or radiance	J mol^{-1}·K^{-1} or W m^{-2} sr^{-1} μm^{-1}
R^2	Coefficient of determination	–
R_{EV} or R_{EV_PATH}	Radiance emitted by the earth in the EV direction or radiance received by the sensor at the EV	W m^{-2} sr^{-1} μm^{-1}

(continued)

(continued)

Variables	Meanings	Units
R_{BKG}	Radiance of the instrument background, excluding the scan mirror's emission	W m^{-2} sr^{-1} μm^{-1}
R_{MIR}, $R_{\mathrm{MIR,b}}$	Radiance of the fire and background pixel in the MIR channels, respectively	W m^{-2} sr^{-1} μm^{-1}
$R_{\mathrm{BB\_PATH}}$, $R_{\mathrm{BCS\_PATH}}$	Radiance of the onboard BB and BCS paths, respectively	W m^{-2} sr^{-1} μm^{-1}
$R_{\mathrm{SV\_PATH}}$, $R_{\mathrm{SVS\_PATH}}$	Radiance of the SV and SVS paths, respectively	W m^{-2} sr^{-1} μm^{-1}
R_i	TOA radiance or reflectance in channel i	W cm^{-2} sr^{-1} μm^{-1} or –
R_{ik}	Radiance of end-member k in channel i	W m^{-2} sr^{-1} μm^{-1}
R_i, $R_{b,i}$	Radiance and background radiance in channel i, respectively	W m^{-2} sr^{-1} μm^{-1}
R_{X}	Total boundary layer resistance of the complete canopy of leaves	s m^{-1}
R_{gi}	Channel radiance observed at ground level	W cm^{-2} sr^{-1} μm^{-1} or W cm^{-2} sr^{-1} cm
R_{gi}^s	Solar irradiance at the ground level in channel i	W m^{-2}
R_{g}	Global solar radiation	W m^{-2}
R_{n}	Surface net radiation	W m^{-2}
R_{nc}	Canopy net radiation	W m^{-2}
R_{ns}	Soil net radiation	W m^{-2}
R_{s}	Soil boundary resistance or at-surface radiance	s m^{-1} or W m^{-2} sr^{-1} μm^{-1}
R_{u}	Upward radiance	W cm^{-2} sr^{-1}
R_{d}	Downward radiance	W cm^{-2} sr^{-1}
$R_{ati}\uparrow$, $R_{ati}\downarrow$	The upward and downward atmospheric thermal radiance in channel i, respectively	W cm^{-2} sr^{-1} μm^{-1} or W cm^{-2} sr^{-1} cm
$R_{sli}\uparrow$, $R_{sli}\downarrow$	The upward and downward solar diffusion radiance resulting from atmospheric scattering of the solar radiance in channel i, respectively	W cm^{-2} sr^{-1} μm^{-1} or W cm^{-2} sr^{-1} cm
R_{HSR}	Measured radiance of the reference HSR	W m^{-2} sr^{-1} μm^{-1}
R_{aic}, R_{bic}	Radiance after and before intercalibration	W m^{-2} sr^{-1} μm^{-1}
S_k	Fractional area of the homogeneous sub-element k in the pixel	–
T_{a}	Air temperature	K
$T_{\mathrm{a}\uparrow}$	Atmospheric equivalent temperature	K
T_{b}	Brightness temperature	K
T_{c}	Canopy temperature	K
T_{f}	Target temperature or temperature of the sub-pixel fire	K

(continued)

(continued)

Variables	Meanings	Units
T_i	Temperature of the *i*th fire component or TOA brightness temperature in channel *i*	K
T_o	Soil temperature or modeled surface temperature when actual ET equals potential ET	K
T_p	Potential temperature	K
T_s	Surface temperature	K
T_v	Mean virtual potential temperature	K
TI	Thermal inertia	$W \cdot s^{1/2}\ m^{-2} \cdot K^{-1}$
T_{gi}	Brightness temperature at ground level in channel *i*	K
T_{sm}	Temperature of the scan mirror	K
T_{sr}	Radiometric temperature	K
T_{st}	Thermodynamic temperature	K
T_{cav}	Equivalent temperature of the instrument scan cavity	K
T_{dew}	Dew point temperature	K
T_{max}	Maximum air temperature or maximum surface temperature at the dry edge	K
T_{min}	Minimum surface temperature at the wet edge	K
T_{pbl}	Average planetary boundary layer temperature	K
T_{ref}	Reference land surface temperature	K
T_{aero}	Aerodynamic temperature	K
$\overline{T}_b$, $\overline{T}_{bf}$	Mean of T for the background and the background fire pixels, respectively	K
T_n, T_f	Brightness temperatures in nadir and forward views, respectively	K
$T_{s,max}$ or $T_{s,dry}$	Surface temperature at the dry limit/pixel	K
$T_{s,min}$ or $T_{s,wet}$	Surface temperature at the wet limit/pixel	K
T_{smax}, T_{smin}	Maximum and minimum land surface temperature of all images, respectively	K
T_{BB}, T_{BCS}	Temperature of the onboard BB and the BCS, respectively	K
T_{MIR}, $T_{MIR,b}$	Brightness temperature of the fire pixel and the background in the MIR channel, respectively	K
T_∞	Modeled surface temperature when evapotranspiration is zero	K
u	Wind speed	$m\ s^{-1}$
u_2	Mean daily or hourly wind speed at 2-m height	$m\ s^{-1}$
u^*	Friction velocity	$m\ s^{-1}$
v	Wavenumber	cm^{-1}
W	Atmospheric total water vapor content	$g\ cm^{-2}$
z_a	Height for the wind speed and humidity measurements	m
z_h	Height of humidity measurement	m
z_m	Height of wind measurement	m
z_{oh}	Roughness length governing transfer of heat and vapor	m
z_{om}	Momentum roughness length	m

(continued)

(continued)

Variables	Meanings	Units
α	Surface albedo or Priestley–Taylor coefficient	–
α_s	Surface shortwave albedo	–
α_k	Relative area of a sub-element	–
β	Bowen ratio or spectral ratio	–
γ	Psychometric constant	kPa °C^{-1}
θ	View zenith angle	Degree
θ_g	Angle between vectors pointing in the surface-to-satellite and specular reflection directions	Degree
θ_i	Incoming zenith angle	Degree
θ_s	Solar zenith angle	Degree
θ_v	Outgoing zenith angle or view zenith angle	Degree
φ	(Relative) azimuth angle	Degree
φ_i	Incoming azimuth angle	Degree
φ_v	Outgoing azimuth angle	Degree
ϕ	A modified Priestley–Taylor coefficient	–
ϕ_{min}, ϕ_{max}	Global minimum and maximum ϕ, respectively	–
$\phi_{min,i}$, $\phi_{max,i}$	Local minimum and maximum ϕ, respectively	–
Φ	Radiant flux	J s^{-1}
ε	Surface emissivity or effective mean emissivity of all fire components over all emitting wavelengths	–
ε_a	Atmospheric emissivity	–
ε_i	Surface emissivity in channel *i* or residual radiance error in channel *i*	– or W m^{-2} sr^{-1} μm^{-1}
ε_n, ε_f	Emissivities in nadir and forward views, respectively	–
ε_s	Surface total emissivity	–
ε_s, ε_v	Soil and vegetation emissivities, respectively	–
$\overline{\varepsilon}$, $\Delta\varepsilon$	Averaged emissivity and difference of emissivity in two channels, respectively	–
ε_{MIR}	Emissivity in the MIR channel	–
δ	Normal optical thickness	–
δ_{ib}, δ_{ibf}	Mean absolute deviation of T_i for the background and background fire pixels, respectively	K
$\delta_{\Delta Tb}$	Mean absolute deviation of ΔT for the background pixels	K
ω	Surface energy flux	W m^{-2}
$d\omega_k$	A small solid angle corresponding to a homogeneous area *k*	sr
Ω	Solid angle	sr
λ	Wavelength	μm
λ_1, λ_2	Lower and upper wavelength limits of the spectral response function, respectively	μm
ρ	Density of the material or air	kg m^{-3}
ρ_0	Bidirectional reflectivity for overhead sun illumination and nadir observation	sr^{-1}

(continued)

(continued)

Variables	Meanings	Units
ρ_{bi}	Bidirectional reflectivity	sr^{-1}
ρ^{sm}_{BCS}, ρ^{sm}_{SVS}, ρ^{sm}_{BB}, ρ^{sm}_{SV}, and ρ^{sm}_{EV}	Reflectances of the scan mirror when sensor views the BCS, SVS, BB, SV, and EV, respectively	–
σ	Stefan–Boltzmann constant	$\mathrm{W\ m^{-2}\ K^{-4}}$
σ_{ib}	Standard deviation of T_i for the background pixels	K
σ_{34b}	Standard deviation of ΔT_{34} for the background pixels	K
$\sigma_{\Delta\mathrm{Tb}}$	Standard deviation of ΔT for the background pixels	K
$\sigma_{\mathrm{T}\varepsilon}$	Covariance of emissivity and temperature within a pixel	K
$\sigma^2{}_{\mathrm{T}}$	Variance of temperature within a pixel	K^2
τ	Atmospheric transmittance	–
ψ_1	Stability correction function for momentum transfer	–
ψ_2	Stability correction functions for heat transfer	–
Δ	Slope of the saturation vapor pressure versus air temperature curve	$\mathrm{kPa\ ^\circ C^{-1}}$
ΔT	Difference of TOA brightness temperature in two channels or the LST error	K
ΔT_{s}	Difference between the afternoon and midnight surface temperatures	K
ΔR_{BB}	Radiance difference between the onboard BB and SV paths	$\mathrm{W\ m^{-2}\ sr^{-1}\ \mu m^{-1}}$
ΔR_{EV}	Radiance difference between the EV and SV paths	$\mathrm{W\ m^{-2}\ sr^{-1}\ \mu m^{-1}}$
ΔR_{BCS}	Radiance difference between the BCS and SVS paths	$\mathrm{W\ m^{-2}\ sr^{-1}\ \mu m^{-1}}$
$\Delta\mathrm{DN}_{\mathrm{EV}}$	Digital number difference between the EV and SV paths	–
$\Delta\mathrm{DN}_{\mathrm{BCS}}$	Digital number difference between the BCS and SVS paths	–
$\overline{\Delta T}_{ij\mathrm{b}}$	Mean of the difference of T between channels i and j for the background pixels	K
Γ_{s}	Fraction of the ratio of soil heat flux to surface net radiation at bare soil surface	–

Acronyms

Variables	Full names
ABL	Atmospheric boundary layer
ALEXI	Atmosphere-Land Exchange Inverse
ANEM	Adjusted normalized-emissivity method
ANN	Artificial neural network
AOI	Area of interest or angle of incidence
AR	Audubon Research Ranch
ASL	Atmospheric surface layer
ASTER	Advanced Spaceborne Thermal Emission and Reflection Radiometer
ATI	Apparent thermal inertia
ATVDI	Advanced temperature-vegetation dryness index
AVHRR	Advanced Very High Resolution Radiometer
BB	Blackbody
BCS	Blackbody calibration source
BR	Bowen ratio
BRDF	Bidirectional reflectance distribution function
CBEM	Classification-based emissivity method
CDIAC	Carbon Dioxide Information Analysis Center
CWSI	Crop water stress index
D/N	Physics-based day/night operational method
DISALEXI	Disaggregated ALEXI
DN	Digital number
DOY	Day of year
DTD	Dual-temperature-difference
EC	Eddy covariance
EF	Evaporative fraction
EF_r	Relative evaporation fraction
ESI	Evaporative stress index
ETM+	Enhanced Thematic Mapper Plus
EV	Earth view

(continued)

(continued)

Variables	Full names
FIFE	First International Satellite Land Surface Climatology Project Field Experiment
FPAR	Fraction of photosynthetically active radiation
FRP	Fire radiative power [W]
GBE	Gray body emissivity
GEWEX	Global energy and water cycle experiment
GMAO	Global Modeling and Assimilation Office
GOES	Geostationary Operational Environmental Satellite
HCMM	Heat capacity mapping mission
IGBP	International Geosphere-Biosphere Programme
ISLSCP	International Satellite Land Surface Climatology Project
ISSTES	Iterative spectrally smooth temperature emissivity separation method
KG	Kendall Grassland
LAI	Leaf Area Index
LAS	Large aperture scintillometer
LECTES	Linear emissivity constraint temperature emissivity separation method
LP DAAC	Land Processes Distributed Active Archive Center
LSE	Land surface emissivity
LST	Land surface temperature
METRIC	Mapping EvapoTranspiration at high Resolution with Internalized Calibration
MIR	Mid-infrared
MMD	Maximum–minimum apparent emissivity difference method
MODIS	Moderate Resolution Imaging Spectroradiometer
MSG	Meteosat Second Generation
MTI	Multispectral Thermal Imager
MW	Microwave
NAALSED	North American Aster Land Surface Emissivity Database
NASA	National Aeronautics and Space Administration
NBEM	NDVI-based emissivity method
NDTI	Normalized difference temperature index
NDVI	Normalized Difference Vegetation Index
$NDVI_{max}$	Maximum NDVI
$NDVI_{min}$	Minimum NDVI
$NDVI_s$	NDVI corresponding to the bare soil
$NDVI_v$	NDVI corresponding to the full vegetation
NEM	Normalization emissivity method
PBL	Planetary boundary layer
PCA	Principle component analysis
RCM	Reference channel method
RE	Residual energy
RH	Relative humidity
RMSD	Root-mean-square difference
RMSE	Root-mean-square error
RTE	Radiative transfer equation
SAVI	Soil-Adjusted Vegetation Index

(continued)

(continued)

Variables	Full names
SEBAL	Surface Energy Balance Algorithm for Land
SEBI	Surface Energy Balance Index
SEBS	Surface Energy Balance System
SI	International System of Units
SM	Spectral smoothness
SMA	Spectral Mixture Analysis
SNR	Signal-to-noise ratio
SR	Spectral ratio
SRB	Surface radiation budget
S-SEBI	Simplified Surface Energy Balance Index
STVDI	Simple temperature-vegetation dryness index
SURFRAD	Surface Radiation Budget Network
SV	Space view
SVAT	Soil-vegetation-atmosphere transfer
SVS	Space view source
TCI	Temperature condition index
TEB	Thermal emissive band
TES	Temperature emissivity separation method
TI	Thermal inertia
TIR	Thermal infrared
TIROS	Television Infrared Observation Satellite
TISI	Temperature-independent spectral index
TM	Thematic Mapper
TOA	Top of the atmosphere
TSRM	Two-step physical retrieval method
TSTIM	Dual-source time integrated model
TTM	Two-temperature method
TVDI	Temperature-vegetation dryness index
TVT	T_s-VI triangle
VCI	Vegetation condition index
VHI	Vegetation Health Index
VI	Vegetation index
VPD	Vapor pressure deficit [kPa or Pa]
VSWI	Vegetation supply water index
VZA	View zenith angle [degree or radian]
VNIR	Visible/near-infrared
WDI	Water deficit index
WS	Wind speed [m/s]
WVS	Water vapor scaling

Chapter 1
Introduction

Remote sensing, by definition, is the acquisition of information about objects or events without making physical contact. In the broad usage, remote sensing refers to the use of both the force fields (such as gravity and magnetism) and electromagnetic spectrum to detect, classify, and identify the objects or events. In the restricted usage, the term remote sensing generally refers to the use of electromagnetic spectrum to acquire information about the earth's surface (land and ocean) and atmosphere using sensors onboard airborne (aircraft, balloons) or spaceborne (satellites, space shuttles) platforms. Remote sensing in its restricted usage is recognized as the only viable means to acquire regional- and mesoscale information of the earth surface and atmosphere in a globally consistent and economically feasible manner. According to the spectral response range of sensors, earth observation by remote sensing can be roughly classified to visible and near-infrared remote sensing (VNIR), thermal infrared (TIR) remote sensing, and microwave remote sensing. Given the broad scope of remote sensing, this book is limited to deal with the TIR remote sensing.

As TIR sensors onboard the spaceborne platforms only measure spectral radiances at the top of the atmosphere, these measured radiances are influenced not only by the surface parameters (emissivity and temperature) but also by the composition and thermal structure of atmosphere along the path between the surface and the sensors. Theoretically, only two surface parameters (land surface emissivity and land surface temperature) can be directly retrieved from TIR data; other surface parameters or variables such as soil moisture and land surface evapotranspiration may be indirectly deduced.

Since the launch of the first infrared observation satellite—the US Television Infrared Operational Satellite (TIROS)—in 1960, great efforts have been made to interpret what the measured quantities from the infrared sensors are and how land/sea surface and atmospheric variables can be retrieved from these measured quantities. As early as in the mid-1980s, we committed ourselves to developing methodologies for retrieving land surface temperature and emissivity from TIR data. After our nearly 30 years of research in TIR remote sensing, it is the right

H. Tang and Z.-L. Li, *Quantitative Remote Sensing in Thermal Infrared: Theory and Applications*, Springer Remote Sensing/Photogrammetry,
DOI 10.1007/978-3-642-42027-6_1,

time to give an in-depth overview of the state of the art in TIR remote sensing. This book therefore provides a comprehensive and in-depth overview of the basic theory of TIR remote sensing and its applications in agriculture, hydrology, and forestry. Specifically, this book highlights the main theory, assumptions, advantages, drawbacks, and perspectives of the current methodologies for retrieving and validating land surface temperature, land surface emissivity, and land surface evapotranspiration from TIR data in the past several decades.

This book is intended to answer the questions what the measured quantities are in TIR remote sensing and what the interesting information the TIR remote sensing can provide is and how this information is produced. It is based in part on three review papers we have published in *Sensors*, *International Journal of Remote Sensing*, and *Remote Sensing of Environment*, respectively. We have also used materials from our published papers in the literature. This book is the point of view of the physicists and contains eight chapters.

The opening chapter (Chap. 2) treats concisely the basic theory of quantitative TIR remote sensing. The first part of this chapter gives various definitions relevant to the radiometry and the land surface evapotranspiration process. The second part of this chapter briefly describes the radiative transfer in the atmosphere in TIR, which reveals physical phenomenon of energy transfer in the form of electromagnetic radiation from land surface to the top of atmosphere and links remotely sensed TIR data with land surface and atmospheric variables. All these definitions and radiative transfer theories are extensively used in the book and serve as the fundamental theories of the subsequent chapters.

Radiometric calibration in TIR is the basis for reliable remote sensing and proper quality of the retrieved parameters and variables. Without radiometric calibration, there is no means to know whether the observed changes are real or are due to sensor error. Quantitative retrieval of land surface parameters and variables from TIR radiometry requires well-calibrated TIR data. Chapter 3 describes in detail four types of radiometric calibration methods (i.e., preflight, in-flight, vicarious, and intercalibration), which are used in different stages and under different conditions for calibrating TIR data, to ensure TIR data to be well calibrated.

As an intrinsic property of natural materials, land surface emissivity is an important surface parameter and can be derived from the emitted radiance measured from TIR data. Besides radiometric calibration, two main problems need to be resolved to obtain land surface emissivity values from space measurements. These problems are often referred to as land surface temperature and emissivity separation from radiance at ground level and atmospheric corrections in the literature. To date, many methods have been proposed with the same goal but different application conditions, advantages, and limitations. Chapter 4 is therefore devoted to reviewing these land surface emissivity retrieval methods and to providing technical assistance for estimating land surface emissivity from space. The first part of this chapter gives the definition of land surface emissivity for heterogeneous and nonisothermal surfaces and illustrates the angular and spectral variation of emissivity for different types of land use and land cover. The second part of this chapter provides a comprehensive and in-depth overview of the current methodologies for land surface emissivity retrieval from space measurements. Validation is an important task because product accuracy is crucial to the scientific communities and because feedback from the validation activity is invaluable to the improvement of quality of the generated products. Without validation, no methods, algorithms, or parameters derived from remotely sensed data can be used with confidence. However, validation of satellite-derived products is often a challenge in itself. The last part of this chapter and the subsequent chapters present the validation methods for satellite-derived land surface emissivity, land surface temperature, and evapotranspiration, which are of great importance in verifying the uncertainty and accuracy of satellite-retrieved parameters and variables.

Land surface temperature is one of the key parameters in the physics of land surface processes. The importance of land surface temperature is being increasingly recognized, and there is a strong interest in developing methodologies to measure land surface temperature from remotely sensed TIR data. However, retrieving land surface temperature is still challenging since land surface temperature retrieval problem is ill-posed. Chapter 5 reviews the current status of the retrieval algorithms for estimating land surface temperature from remotely sensed TIR data. A brief theoretical background of the subject is presented along with a survey of the algorithms employed for obtaining land surface temperature from remotely sensed TIR data. The discussion focuses on TIR data acquired from polar-orbiting satellites because of their widespread use, global applicability, and higher spatial resolution compared to geostationary satellites. The theoretical framework and methodologies used to derive the land surface temperature from TIR data are reviewed, followed by the methodologies for validating satellite-derived land surface temperature.

Chapter 6 provides an overview of a variety of remote sensing evapotranspiration (ET) estimation methods and models that have been developed at field, regional, and continental scales. In the first part of this chapter, the empirical regression method, the single- and dual-source energy balance method, the Penman–Monteith equation, the modified Priestley–Taylor equation, and the data assimilation method are reviewed. In the single-source energy balance method, several representative end-member-based models are highlighted. Representative methods of upscaling instantaneous ET to daily values are also presented. For each specific ET retrieval and upscaling method or model, the main theory and assumption are given, and the advantages, limitations, and potentialities are highlighted. The extent of application of each of the upscaling method is defined. A comparison of the representative remote sensing ET estimation models and ET upscaling methods is performed. In the second part of this chapter, several exercises are outlined in validating remote sensing estimates of ET. In this part, surface energy components measured by eddy covariance system and large aperture scintillometer in a number of field experiments are used as ground-truth to validate the instantaneous and daily surface energy components and ET estimated by the various models and upscaling methods at the satellite pixel scale. At the end, problems that remain unresolved at present and need further investigation are addressed for the remote sensing ET retrieval and validation.

Chapter 7 is devoted to the applications of TIR remote sensing in agricultural drought monitoring and thermal anomaly detection. For the agricultural drought monitoring, several commonly applied drought indices that have been developed based on the TIR remote sensing are introduced, and applications of these indices are given. For the thermal anomaly detection, the background and theoretical basis for the fire detection are presented. Two types of algorithms for active fire detection, namely, the fixed-threshold algorithm and the contextual algorithm, and methods for fire characterization are described. Finally, the use of TIR remote sensing to study the urban heat island effect is briefly discussed.

Chapter 8 provides an in-depth perspective to the future trend of the TIR remote sensing, especially giving a perspective of what should be focused on in the future in the remote sensing retrieval and validation of land surface emissivity, land surface temperature, and evapotranspiration.

This book is in fact the fruits of research that we have carried out for decades. It is also the result of reflections and many discussions we have had with the colleagues with whom we had the privilege of working, both in China, France, and abroad.

This book is of particular interest to students, researchers, experts, and decision-makers whose interest focuses on the retrieval and validation of land surface temperature, land surface emissivity, the estimation and validation of evapotranspiration at satellite pixel scale, and the application of TIR remote sensing in agriculture, hydrology, and forestry.

Like any human work, this book certainly contains errors that have escaped to us despite the care we have taken in its drafting. We are grateful to all those who have helped us in writing this book, particularly our colleagues Dr. B-H. Tang for his contribution to Chap. 5, Dr. H. Wu for Chaps. 2 and 4, Dr. R-L. Tang for Chaps. 2 and 6, Mr. S-B. Duan for Chaps. 3 and 7, and Dr. L. Sun for Chap. 7.

We especially thank Dr. R-L. Tang for editing and preparing the manuscript for transmittal to the publisher.

We are grateful to the various people who have supplied us with materials for illustrations and to the holders of the copyrights of material that we have used.

Chapter 2
Basic Theory of Quantitative Remote Sensing

2.1 Basic Concepts

2.1.1 Definitions Relevant to Radiometry

2.1.1.1 Digital Number (DN)

Digital numbers (DNs) are what we get after purchasing data from the data providers (Colwell et al. 1983; Liang 2004). DNs are the scaled integers from quantification that is not a physical quantity. Most quantization systems in remote sensing are linear having 6–12 bits. The DN can be any integer in this set:

$$\mathrm{DN} \in [1, 2^q] \tag{2.1}$$

where q is an integer representing the bits. For example, an 8-bit linear quantization system equally divides a given range of response of the sensor into 255 steps. The same response of a sensor will generate totally different DN value by different quantization system. The larger the q is, the more accurate the radiometric measurement is.

2.1.1.2 Solid Angle (Ω)

The solid angle (Ω) subtended by an area (A) on a spherical surface is equal to the area divided by the square of the radius of the sphere (Colwell et al. 1983; Liang 2004) (Fig. 2.1), namely,

$$\Omega = \frac{A}{r^2} \tag{2.2}$$

H. Tang and Z.-L. Li, *Quantitative Remote Sensing in Thermal Infrared: Theory and Applications*, Springer Remote Sensing/Photogrammetry,
DOI 10.1007/978-3-642-42027-6_2,

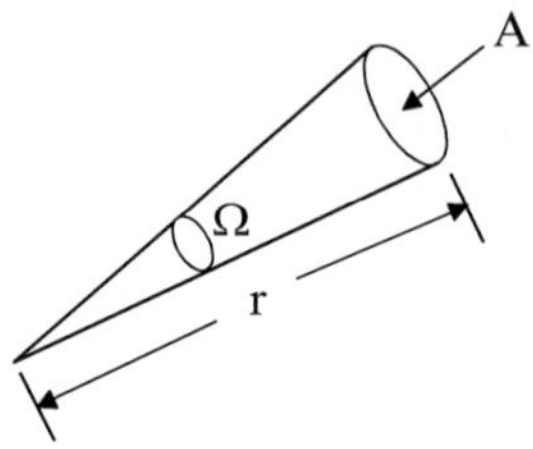

Fig. 2.1 Illustration of a solid angle (Ω). *A* is the spherical area of the sphere, and *r* is the radius of this sphere

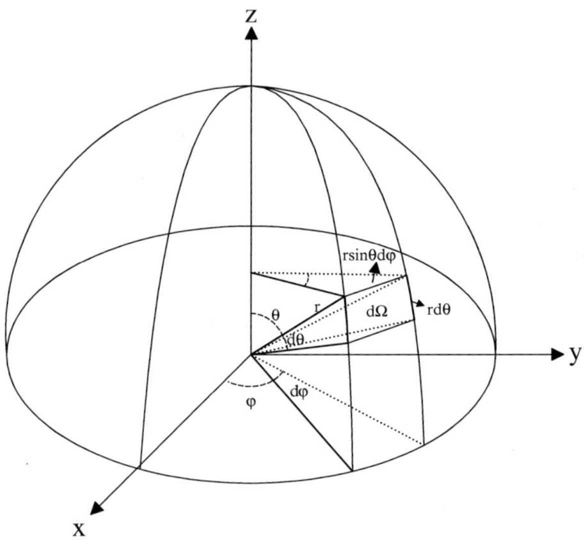

Fig. 2.2 Illustration of a differential solid angle and its representation. θ and φ are zenith angle and azimuth angle, respectively. *r* is the radius of the sphere. dθ, dφ, and dΩ are differential forms of zenith angle, azimuth angle, and solid angle, respectively

In the International System of Units (SI), a solid angle is a dimensionless unit of measurement called a steradian (symbol: sr). For a sphere whose surface area is $4\pi r^2$, its solid angle is 4π. Thus, the solid angle of the upper or lower hemisphere is 2π.

A solid angle is often represented by a zenith angle (θ) and an azimuth angle (φ) in polar coordinates (Fig. 2.2).

For a given zenith angle θ and azimuth angle φ, a differential element of solid angle is represented as

$$\mathrm{d}\Omega = \frac{\mathrm{d}A}{r^2} = \frac{(r\mathrm{d}\theta)\,(r\sin\theta\mathrm{d}\varphi)}{r^2} = \sin\theta\;\mathrm{d}\theta\mathrm{d}\varphi \tag{2.3}$$

Note that the zenith angle θ ranges from 0 to 180°. It measures from the normal direction. Usually the range of 0–90° represents the upwelling hemisphere and 90–180° the downward hemisphere. The azimuth angle φ ranges from 0 to 360° (clockwise positive from north).

It is important to mention that the viewing zenith angle is not equivalent to the sensor scan angle although they are closely related. The viewing zenith angle θ at a specific location of the earth surface is always larger than the sensor's scan angle because of the spherical earth surface. The larger the sensor scan angle is, the larger their difference is (Liang 2004).

2.1.1.3 Radiant Flux (Φ)

In radiometry, radiant flux, also called radiant power, is the measure of the total power of electromagnetic radiation, including infrared, ultraviolet, and visible light. Radiant flux (Φ) is defined as the total radiant energy Q that passes a spatial position during time period t (Slater 1980; Colwell et al. 1983), namely,

$$\Phi = \frac{\mathrm{d}Q}{\mathrm{d}t} \tag{2.4}$$

The unit of radiant flux is either joule/second [J/s] or, most commonly, the watt [W].

2.1.1.4 Radiant Intensity (I)

The radiant intensity (I) is a measure of the intensity of electromagnetic radiation. It is defined as radiant flux per unit solid angle with unit W sr^{-1} (Colwell et al. 1983):

$$I = \frac{\mathrm{d}\Phi}{\mathrm{d}\Omega}. \tag{2.5}$$

2.1.1.5 Radiance (R) and Spectral Radiance (R_λ)

Radiance (R) is a measure of the quantity of radiation that passes through or is emitted from a surface and falls within a given solid angle in a specified direction. It is defined as the radiant flux per unit solid angle per unit projected source area in a given direction (Colwell et al. 1983), namely,

$$R = \frac{\mathrm{d}^2\Phi}{\mathrm{d}A\cos\theta\,\mathrm{d}\Omega} \tag{2.6}$$

The unit of radiance is W cm^{-2} sr^{-1}. The projected area of an elementary plane segment of area A, in the direction at zenith angle θ, is $A\cos\theta$.

Spectral radiance (R_λ) is a physical quantity commonly used in remote sensing. It represents the radiant flux per unit solid angle per unit projected source area per unit wavelength (or wavenumber). A typical unit could be W cm^{-2} sr^{-1} μm^{-1} for wavelength λ (unit: μm) or W cm^{-2} sr^{-1} cm for wavenumber ν (unit: cm^{-1}). To convert radiance in W cm^{-2} sr^{-1} μm^{-1} to W cm^{-2} sr^{-1} cm, one must multiply by $\lambda^2/10^4$. Similarly, in converting radiance in W cm^{-2} sr^{-1} cm to W cm^{-2} sr^{-1} μm^{-1}, one must multiply by $\nu^2/10^4$.

2.1.1.6 Irradiance (*E*) and Radiant Exitance (*M*)

The irradiance (*E*) is defined as the total incident radiant power (flux) per unit area on the surface, and its unit is W m^{-2} (Colwell et al. 1983)

$$E = \frac{\mathrm{d}\Phi}{\mathrm{d}A} \tag{2.7}$$

Thus, the irradiance on the surface can be estimated by integrating incident radiance *R* over the whole 2π hemisphere, namely,

$$E = \int_0^{2\pi} \int_0^{\pi/2} R(\theta, \phi) \cos\theta \sin\theta \, \mathrm{d}\theta \mathrm{d}\phi \tag{2.8}$$

It is worth noting that the radiance is a function of direction; however, the irradiance clearly is not. For isotropic radiation where radiance is independent of the direction, the irradiance *E* becomes πR.

Compared with the irradiance that refers to the density of radiant flux incident on the surface, radiant exitance (*M*) is defined as the density of radiant flux leaving a surface in all directions in a hemisphere over the surface. The unit of radiant exitance is also W m^{-2}.

2.1.1.7 Bidirectional Reflectance Distribution Function (BRDF)

Bidirectional reflectance distribution function (BRDF) is a four-dimensional function that reflects the directional reflectance properties of a surface. BRDF is defined as the ratio of the reflected spectral radiance from the surface in the outgoing direction at zenith angle θ_v and azimuth angle φ_v to the directional spectral irradiance on the surface in the incoming direction at zenith angle θ_i and azimuth angle φ_i (Nicodemus 1965; Martonchik et al. 2000) (Fig. 2.3), namely,

$$\mathrm{BRDF}(\theta_i, \varphi_i, \theta_v, \varphi_v) = \frac{\mathrm{d}R(\theta_i, \phi_i, \theta_v, \varphi_v)}{\mathrm{d}E(\theta_i, \varphi_i)} \tag{2.9}$$

where *E* is the incoming irradiance. The unit of BRDF is sr^{-1}.

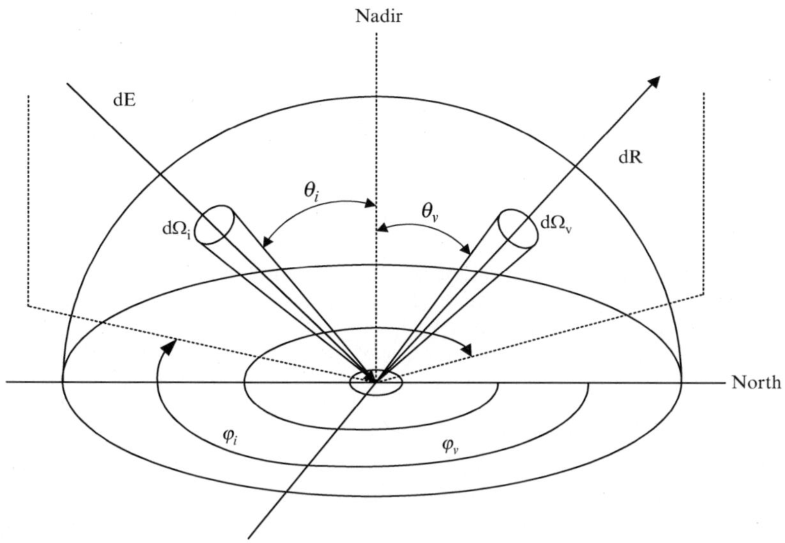

Fig. 2.3 Illustration of a BRDF. θ_v and φ_v are the zenith angle and azimuth angle, respectively, for the outgoing direction. θ_i and φ_i are the zenith angle and azimuth angle, respectively, for the incoming direction. $d\Omega_v$ and $d\Omega_i$ are the differential solid angles for the outgoing and incoming direction, respectively. dE is the differential incoming irradiance, and dR is the differential outgoing radiance

2.1.1.8 Albedo (α)

Albedo (α) is the diffuse reflectivity or the reflecting power of a surface. It is defined as the ratio of radiant exitance from the surface to incident irradiation upon it (Lambert 1760; Stephens 1994; Schaepman-Strub et al. 2006), namely,

$$\alpha = \frac{M}{E} \tag{2.10}$$

Being a dimensionless fraction, it is measured on a scale from 0, for no reflecting power of a perfectly black surface, to 1, for perfect reflection of a white surface.

2.1.1.9 Blackbody

A blackbody is an ideal thermal emitter that transforms heat energy into radiant energy with maximum rate permitted by thermodynamic laws. Any real material at the same temperature cannot emit thermally at a rate exceeding that of a blackbody. Likewise, a blackbody must absorb and convert all incident radiant energy into heat energy regardless of the spectral band of the radiant energy (Colwell et al. 1983).

2.1.1.10 Planck's Law

Planck's law describes the spectral distribution of the radiation from a blackbody as (Planck 1914)

$$B_\lambda(T) = \frac{C_1}{\lambda^5\left[\exp\left(\frac{C_2}{\lambda T}\right) - 1\right]} \tag{2.11}$$

where $B_\lambda(T)$ is the spectral radiance [W m^{-2} μm^{-1} sr^{-1}] of a blackbody at temperature T [K] and wavelength λ [μm]; C_1 and C_2 are the physical constants ($C_1 = 1.191 \times 10^8$ W μm^4 sr^{-1} m^{-2}, $C_2 = 1.439 \times 10^4$ μm · K).

2.1.1.11 Stefan–Boltzmann's Law

Integrating the emitted radiance over the whole spectrum gives an expression for the total flux emitted by an isotropic blackbody of unit area. This is known as the Stefan–Boltzmann's law (Stefan 1879; Boltzmann 1884):

$$M(T) = \pi \int_0^\infty B_\lambda(T)\mathrm{d}\lambda = \sigma T^4 \tag{2.12}$$

where M is the emitted radiant exitance, and σ is the Stefan–Boltzmann constant ($\sigma = 5.67 \times 10^{-8}$ W m^{-2} K^{-4}).

2.1.1.12 Wien's Law

Differentiating $B_\lambda(T)$ with respect to the wavelength and solving for the maximum gives Wien's law, which gives the wavelength of maximum emitted radiance at a given temperature (Wannier 1987):

$$\lambda_m = \frac{2{,}898\ [\mu\mathrm{m\ K}]}{T} \tag{2.13}$$

Thus, the sun, with a temperature of 6,000 K, will have a maximum emitted radiance at $\lambda_m = 0.48$ μm. A surface at temperature 300 K will have a maximum emitted radiance at $\lambda_m = 9.66$ μm, that is, in the infrared region.

2.1.1.13 Spectral Emissivity (ε)

The land surface emissivity (ε) is a measure of the ability of surface to emit thermal radiation. The spectral emissivity, ε_λ, is defined as the ratio of the spectral radiance, $R_\lambda(T_s)$, emitted by a body at the temperature, T_s, and the spectral radiance, $B_\lambda(T_s)$, emitted by a blackbody at the same temperature, T_s (Becker and Li 1995; Rees 2001):

$$\varepsilon_\lambda = \frac{R_\lambda(T_s)}{B_\lambda(T_s)} \tag{2.14}$$

where $B_\lambda(T_s)$ is the Planck's law.

2.1.1.14 Gray Body

A gray body is a hypothetical source that would radiate as a blackbody but with an emissivity lower than unity and constant with wavelength. The gray body assumption is more useful as a first approximation to the actual emission of a real object or medium than is the ideal blackbody radiation. It is used in radiative transfer problem, for instance, in planetary atmospheres (Colwell et al. 1983).

2.1.1.15 Thermodynamic Temperature (T_{st})

Thermodynamic temperature (T_{st}) is defined for a medium in thermal equilibrium from the second principle of the thermodynamics. It can be measured directly by a thermometer (Norman and Becker 1995).

2.1.1.16 Radiometric Temperature or Skin Temperature (T_{sr})

Radiometric temperature (T_{sr}) is defined from the radiance, R_λ, emitted by a surface (Norman and Becker 1995). If ε_λ is the emissivity of this surface, the radiometric temperature is the temperature of a blackbody with radiance, $R_\lambda/\varepsilon_\lambda$, namely,

$$T_{sr} = B^{-1}\left[\frac{R_\lambda}{\varepsilon_\lambda}\right] \tag{2.15}$$

where B^{-1} represents the inverse function of Planck's law. This radiometric temperature is also called skin temperature since it corresponds to the radiation emitted from a depth of the order of the penetration depth, namely, of the order of the wavelength. Radiometric and thermodynamic temperatures are equivalent to measure a surface temperature that can be used in the heat flux equation with appropriate resistance. However, for heterogeneous and nonisothermal surfaces, those temperatures are different, and the skin temperature measured by a radiometer is no longer equal to the bulk temperature measured by a thermometer. This leads to great difficulties in relating large-scale radiometric temperature to thermometric measurement and therefore to the heat fluxes.

2.1.1.17 Brightness Temperature (T_b)

Brightness temperature (T_b) is a directional temperature obtained by equating the measured radiance with the integral over wavelength of the Planck's law times the sensor response. It is the temperature of a blackbody that would have the same radiance as the radiance actually observed with the sensor (Norman and Becker 1995), namely,

$$T_b = B^{-1}\left[\frac{\int_{\lambda_1}^{\lambda_2} f_i(\lambda) R_\lambda \mathrm{d}\lambda}{\int_{\lambda_1}^{\lambda_2} f_i(\lambda)\mathrm{d}\lambda}\right] \tag{2.16}$$

where B^{-1} represents the inverse function of Planck's law, λ is the wavelength, λ_1 and λ_2 are the lower and upper wavelengths of sensitivity of the sensor in channel i, and R_λ is the measured radiance at wavelength λ, $f_i(\lambda)$ is the spectral response function of the sensor in channel i.

2.1.1.18 Kirchhoff's Law

In general, a medium that absorbs radiation may also emit radiation at the same wavelength. Kirchhoff's law states that the absorptivity of a body is equal to its emissivity at the same temperature (Kirchhoff 1860), namely,

$$a_k = \varepsilon \tag{2.17}$$

where a_k is the absorptivity of a body. A body that is a good radiator (or emitter) is also a good absorber. Therefore, an ability to emit is oppositely related to its ability to reflect. Kirchhoff's law requires thermodynamic equilibrium, implying that the surface is isothermal and radiates to an isothermal background at the same temperature (Salisbury et al. 1994). This relationship is often used to measure emissivity of materials by reflectance in laboratory.

2.1.1.19 Atmospheric Window

Atmospheric window is the spectral intervals where the electromagnetic spectrum can be transmitted through the atmosphere with relatively little interference. Only sensor channels in atmospheric windows are used for land applications (Colwell et al. 1983).

2.1.1.20 Atmospheric Transmittance (τ)

For a horizontal atmospheric layer, atmospheric transmittance (τ) is defined by the ratio of the transmitted radiance, R_T, and the incoming radiance, R_0, of a direction (Colwell et al. 1983), namely,

$$\tau = \frac{R_T}{R_0} \tag{2.18}$$

2.1.2 Definitions Relevant to Land Surface Evapotranspiration Process

2.1.2.1 Evapotranspiration (ET)

The phenomenon by which a substance is converted from the liquid or solid state into vapor is defined as evaporation (Brutsaert 1982). In the case of a solid substance, the phenomenon is often referred to as sublimation. The vaporization of water through the stomata of living plants is defined as transpiration. The sum of water evaporation and canopy transpiration from the earth's land surface to atmosphere is called evapotranspiration.

2.1.2.2 Latent Heat of Vaporization (L)

The energy required to change a unit mass of water from liquid to water vapor in a constant pressure and constant temperature process, approximating 2.45 MJ kg^{-1} at 20 °C (Allen et al. 1998).

2.1.2.3 Saturation Vapor Pressure (e_s)

The vapor pressure at which a change in phase can occur at constant temperature (Allen et al. 1998), expressed as

$$e_s = 0.6108 \exp\left[\frac{17.27T_a}{T_a + 237.3}\right] \tag{2.19}$$

where e_s is in kPa and T_a is the air temperature [°C].

2.1.2.4 Dew Point Temperature (T_{dew})

The temperature to which the air needs to be cooled to make the air saturated (Allen et al. 1998).

2.1.2.5 Actual Vapor Pressure (e_a)

The saturation vapor pressure at the dew point temperature (Allen et al. 1998), expressed as

$$e_a = e_s(T_{dew}) = 0.6108 \exp\left[\frac{17.27T_{dew}}{T_{dew} + 237.3}\right] \tag{2.20}$$

where T_{dew} is in °C.

2.1.2.6 Relative Humidity (RH)

The ratio of the actual vapor pressure and the saturation vapor pressure at the same temperature and pressure (Allen et al. 1998), expressed as

$$\mathrm{RH} = 100\frac{e_a}{e_s} \tag{2.21}$$

where RH is the relative humidity.

2.1.2.7 Specific Heat at Constant Pressure (c_p)

The amount of energy required to increase the temperature of a unit mass of air by one degree at constant pressure (Allen et al. 1998). For average atmospheric conditions, a value $c_p = 1.013$ J/(g · K) can be used.

2.1.2.8 Potential Temperature (T_p)

The temperature that would result if air were brought adiabatically to a standard pressure level (Brutsaert 1982), expressed as

$$T_p = (T_a + 273.15)\left(\frac{P_0}{P}\right)^{R/(28.97c_p)} \tag{2.22}$$

where T_p is the potential temperature [K]; T_a is the current temperature of the air parcel [°C]; R is the gas constant of air, 8.3144621 J/(mol · K); P_0 is the standard reference pressure [Pa]; P is the current pressure of the air parcel [Pa]; and c_p is the specific heat capacity at a constant pressure.

2.1.2.9 Momentum Roughness Length (z_{om})

The height at which the wind speed theoretically becomes zero when extrapolating the logarithmic wind speed profile downward through the surface layer (Carlson et al. 1981), empirically expressed as

$$z_{om} = 0.13h \tag{2.23}$$

where z_{om} is the momentum roughness length [m], and h is the vegetation height [m].

2.1.2.10 Zero Plane Displacement Height (d)

The height above the ground at which zero wind speed is achieved as a result of flow obstacles such as trees or buildings, empirically expressed as

$$d = 0.67h \tag{2.24}$$

where d is the zero plane displacement height [m], and h is the height of the vegetation [m].

2.1.2.11 Friction Velocity (u^*)

The square root of the surface shear stress divided by the air density (Brutsaert 1982), expressed as

$$u^* = \sqrt{\frac{N_s}{\rho}} \tag{2.25}$$

where u^* is the friction velocity [m/s], N_s is the shear stress [kg/(m · s^2)], and ρ is the density of the air [kg/m^3].

2.1.2.12 Aerodynamic Resistance (r_a)

The transfer of heat and water vapor from the evaporating surface into the air above the canopy (Brutsaert 1982), under the neutral atmospheric conditions expressed as

$$r_a = \frac{\ln\left[\frac{z_m - d}{z_{om}}\right] \ln\left[\frac{z_h - d}{z_{oh}}\right]}{k^2 u_z} \tag{2.26}$$

where r_a is the aerodynamic resistance [s/m]; z_m is the height of wind measurements [m]; z_h is the height of humidity measurements [m]; d is the zero plane displacement height [m]; z_{om} is the roughness length governing momentum transfer [m]; z_{oh} is the roughness length governing transfer of heat and vapor [m]; k is the von Karman's constant, 0.4; and u_z is the wind speed at height z [m/s]. The aerodynamic resistance is a reciprocal of the aerodynamic conductance.

2.1.2.13 Soil Boundary Resistance (R_s)

The resistance to heat flow in the boundary layer immediately above the soil surface (Norman et al. 1995).

2.1.2.14 Bulk Surface Resistance (r_s)

The resistance of vapor flow through the transpiring canopy and evaporating soil surface (Allen et al. 1998).

2.1.2.15 Aerodynamic Temperature

The average temperature of all the canopy elements weighted by the relative contribution of each element to the overall aerodynamic conductance and is recognized as the temperature of the apparent sources or sinks of sensible heat (Moran et al. 1989).

2.1.2.16 Monin–Obukhov Length (L_{MO})

The height at which turbulence is generated more by buoyancy than by wind shear (Monin and Obukhov 1954), expressed as

$$L_{MO} = \frac{u{*}^3 \rho c_p T_v}{kgH} \tag{2.27}$$

where L_{MO} is the Monin–Obukhov length [m], u^* is the surface friction velocity [m/s], c_p is the specific heat at constant pressure [J/(kg · K)], ρ is the air density [kg/m^3], T_v is the mean virtual potential temperature [K], k is the von Karman's constant, g is the acceleration of gravity [m/s^2], and H is the sensible heat flux [W/m^2].

2.1.2.17 Latent Heat Flux (LE)

The flux of heat from the earth's surface to the atmosphere that is associated with evaporation or transpiration of water at the surface and subsequent condensation of water vapor in the troposphere, expressed as

$$\mathrm{LE} = \frac{\rho c_p}{\gamma} \frac{e_\mathrm{s} - e_\mathrm{a}}{r_\mathrm{w}} \tag{2.28}$$

where LE is the latent heat flux [W/m^2], ρ is the air density [kg/m^3], c_p is the specific heat of air at constant pressure [J/(g $\cdot$ K)], γ is the psychometric constant [kPa/°C], e_s is the saturated vapor pressure near the surface [kPa], e_a is the actual vapor pressure at the reference height [kPa], and r_w is the resistance to water transfer [s/m].

2.1.2.18 Bowen Ratio (β)

The ratio of energy fluxes from one state to another by sensible and latent heating, respectively (Fritschen 1965), expressed as

$$\beta = \frac{H}{\mathrm{LE}} \tag{2.29}$$

where β is the Bowen ratio [–], H is the sensible heat flux [W/m^2], and LE is the latent heat flux [W/m^2].

2.1.2.19 Potential Evaporation ($\mathrm{ET_p}$)

The evaporation that would occur if water were plentiful. It is a measure of the ability of the atmosphere to remove water from the surface through the processes of evaporation and transpiration assuming no control on water supply (Allen et al. 1998).

2.1.2.20 Equilibrium Evaporation ($\mathrm{LE_e}$)

The evaporation under conditions of zero advection and no boundary layer growth when the air above a free water surface of large spatial scale becomes saturated (Kim and Entekhabi 1997), expressed as

$$\mathrm{LE_e} = \frac{\Delta}{\Delta + \gamma} (R_\mathrm{n} - G) \tag{2.30}$$

where Δ represents the slope of the saturation vapor pressure versus air temperature curve [kPa/°C], γ is the psychrometric constant [kPa/°C], R_n is the net radiation flux [W/m^2], and G is the soil heat flux [W/m^2].

2.1.2.21 Reference Evaporation (ET$_r$)

The evaporation from a saturated reference surface (e.g., grass) with an assumed height of 0.12 m having a surface resistance of 50 s/m during daytime and 200 s/m during nighttime and an albedo of 0.23 (ASCE-EWRI 2005), expressed as

$$\mathrm{ET_r} = \frac{0.408\Delta\left(R_n - G\right) + \gamma \frac{C_n}{T_a + 273} u_2 \left(e_s - e_a\right)}{\Delta + \gamma\left(1 + C_d u_2\right)} \tag{2.31}$$

where ET$_r$ is the standardized reference crop ET [mm/day], R_n is the net radiation at the crop surface [MJ/(m$^2\cdot$day)], G is the soil heat flux at the soil surface [MJ/(m$^2\cdot$day)], T_a is the mean daily or hourly air temperature at 1.5–2.5 m height [°C], u_2 is the mean daily or hourly wind speed at 2-m height [m/s], e_s is the mean saturation vapor pressure at 1.5–2.5 m height [kPa], e_a is the mean actual vapor pressure at 1.5–2.5 m height [kPa], Δ is the slope of the saturation vapor pressure–temperature curve [kPa/°C], γ is the psychrometric constant [kPa/°C], C_n is the numerator constant that changes for reference type and calculation time step, and C_d is the denominator constant that changes for reference type and calculation time step.

2.2 Radiative Transfer in the Atmosphere

2.2.1 A General Form of Radiative Transfer Equation

The quantity of radiance will change when radiation traverses a medium (Fig. 2.4). Generally, there are two effects on radiation. On the one hand, the radiance will be

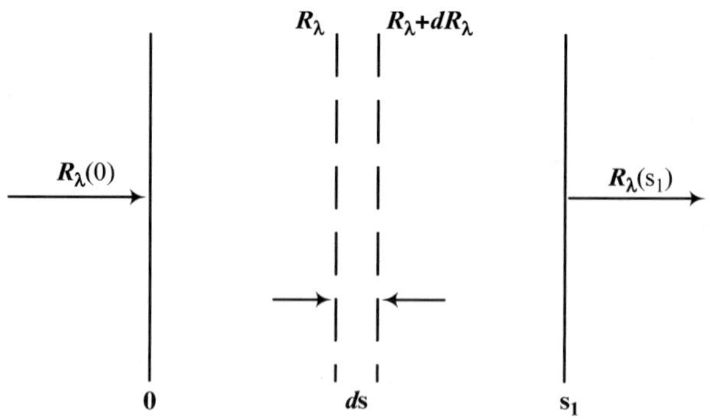

Fig. 2.4 Description of the radiance change when it traverses a medium. $R_\lambda(0)$ and $R_\lambda(s_1)$ are the radiances at thickness 0 and s_1, respectively. ds is a differential thickness, and dR_λ is the increment of radiance after traversing a thickness ds

weakened by its interaction with the medium. The incremental radiance (dR_λ) after traversing a thickness ds in the direction of its propagation due to absorption and scattering can be expressed as

$$dR_\lambda = -k_\lambda \rho R_\lambda ds \tag{2.32}$$

where ρ is the density of the material and k_λ denotes the mass extinction cross section (the sum of the mass absorption and scattering cross sections in units of area per mass) for radiation at wavelength λ. Thus, the reduction in radiance is the combination of absorption and scattering of the materials.

On the other hand, the radiance may be strengthened by emission from the material plus multiple scattering from all other directions into the solid angle at the same wavelength. Therefore, the incremental radiance (dR_λ) due to emission and multiple scattering can be given as

$$dR_\lambda = j_\lambda \rho ds \tag{2.33}$$

where the source function coefficient, j_λ, has the same physical meaning as the mass extinction cross section, k_λ.

Combining Eqs. (2.32) and (2.33), the final incremental radiance (dR_λ) is

$$dR_\lambda = -k_\lambda \rho R_\lambda ds + j_\lambda \rho ds \tag{2.34}$$

If the source function J_λ is defined in unit of radiance by

$$J_\lambda = j_\lambda / k_\lambda \tag{2.35}$$

then Eq. (2.34) can be rewritten as

$$\frac{dR_\lambda}{k_\lambda \rho ds} = -R_\lambda + J_\lambda \tag{2.36}$$

This general radiative transfer equation is fundamental to any radiative transfer process below. A more detailed description of the atmospheric radiation can be found in Liou (2002).

2.2.2 The Radiative Transfer Equation for Plane-Parallel Atmospheres

As discussed in Sect. 2.2.1, the radiance passing through a medium will undergo the absorption, scattering, and emission processes simultaneously. For various applications, it is required to introduce some reasonable hypothesis to simplify the radiative transfer equation described earlier. It is physically appropriate to assume that the atmosphere is plane-parallel in localized circumstance, where its variations exist only in the vertical direction (Liou 2002). In this case, the general equation of radiative transfer defined in Eq. (2.36) becomes

$$\frac{\mathrm{d}R\left(z,\theta,\varphi\right)}{k\rho\mathrm{d}z/\cos\theta} = -R\left(z,\theta,\varphi\right) + J\left(z,\theta,\varphi\right) \tag{2.37}$$

where θ and φ denote the zenith and azimuthal angle, respectively, and z denotes the normal distance with respect to the plane of stratification, that is, $\mathrm{d}z = \mathrm{d}s\cos\theta$. For simplicity, the subscript λ in radiative transfer equation is omitted. The first term in the right-hand side of Eq. (2.37) denotes the reduction of the radiant intensity due to absorption and scattering, and the second term represents the increase in the radiance from emission and scattering. This is the basic equation for plane-parallel atmospheres.

To seek a solution for Eq. (2.37), the normal optical thickness (δ) is introduced as

$$\delta = \int_{z}^{\infty} k\rho\mathrm{d}z' \tag{2.38}$$

The differential form of Eq. (2.38) is

$$\mathrm{d}\delta = -k\rho\mathrm{d}z \tag{2.39}$$

Let $\mu = \cos\theta$ and multiply by a factor $-(e^{-\delta/u}k\rho\mathrm{d}z)/u$, Eq. (2.37) can be rewritten as

$$\mathrm{d}\left[R\left(z,u,\varphi\right)e^{-\delta/u}\right] = J\left(z,u,\varphi\right)e^{-\delta/u}\frac{1}{u}k\rho\mathrm{d}z \tag{2.40}$$

Consequently, the upward radiance (R_{u}) for a plane-parallel atmosphere as illustrated in Fig. 2.5 at a given height z can be estimated by integrating the normal distance from 0 to z upon Eq. (2.40)

$$R_{\mathrm{u}}\left(z;u,\varphi\right) = R\left(0;u,\varphi\right)e^{-(\delta_0-\delta_z)/u} + \int_0^z J\left(z';u,\varphi\right)e^{-(\delta'-\delta_z)/u}\frac{1}{u}k\rho\mathrm{d}z' \tag{2.41}$$

where δ_0 and δ_z are the normal optical thicknesses at height $z = 0$ and the given height z, respectively.

Similarly, the downward radiance (R_{d}) at a given height z can be estimated by integrating the normal distance from ∞ to z upon Eq. (2.40) and by replacing u by $-u$:

$$R_{\mathrm{d}}\left(z;-u,\varphi\right) = R\left(\infty;-u,\varphi\right)e^{-\delta_z/u} - \int_{\infty}^{z} J\left(z';-u,\varphi\right)e^{-(\delta_z-\delta')/u}\frac{1}{u}k\rho\mathrm{d}z' \tag{2.42}$$

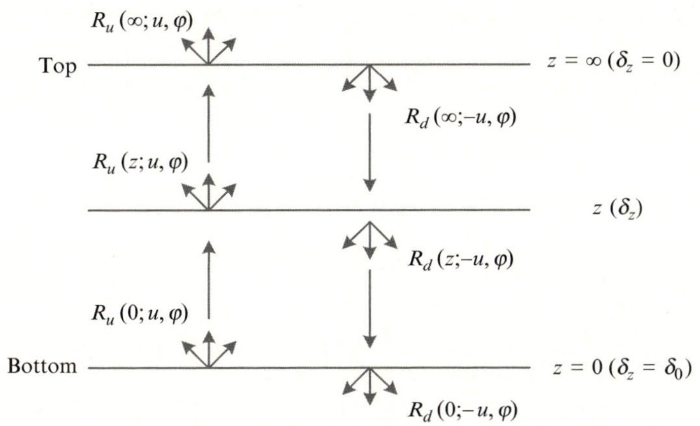

Fig. 2.5 Illustration of the upward and downward radiance at a given height z in a plane-parallel atmosphere. $R_u(z;u,\varphi)$ and $R_d(z;-u,\varphi)$ are upward and downward radiances, respectively, at height z. δ_z is the corresponding normal optical thickness at height z

It is worth noting that the source function J represents the combined effects of emission and scattering, namely,

$$J(z,\theta,\varphi) = B(T(z)) + S(z,\theta,\varphi,P,r,\ldots) \tag{2.43}$$

The first term in the right-hand side of Eq. (2.43) denotes the increase in the radiance from blackbody emission of the atmosphere, and the second term represents the increase in the radiance from multiple scattering of the atmosphere. This scattering source function is characterized by many factors, such as P (the phase function) and r (particle radius).

For the applications of retrieving land surface parameters from remotely sensed data, it is desirable to get radiance at the top and bottom of the atmosphere directly. Upon setting $z=\infty$ ($\delta_z=0$) in Eq. (2.41) and combining Eq. (2.43), we obtain

$$\begin{aligned} R_u(\infty;u,\varphi) &= R(0;u,\varphi)\,e^{-\delta_0/u} + \int_0^\infty B(z';u,\varphi)\,e^{-\delta'/u}\frac{1}{u}k\rho dz' \\ &\quad + \int_0^\infty S(z';u,\varphi,P,r,\ldots)\,e^{-\delta'/u}\frac{1}{u}k\rho dz' \\ &= R_{\text{bottom}\uparrow}\tau + R_{at\uparrow} + R_{sl\uparrow} \end{aligned} \tag{2.44}$$

The first term ($R_{\text{bottom}\uparrow}$) represents the bottom radiance contribution before passing through the atmosphere (attenuated to the top). Similar to Eq. (2.39), τ is the transmittance of atmosphere. The second term ($R_{at\uparrow}$) is the radiance contributed from upward emission of atmosphere, and the third term ($R_{sl\uparrow}$) results from upward atmospheric scattering radiance.

Similarly, setting $z=0$ $(\delta_z=\delta_0)$ in Eq. (2.42) and combining Eq. (2.43), we obtain

$$\begin{aligned} R_{\mathrm{d}}(0;-u,\varphi) &= R(\infty;-u,\varphi)\,e^{-\delta_0/u} - \int_{\infty}^{0} B\left(z';-u,\varphi\right) e^{-(\delta_0-\delta')/u}\frac{1}{u}k\rho \mathrm{d}z' \\ &\quad - \int_{\infty}^{0} S\left(z';-u,\varphi,P,r,\ldots\right) e^{-(\delta_0-\delta')/u}\frac{1}{u}k\rho \mathrm{d}z' \\ &= R_{\mathrm{top}\downarrow}\tau + R_{at\downarrow} + R_{sl\downarrow} \end{aligned} \tag{2.45}$$

The first term ($R_{\mathrm{top}\downarrow}$) represents the top radiance contribution before passing through the atmosphere (attenuated to the bottom). Similar to Eq. (2.39), τ is the transmittance of atmosphere. The second term ($R_{at\downarrow}$) is the radiance contributed from downward emission of atmosphere, and the third term ($R_{sl\downarrow}$) results from downward atmospheric scattering radiance.

2.2.3 Approximations for Radiative Transfer Equation

An infrared sensor onboard a satellite viewing the earth's surface measures the radiation from earth and its atmosphere along the line of sight within a finite bandwidth. Using the radiative transfer equation analyzed above and assuming a cloud-free atmosphere under local thermodynamic equilibrium, as illustrated in Fig. 2.6, the channel infrared radiance R_i received by a sensor at the top of the atmosphere (TOA) can be approximately written as (Chandrasekhar 1960)

$$\begin{aligned} R_i(\theta,\varphi) = &\underbrace{R_{gi}(\theta,\varphi)\;\tau_i(\theta,\varphi)}_{\substack{\text{Surface outgoing radiation term}\\ \text{attenuated by the atmosphere}}} + \underbrace{R_{at_i\uparrow}(\theta,\varphi)}_{\text{Atmospheric emission term}} \\ &+ \underbrace{R_{sl_i\uparrow}(\theta,\varphi)}_{\text{Atmospheric scattering term}} \end{aligned} \tag{2.46}$$

with R_{gi} being the channel radiance observed in channel i at ground level given by

$$\begin{aligned} R_{gi}(\theta,\varphi) = &\underbrace{\varepsilon_i(\theta,\varphi)\,B_i(T_s)}_{\text{Surface emission term}} + \underbrace{[1-\varepsilon_i(\theta,\varphi)]\,R_{at_i\downarrow}}_{\substack{\text{Surface reflected downwelling}\\ \text{atmospheric emission term}}} \\ &+ \underbrace{[1-\varepsilon_i(\theta,\varphi)]\,R_{sl_i\downarrow}}_{\substack{\text{Surface reflected downwelling}\\ \text{atmospheric scattering term}}} + \underbrace{\rho_{b_i}(\theta,\varphi,\theta_{\mathrm{s}},\varphi_{\mathrm{s}})\,E_i\cos(\theta_{\mathrm{s}})\,\tau_i(\theta_{\mathrm{s}},\varphi_{\mathrm{s}})}_{\substack{\text{Surface reflected downwelling}\\ \text{solar beam term}}} \end{aligned} \tag{2.47}$$

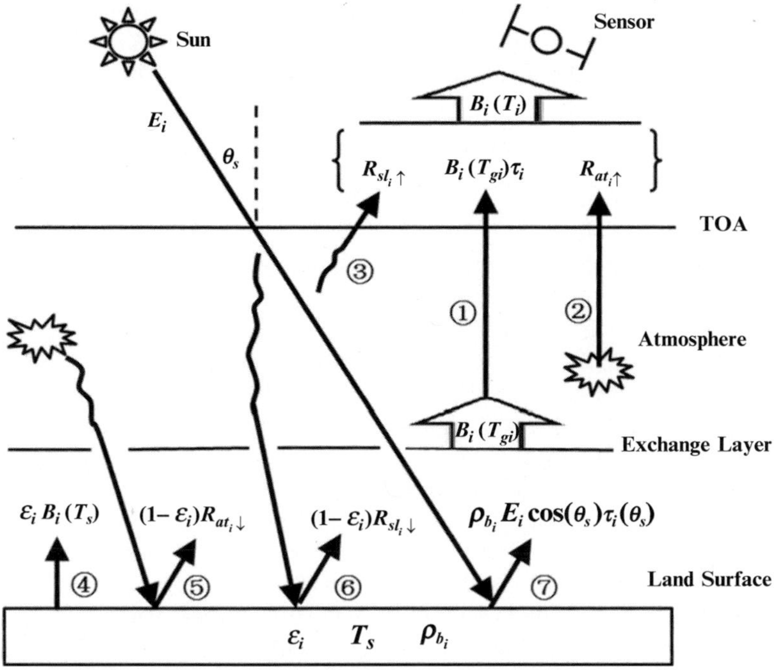

Fig. 2.6 Illustration of radiative transfer equation in infrared regions. θ_s and φ_s are the solar zenith and azimuthal angles, respectively. ε_i and T_s are the effective surface emissivity and surface temperature, respectively, in channel i. ρ_{b_i} is the bidirectional reflectivity of the surface. E_i is the channel solar irradiance at the TOA. $B_i(T)$ is the channel radiance of a blackbody at temperature T. τ_i is the effective channel transmittance of the atmosphere. $R_{at_i}\uparrow$ and $R_{at_i}\downarrow$ are the upward and downward atmospheric thermal radiances, respectively. $R_{sl_i}\uparrow$ and $R_{sl_i}\downarrow$ are the upward and downward solar diffusion radiances, respectively, resulting from atmospheric scattering of the solar radiances (Adopted from Li et al. (2013), with permission from Elsevier)

where θ and φ represent the zenithal and azimuthal viewing angles, respectively. For simplicity, the zenithal and azimuthal viewing angles are ignored in the following expressions. τ_i is the effective transmittance of the atmosphere in channel i. $R_{gi}\tau_i$ is the radiance observed at ground level attenuated by the atmosphere (path ①). $R_{at_i}\uparrow$ is the upward atmospheric thermal radiance (path ②). $R_{sl_i}\uparrow$ is the upward solar diffusion radiance resulting from atmospheric scattering of the solar radiance (path ③). ε_i and T_s are the effective surface emissivity and surface temperature, respectively, in channel i. $\varepsilon_i B_i(T_s)$ represents the radiance emitted directly by surface (path ④). $R_{at_i}\downarrow$ is the downward atmospheric thermal radiance. $R_{sl_i}\downarrow$ is the downward solar diffusion radiance. $(1-\varepsilon_i)\ R_{at_i}\downarrow$ and $(1-\varepsilon_i)\ R_{sl_i}\downarrow$ represent the downward atmospheric thermal radiance and solar diffusion radiance reflected by the surface (paths ⑤ and ⑥), respectively. ρ_{b_i} is the bidirectional reflectivity of the surface. E_i is the solar irradiance at the TOA. θ_s and φ_s are the solar zenithal and azimuthal angles, respectively. $\rho_{b_i} E_i \cos(\theta_s)\ \tau_i\ (\theta_s, \varphi_s)$ is the direct solar radiance reflected by the surface (path ⑦). Because the contribution of solar radiation at the TOA is negligible in the 8–14 μm window during both day and night and in the 3–5 μm window at night, the solar-related items (paths ③, ⑥, and ⑦) can be neglected without loss of accuracy.

In this approximated radiative transfer equation, path ① represents the radiance observed at ground level attenuated by the atmosphere. Paths ② and ③ represent the upward atmospheric thermal radiance and the upward solar diffusion radiance, respectively. Path ④ represents the radiance emitted directly by the surface. Paths ⑤ and ⑥ represent the downward atmospheric thermal radiance and solar diffusion radiance reflected by the surface, respectively. Path ⑦ represents the direct solar radiance reflected by the surface.

For convenience and mathematical shorthand, the radiances R_i and R_{gi} measured at the TOA and at ground level, respectively, are generally expressed in terms of the brightness temperatures. The TOA and ground level brightness temperatures T_i and T_{gi} are defined respectively by

$$B_i\left(T_i\right) = R_i \text{ and } B_i\left(T_{gi}\right) = R_{gi} \tag{2.48}$$

It is worth noting that all variables/parameters in Eqs. (2.46) and (2.47), except for the angles (θ, ϕ, θ_s, and ϕ_s), are channel-effective values. Most satellite sensors measure the outgoing radiation with a finite spectral bandwidth, and the channel-effective quantities of interest are therefore a weighted average, expressed by

$$X_i = \frac{\int_{\lambda_1}^{\lambda_2} g_i\left(\lambda\right) X_\lambda \mathrm{d}\lambda}{\int_{\lambda_1}^{\lambda_2} g_i\left(\lambda\right) \mathrm{d}\lambda} \tag{2.49}$$

where $g_i(\lambda)$ is the spectral response function of channel i; λ_1 and λ_2 are the lower and upper boundaries of the wavelength, respectively, in channel i; and X stands for $B(T)$, R, R_g, $R_{at\uparrow}$, $R_{sl\uparrow}$, $R_{at\downarrow}$, $R_{sl\downarrow}$, E, ε, τ, or ρ_b.

Equations (2.46) and (2.47) are actually approximations to the theoretical radiative transfer equation in which monochromatic quantities are replaced with channel-effective values, but these approximations or simplifications require several important preconditions.

First, the integral of a product is assumed to be equal to the product of the integrals. This assumption is true only if the variables are constant within the integration limits. In practice, this situation rarely happens. Fortunately, the bandwidth of the channel is generally narrow, and the various spectral quantities X_λ involved in Eq. (2.49) do not change rapidly. Therefore, the use of the weighted averages as defined by Eq. (2.49) in Eqs. (2.46) and (2.47) is a good approximation to the radiative transfer equation with monochromatic quantities.

Second, in the calculation of the downward radiations reflected by the surface, the surface is assumed to be Lambertian. Furthermore, the downward atmospheric and solar diffuse radiations are assumed to be isotropic. Although these conditions are never fulfilled, this simplification of Eqs. (2.46) and (2.47) is reasonable and does not introduce significant errors due to the reason that the surface-reflected downward atmospheric thermal radiation term is much smaller than the surface thermal emission, and the surface-reflected diffuse solar radiation term is much smaller than the surface-reflected direct solar term.

References

Allen, R. G., Pereira, L. S., Raes, D., & Smith, M. (1998). Crop evapotranspiration-guidelines for computing crop water requirements (FAO Irrigation and drainage paper 56). Rome: FAO, *300*, 6541.

ASCE-EWRI. (2005). *The ASCE standardized reference evapotranspiration equation*. Technical committee report to the environmental and water resources Institute of the American Society of Civil Engineers from the Task Committee on Standardization of Reference Evapotranspiration. ASCE-EWRI, 1801 Alexander Bell Drive, Reston, VA 20191-4400, 173 pp.

Becker, F., & Li, Z. L. (1995). Surface temperature and emissivity at various scales: Definition, measurement and related problems. *Remote Sensing Reviews, 12*(3–4), 225–253.

Boltzmann, L. (1884). Ableitung des Stefan'schen Gesetzes, betreffend die Abhängigkeit der Wärmestrahlung von der Temperatur aus der electromagnetischen Lichttheorie. In *Annalen der Physik und Chemie*, Bd. 22, S. 291–294.

Brutsaert, W. (1982). *Evaporation into the atmosphere: Theory, history, and applications* (p. 299). Dordrecht: Reidel Publishing Co.

Carlson, T. N., Dodd, J. K., Benjamin, S. G., & Cooper, J. N. (1981). Satellite estimation of the surface energy balance, moisture availability and thermal inertia. *Journal of Applied Meteorology, 20*, 67–87.

Chandrasekhar, S. (1960). *Radiative transfer* (Rev. reprint ed.). New York: Dover Publications. ISBN 978-0-486-60590-6.

Colwell, R. N., Ulaby, F. T., Simonett, D. S., Estes, J. E., & Thorley, G. A. (1983). *Manual of remote sensing* (Theory, instruments and techniques, Vol. 1). Falls Church: American Society of Photogrammetry.

Fritschen, L. J. (1965). Accuracy of evapotranspiration determinations by the Bowen ratio method. *Hydrological Sciences Journal, 10*(2), 38–48.

Kim, C. P., & Entekhabi, D. (1997). Examination of two methods for estimating regional evaporation using a coupled mixed layer and land surface model. *Water Resources Research, 33*(9), 2109–2116.

Kirchhoff, G. (1860). Ueber das Verhältniss zwischen dem Emissionsvermögen und dem Absorptionsvermögen der Körper für Wärme and Licht. *Annalen der Physik und Chemie, 109*, 275–301. (Translated by Guthrie, F. as Kirchhoff, G. (1860). On the relation between the radiating and absorbing powers of different bodies for light and heat. *Philosophical Magazine*. Series 4, *20*: 1–21.)

Lambert, J. H. (1760). *Photometria, sive de Mensura et Gradibus Luminis, Colorum et Umbrae*. Augsburg: Vidvae Eberhardi Klett.

Li, Z. L., Tang, B. H., Wu, H., Ren, H., Yan, G., Wan, Z., Trigo, I., & Sobrino, J. A. (2013). Satellite-derived land surface temperature: Current status and perspectives. *Remote Sensing of Environment, 131*, 14–37.

Liang, S. (2004). *Quantitative remote sensing of land surfaces*. Hoboken: Wiley-Interscience.

Liou, K. N. (2002). *An introduction to atmospheric radiation* (Vol. 84). Boston: Academic Press.

Martonchik, J. V., Bruegge, C. J., & Strahler, A. H. (2000). A review of reflectance nomenclature used in remote sensing. *Remote Sensing Reviews, 19*, 9–20.

Monin, A. S., & Obukhov, A. (1954). Basic laws of turbulent mixing in the surface layer of the atmosphere. Trudy Geofizicheskogo Instituta, Akademiya Nauk SSSR, 24, 163–187.

Moran, M. S., Jackson, R. D., Raymond, L. H., Gay, L. W., & Slater, P. N. (1989). Mapping surface energy balance components by combining Landsat Thematic Mapper and ground-based meteorological data. *Remote Sensing of Environment, 30*(1), 77–87.

Nicodemus, F. E. (1965). Directional reflectance and emissivity of an opaque surface. *Applied Optics, 4*(7), 767–773.

Norman, J. M., & Becker, F. (1995). Terminology in thermal infrared remote sensing of natural surfaces. *Agricultural and Forest Meteorology, 77*(3), 153–166.

Norman, J. M., Kustas, W. P., & Humes, K. S. (1995). Source approach for estimating soil and vegetation energy fluxes in observations of directional radiometric surface temperature. *Agricultural and Forest Meteorology, 77*(3), 263–293.

Planck, M. (1914). *The theory of heat radiation* (M. Masius, Trans.) (2nd ed.). P. Blakiston's Son & Co. OL 7154661M.

Rees, W. G. (2001). *Physical principles of remote sensing*. Cambridge: Cambridge University Press.

Salisbury, J. W., Wald, A., & D'Aria, D. M. (1994). Thermal-infrared remote sensing and Kirchhoff's Law: 1. Laboratory measurements. *Journal of Geophysical Research, 99*(B6), 11897–11911.

Schaepman-Strub, G., Schaepman, M. E., Painter, T. H., Dangel, S., & Martonchik, J. V. (2006). Reflectance quantities in optical remote sensing-definitions and case studies. *Remote Sensing of Environment, 103*, 27–42.

Slater, P. N. (1980). *Remote sensing: Optics and optical systems*. Reading: Addison-Wesley Pub. Co.

Stefan, J. (1879). Über die Beziehung zwischen der Wärmestrahlung und der Temperatur. In Sitzungsberichte der mathematisch-naturwissenschaftlichen Classe der kaiserlichen Akademie der Wissenschaften, Bd. 79 (Wien 1879), S. 391–428.

Stephens, G. L. (1994). *Remote sensing of the lower atmosphere* (Vol. 515). New York: Oxford University Press.

Wannier, G. H. (1987). *Statistical physics*. New York: Dover Publications. Chapter 10.2. ISBN 978-0-486-65401-0. OCLC 15520414.

Chapter 3
Radiometric Calibration in Thermal Infrared

3.1 Background

Calibration is defined by the Working Group on Calibration and Validation (WGCV) of the International Committee on Earth Observation Satellites (CEOS) as the process of quantitatively defining the system response to known, controlled signal inputs (Belward 1999). The main fundamental aspects that need to be calibrated are the sensor system's response to electromagnetic radiation as a function of (1) wavelength and/or spectral band (spectral response), (2) the intensity of the input signals (radiometric response), (3) different locations across the instantaneous field of view and/or the overall scene (spatial response or uniformity), (4) different integration times and lens or aperture setting, and (5) unwanted signals such as stray light and leakage from other spectral bands (Liang 2004). Estimating land surface bio-/geophysical variables accurately from remotely sensed data relies largely on the accuracy of radiometric calibration. Radiometric calibration is a process that coverts the observed digital numbers (DN) to physical quantities of radiance. Preflight calibration, in-flight calibration, vicarious calibration, and intercalibration are critical components of a calibration system.

3.2 Preflight Calibration

The preflight calibration of satellite instrument is performed at the instrument builders' facilities prior to launch (Sakuma and Ono 1993; Ono et al. 1996; Thome et al. 1998). Preflight calibration involves determining the relationship between instrument output and radiant input, while the instrument views a calibrated, well-characterized radiant source (Butler and Barnes 1998). There are two reasons to perform preflight calibration (Thome et al. 1997). First, preflight calibration allows the instrument to be tested to ensure that it operates properly before being integrated into the launch vehicle. Second, it can also be easier to control and

H. Tang and Z.-L. Li, *Quantitative Remote Sensing in Thermal Infrared: Theory and Applications*, Springer Remote Sensing/Photogrammetry,
DOI 10.1007/978-3-642-42027-6_3, © Springer-Verlag Berlin Heidelberg 2014

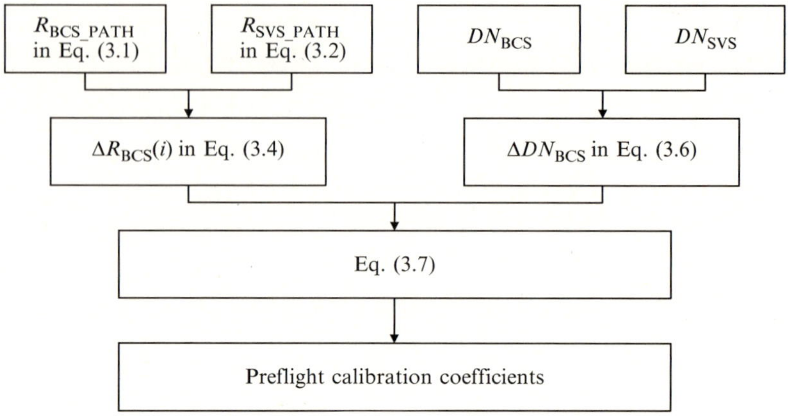

Fig. 3.1 Flowchart of the preflight calibration of the MODIS thermal emissive bands

perform than methods used after launch. For the preflight calibration of the infrared bands, it is essential that the thermal environment be well controlled and accurately monitored. This is achieved by performing tests in a thermal vacuum chamber with the instrument under test surrounded by temperature controlled panels to allow it operating at temperatures close to those expected from the thermal model at flight conditions (Smith et al. 2001, 2012). As an example, the MODIS thermal emissive bands (TEB) are used to describe the preflight calibration of infrared bands.

The preflight calibration of the MODIS TEB is performed in a thermal vacuum chamber at cold, nominal, and hot instrument temperature plateaus using both primary and redundant electronics configurations (Xiong and Barnes 2006a). The preflight calibration source used for the TEB is a large aperture blackbody (BB) calibration source (BCS) with an emissivity better than 0.9995 over the TEB spectral range. A space view source (SVS), which is similar to the BCS and operates at an extremely low temperature, is used to simulate the deep space view. Both the BCS and SVS are inside the thermal vacuum chamber during the preflight calibration of the MODIS TEB. The BCS is located at −45° scan angle, which corresponds to a low angle of incidence (AOI) to the scan mirror to minimize the effects of reflectance variations. The scan mirror exhibits a significant variation in average reflectance as a function of wavelength and AOI.

The preflight calibration algorithm of the MODIS TEB is based on a quadratic relationship between the detector's digital response and the input radiance (Guenther et al. 1998). It is implemented on a scan-by-scan basis to minimize the effects of the instrument background. Figure 3.1 shows the flowchart of the preflight calibration of the MODIS TEB. When the MODIS views the BCS, the spectral radiance of the BCS path $R_{\mathrm{BCS\_PATH}}$ is given by

$$R_{\mathrm{BCS\_PATH}} = \rho_{\mathrm{BCS}}^{\mathrm{sm}} \varepsilon_{\mathrm{BCS}} B(T_{\mathrm{BCS}}) + \left(1 - \rho_{\mathrm{BCS}}^{\mathrm{sm}}\right) B(T_{\mathrm{sm}}) + R_{\mathrm{BKG}} \tag{3.1}$$

where $\rho_{\mathrm{BCS}}^{\mathrm{sm}}$ is the reflectance of the scan mirror, which integrates over the spectral response functions at the BCS AOI to the scan mirror; $\varepsilon_{\mathrm{BCS}}$ is the emissivity of the BCS; $B(T_{\mathrm{BCS}})$ and $B(T_{\mathrm{sm}})$ are the radiances calculated from the Planck functions at the temperatures of the BCS and the scan mirror, respectively; and R_{BKG} is the radiance of the instrument background, which excludes the emission from the scan mirror.

The first term in the right-hand side of Eq. (3.1) is the emission from the BCS after reflected by the scan mirror. The second term is the emission from the scan mirror. This term is explicitly separated from the total radiance of the instrument background to indicate the AOI dependence of the scan mirror. The third term is the radiance of the instrument background, which excludes the emission from the scan mirror.

Similarly, when the MODIS views the SVS, the spectral radiance of the SVS path $R_{\mathrm{SVS\_PATH}}$ is given by

$$R_{\mathrm{SVS\_PATH}} = \left(1 - \rho_{\mathrm{SVS}}^{\mathrm{sm}}\right) B(T_{\mathrm{sm}}) + R_{\mathrm{BKG}} \tag{3.2}$$

where $\rho_{\mathrm{SVS}}^{\mathrm{sm}}$ is the reflectance of the scan mirror, which integrates over the spectral response functions at the SVS AOI to the scan mirror.

To remove the effects of the instrument background, the radiance of the SVS path $R_{\mathrm{SVS\_PATH}}$ is subtracted from the radiance of the BCS path $R_{\mathrm{BCS\_PATH}}$. The spectral radiance difference ΔR_{BCS} between the radiances of the BCS and SVS paths is thus given by

$$\begin{aligned} \Delta R_{\mathrm{BCS}} &= R_{\mathrm{BCS\_PATH}} - R_{\mathrm{SVS\_PATH}} \\ &= \rho_{\mathrm{BCS}}^{\mathrm{sm}} \varepsilon_{\mathrm{BCS}} B(T_{\mathrm{BCS}}) + \left(\rho_{\mathrm{SVS}}^{\mathrm{sm}} - \rho_{\mathrm{BCS}}^{\mathrm{sm}}\right) B(T_{\mathrm{sm}})\,. \end{aligned} \tag{3.3}$$

For a specific MODIS band i, the band averaged radiance difference $\Delta R_{\mathrm{BCS}}(i)$ between the radiances of the BCS and SVS paths can be expressed as

$$\Delta R_{\mathrm{BCS}}(i) = \rho_{\mathrm{BCS}}^{\mathrm{sm}} \varepsilon_{\mathrm{BCS}} \overline{B(T_{\mathrm{BCS}})} + \left(\rho_{\mathrm{SVS}}^{\mathrm{sm}} - \rho_{\mathrm{BCS}}^{\mathrm{sm}}\right) \overline{B(T_{\mathrm{sm}})} \tag{3.4}$$

with

$$\overline{B(T)} = \frac{\int B(\lambda, T)\, f_i(\lambda)\, \mathrm{d}\lambda}{\int f_i(\lambda)\, \mathrm{d}\lambda} \tag{3.5}$$

where i is the band number, λ is the wavelength, and $f_i(\lambda)$ is the spectral response function.

The digital number difference ΔDN_{BCS} between the digital numbers of the BCS and SVS paths is given by

$$\Delta DN_{BCS} = \frac{1}{M_{BCS}} \sum_{m=1}^{M_{BCS}} DN_{BCS}(m) - \frac{1}{N_{SVS}} \sum_{n=1}^{N_{SVS}} DN_{SVS}(n) \tag{3.6}$$

where DN_{BCS} and DN_{SVS} are the digital numbers of the BCS and SVS paths, respectively, and M_{BCS} and N_{SVS} are the number of the BCS and SVS frames, respectively.

The band-averaged radiance difference $\Delta R_{BCS}(i)$ can be expressed as a function of the digital number difference ΔDN_{BCS}:

$$\Delta R_{BCS}(i) = a_0(i) + a_1(i)\Delta DN_{BCS} + a_2(i)(\Delta DN_{BCS})^2 \tag{3.7}$$

where a_0, a_1, and a_2 are the offset, linear, and nonlinear calibration coefficients, respectively.

The offset and nonlinear calibration coefficients a_0 and a_2 are initially derived from preflight calibration and updated on-orbit using data collected from the onboard BB warm-up and cooldown cycles. The linear calibration coefficient a_1 is determined on a scan-by-scan basis during the in-flight calibration.

3.3 In-flight Calibration

Preflight calibration may change in space due to the variations in the environment surrounding the sensor in a spaceborne environment. Therefore, it is necessary to perform in-flight or on-orbit calibration. In-flight calibration is usually performed on a routine basis with onboard calibration systems. Same as the preflight calibration, the MODIS TEB is used as an example to describe the in-flight calibration of infrared bands.

An onboard BB and a space view (SV) port comprise an onboard calibration system for the MODIS TEB. The onboard BB shown in Fig. 3.2 is a full aperture V-grooved flat panel with 12 embedded thermistors. The SV port is designed to provide scan-to-scan measurements of instrument background and electronic offsets.

The in-flight calibration of the MODIS TEB uses a quadratic algorithm on a scan-by-scan basis for each band, detector, and mirror side (Xiong et al. 2009; Chang and Xiong 2011). Figure 3.3 shows the flowchart of the in-flight calibration of the MODIS TEB. The preflight calibration of the MODIS TEB is based on the BCS and SVS, whereas the in-flight calibration of the MODIS TEB is based on the onboard BB and SV ports. For each scan, the MODIS views the onboard BB and deep space through the SV port for calibration and then views the earth for scene observations.

Fig. 3.2 MODIS onboard blackbody used for the thermal emissive bands calibration (Reproduced from Xiong et al. (2008), with permission from IEEE publisher)

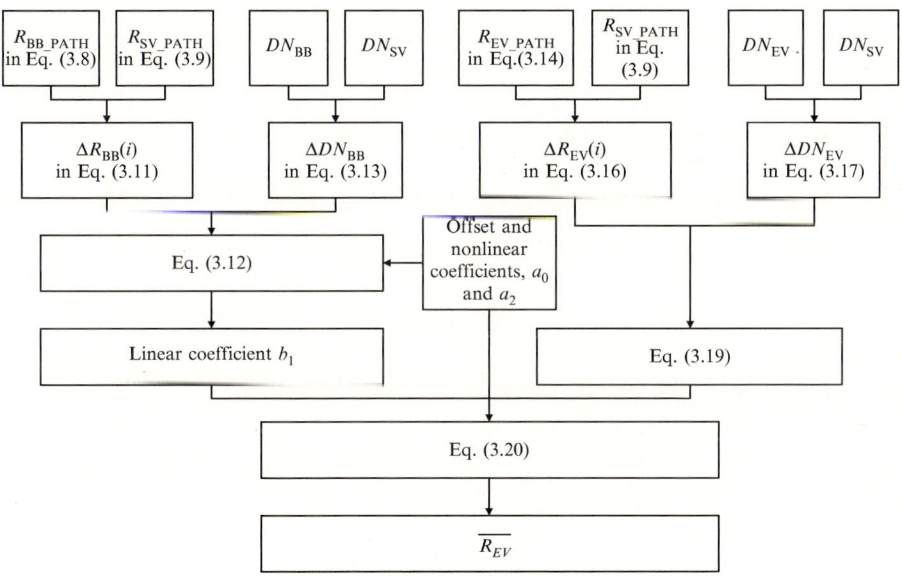

Fig. 3.3 Flowchart of the in-flight calibration of the MODIS thermal emissive bands

When the MODIS views the onboard BB, the spectral radiance of the onboard BB path $R_{\mathrm{BB\_PATH}}$ is given by

$$R_{\mathrm{BB\_PATH}} = \rho_{\mathrm{BB}}^{\mathrm{sm}} \varepsilon_{\mathrm{BB}} B(T_{\mathrm{BB}}) + \rho_{\mathrm{BB}}^{\mathrm{sm}} (1 - \varepsilon_{\mathrm{BB}}) \varepsilon_{\mathrm{cav}} B(T_{\mathrm{cav}}) + \left(1 - \rho_{\mathrm{BB}}^{\mathrm{sm}}\right) B(T_{\mathrm{sm}}) + R_{\mathrm{BKG}} \tag{3.8}$$

where $\rho_{\mathrm{BB}}^{\mathrm{sm}}$ is the reflectance of the scan mirror, which integrates over the spectral response functions at the onboard BB AOI to the scan mirror; $\varepsilon_{\mathrm{BB}}$ and $\varepsilon_{\mathrm{cav}}$ are the emissivities of the onboard BB and instrument scan cavity, respectively; and $B(T_{\mathrm{BB}})$ and $B(T_{\mathrm{cav}})$ are the radiances calculated from the Planck functions at the temperatures of the onboard BB and instrument scan cavity, respectively.

The first term in the right-hand side of Eq. (3.8) is the emission from the onboard BB after reflected by the scan mirror. The second term represents the emission from the instrument scan cavity after reflected by the onboard BB and scan mirror. The third term is the emission from the scan mirror. The fourth term is the radiance of the instrument background, which excludes the emission from the scan mirror. If the onboard BB has perfect emissivity (i.e., $\varepsilon_{\mathrm{BB}} = 1$), the contribution from the instrument scan cavity will disappear. The emissivities of the onboard BB and instrument scan cavity, $\varepsilon_{\mathrm{BB}}$ and $\varepsilon_{\mathrm{cav}}$, are determined from preflight calibration and characterization. The emissivity of the onboard BB is characterized using detector responses to both the BCS (reference source) and the onboard BB under the same thermal vacuum conditions and at the same time (Xiong et al. 2005; Xiong and Barnes 2006b). Results from all detectors in a band are averaged to determine the in-band spectral emissivity.

Similarly, when the MODIS views the deep space through the SV port, the spectral radiance of the SV path $R_{\mathrm{SV\_PATH}}$ is given by

$$R_{\mathrm{SV\_PATH}} = \left(1 - \rho_{\mathrm{SV}}^{\mathrm{sm}}\right) B(T_{\mathrm{sm}}) + R_{\mathrm{BKG}} \tag{3.9}$$

where $\rho_{\mathrm{SV}}^{\mathrm{sm}}$ is the reflectance of the scan mirror, which integrates over the spectral response functions at the SV AOI to the scan mirror.

The spectral radiance difference ΔR_{BB} between the radiances of the onboard BB and SV paths can be expressed as

$$\begin{aligned}\Delta R_{\mathrm{BB}} &= R_{\mathrm{BB\_PATH}} - R_{\mathrm{SV\_PATH}} \\ &= \rho_{\mathrm{BB}}^{\mathrm{sm}} \varepsilon_{\mathrm{BB}} B(T_{\mathrm{BB}}) + \left(\rho_{\mathrm{SV}}^{\mathrm{sm}} - \rho_{\mathrm{BB}}^{\mathrm{sm}}\right) B(T_{\mathrm{sm}}) + \rho_{\mathrm{BB}}^{\mathrm{sm}} (1 - \varepsilon_{\mathrm{BB}}) \varepsilon_{\mathrm{cav}} B(T_{\mathrm{cav}}) .\end{aligned} \tag{3.10}$$

For a specific MODIS band, the band-averaged radiance difference $\Delta R_{\mathrm{BB}}(i)$ between the radiances of the onboard BB and SV paths can be expressed as

$$\Delta R_{\mathrm{BB}}(i) = \rho_{\mathrm{BB}}^{\mathrm{sm}} \varepsilon_{\mathrm{BB}} \overline{B(T_{\mathrm{BB}})} + \left(\rho_{\mathrm{SV}}^{\mathrm{sm}} - \rho_{\mathrm{BB}}^{\mathrm{sm}}\right) \overline{B(T_{\mathrm{sm}})} + \rho_{\mathrm{BB}}^{\mathrm{sm}} (1 - \varepsilon_{\mathrm{BB}}) \varepsilon_{\mathrm{cav}} \overline{B(T_{\mathrm{cav}})} \tag{3.11}$$

where $\overline{B(T_{BB})}$ and $\overline{B(T_{cav})}$ are the Planck emission terms determined using Eq. (3.5).

As mentioned previously, the offset and nonlinear calibration coefficients a_0 and a_2 in the quadratic algorithm are initially derived from preflight calibration and updated on-orbit using data collected from the onboard BB warm-up and cooldown cycles (Xiong et al. 2003). The onboard BB cycle provides the TEB responses over the temperature of the onboard BB from 270 to 315 K. The linear calibration coefficient b_1 (instead of a_1 to emphasize its on-orbit computation) is determined on a scan-by-scan basis by (Xiong et al. 2008):

$$b_1(i) = \frac{\Delta R_{BB}(i) - a_0(i) - a_2(i)(\Delta DN_{BB})^2}{\Delta DN_{BB}} \tag{3.12}$$

with

$$\Delta DN_{BB} = \frac{1}{M_{BB}} \sum_{m=1}^{M_{BB}} DN_{BB}(m) - \frac{1}{N_{SV}} \sum_{n=1}^{N_{SV}} DN_{SV}(n) \tag{3.13}$$

where DN_{BB} and DN_{SV} are the digital numbers of the onboard BB and SV paths, respectively, and M_{BB} and M_{SV} are the number of the onboard BB and SV frames, respectively.

When the MODIS views the earth, the spectral radiance of the earth view (EV) path R_{EV_PATH} is given by

$$R_{EV\_PATH} = \rho_{EV}^{sm} R_{EV} + \left(1 - \rho_{EV}^{sm}\right) B(T_{sm}) + R_{BKG} \tag{3.14}$$

where ρ_{EV}^{sm} is the reflectance of the scan mirror, which integrates over the spectral response functions at the EV AOI to the scan mirror, and R_{EV} is the spectral radiance from the EV.

The spectral radiance difference ΔR_{EV} between the radiances of the EV and SV paths can be expressed by

$$\begin{aligned} \Delta R_{EV} &= R_{EV\_PATH} - R_{SV\_PATH} \\ &= \rho_{EV}^{sm} R_{EV} + \left(\rho_{SV}^{sm} - \rho_{EV}^{sm}\right) B(T_{sm}). \end{aligned} \tag{3.15}$$

For a specific MODIS band, the band-averaged radiance difference $\Delta R_{EV}(i)$ between the radiances of the EV and SV paths can be expressed as

$$\Delta R_{EV}(i) = \rho_{EV}^{sm} \overline{R_{EV}} + \left(\rho_{SV}^{sm} - \rho_{EV}^{sm}\right) \overline{B(T_{sm})}. \tag{3.16}$$

The digital number difference ΔDN_{EV} between the digital numbers of the EV and SV paths is given by

$$\Delta \mathrm{DN}_{\mathrm{EV}} = \mathrm{DN}_{\mathrm{EV}} - \frac{1}{N_{\mathrm{SV}}} \sum_{n=1}^{N_{\mathrm{SV}}} \mathrm{DN}_{\mathrm{SV}}(n) \tag{3.17}$$

where $\mathrm{DN}_{\mathrm{EV}}$ is the digital number of the EV path.

The band-averaged radiance difference $\Delta R_{\mathrm{EV}}(i)$ can be expressed as a function of the digital number difference $\Delta \mathrm{DN}_{\mathrm{EV}}$:

$$\Delta R_{\mathrm{EV}}(i) = a_0(i) + b_1(i)\Delta \mathrm{DN}_{\mathrm{EV}} + a_2(i)(\Delta \mathrm{DN}_{\mathrm{EV}})^2 \tag{3.18}$$

then

$$\rho_{\mathrm{EV}}^{\mathrm{sm}} \overline{R_{\mathrm{EV}}} + \left(\rho_{\mathrm{SV}}^{\mathrm{sm}} - \rho_{\mathrm{EV}}^{\mathrm{sm}}\right) \overline{B(T_{\mathrm{sm}})} = a_0(i) + b_1(i)\Delta \mathrm{DN}_{\mathrm{EV}} + a_2(i)(\Delta \mathrm{DN}_{\mathrm{EV}})^2. \tag{3.19}$$

From Eq. (3.19), the band-averaged radiance from the EV $\overline{R_{\mathrm{EV}}}$ can be derived as

$$\overline{R_{\mathrm{EV}}} = \frac{1}{\rho_{\mathrm{EV}}^{\mathrm{sm}}} \left[a_0(i) + b_1(i)\Delta \mathrm{DN}_{\mathrm{EV}} + a_2(i)(\Delta \mathrm{DN}_{\mathrm{EV}})^2\right] - \frac{\rho_{\mathrm{SV}}^{\mathrm{sm}} - \rho_{\mathrm{EV}}^{\mathrm{sm}}}{\rho_{\mathrm{EV}}^{\mathrm{sm}}} \overline{B(T_{\mathrm{sm}})}. \tag{3.20}$$

3.4 Vicarious Calibration

Vicarious calibration refers to a technique that makes use of natural or artificial sites on the surface of the earth for the postlaunch calibration of sensors. It is a technique in which calibrated ground-based or airborne radiometers deployed on or above a spectrally and spatially homogeneous target take simultaneous measurements during periods of satellite instrument overpasses (Wan et al. 1999). Vicarious calibration is performed in the thermal infrared (TIR) spectral region using either a temperature-based method or a radiance-based method (Thome et al. 1998).

3.4.1 Temperature-Based Method

In the temperature-based method, at-sensor radiance is estimated using a radiative transfer model with surface temperature and emissivity measured on site, and atmospheric profiles measured by a radiosonde launched from the site, or obtained from other sources such as a numerical forecast model (Thome et al. 1998). The at-sensor radiance can be expressed as

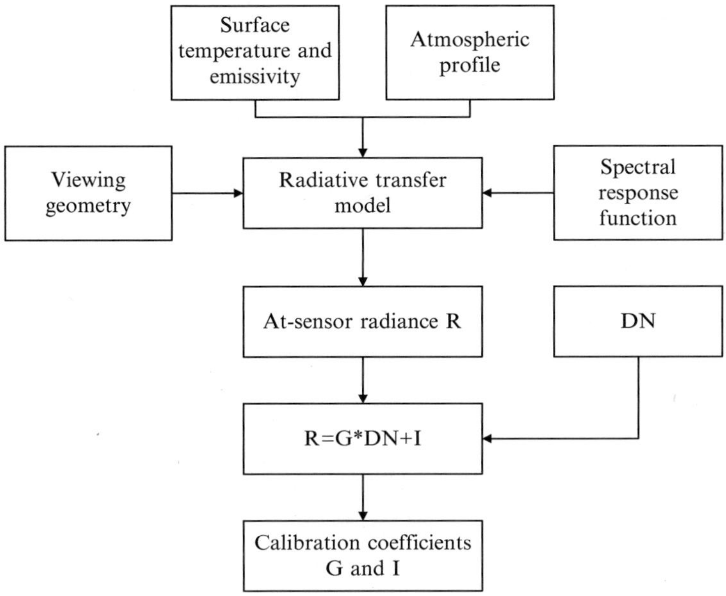

Fig. 3.4 Flowchart of the temperature-based vicarious calibration method

$$R_k = \left[\varepsilon_k B_k\left(T_s\right) + \left(1 - \varepsilon_k\right) R_{atk\downarrow}\right] \tau_k + R_{atk\uparrow} \tag{3.21}$$

where k is the band number, R is the at-sensor radiance, ε is the surface emissivity, T_s is the surface temperature, B is the Plank function, $R_{at\downarrow}$ is the atmospheric downwelling radiance, τ is the atmospheric transmittance, and $R_{at\uparrow}$ is the atmospheric upwelling radiance.

In the temperature-based method, the at sensor radiance R is derived from Eq. (3.21) with surface parameters (T_s and ε) measured on site and atmospheric parameters (τ, $R_{at\downarrow}$, and $R_{at\uparrow}$) calculated from atmospheric profiles using a radiative transfer model. The flowchart of the temperature-based method is shown in Fig. 3.4. To obtain accurate surface-leaving radiance from ground-based measurements, it needs to select suitable test sites and to rigorously calibrate the ground-based instruments before, during, and after field campaigns. Large homogeneous test sites such as silt playas and inland lakes should be chosen because their surface emissivities can be measured or modeled and their in situ surface temperatures can be measured more accurately. Because the field of view of ground-based instruments is much smaller than those of satellite sensors, the viewing angle and measurement time may not be exactly the same, in situ measurements must be made at multiple locations, multiple viewing angles, and multiple times (before and after the overpasses of satellite sensors) to reduce the uncertainties in temporal interpolation, spatial sampling and averaging, and geometric co-registration (Wan et al. 1999).

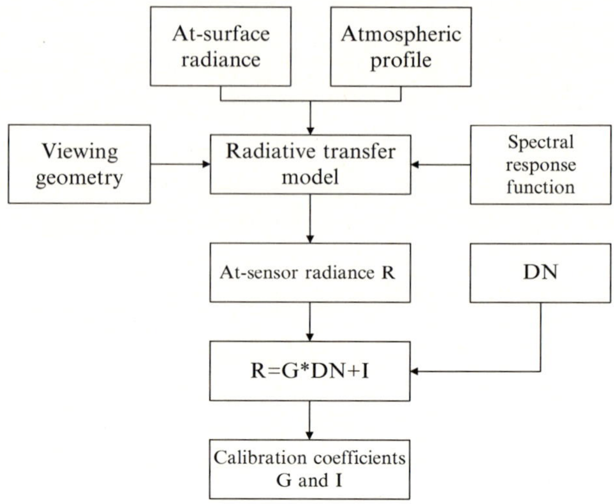

Fig. 3.5 Flowchart of the radiance-based vicarious calibration method

3.4.2 *Radiance-Based Method*

In the radiance-based method, a well-calibrated radiometer is flown above a test site. The radiometer measures the upwelling radiance from the test site at the time of satellite overpass (Slater et al. 1996). The at-sensor radiance can be expressed as

$$R_k = R_{s,k}\tau_k + R_{atk\uparrow} \tag{3.22}$$

where R_s is the at-surface radiance defined by

$$R_{s,k} = \varepsilon_k B_k (T_s) + (1 - \varepsilon_k) R_{atk\downarrow} \tag{3.23}$$

In the radiance-based method, the at-sensor radiance R is calculated from Eq. (3.22) with at-surface radiance R_s measured on site and atmospheric parameters (τ and $R_{at\uparrow}$) calculated from atmospheric profiles by using a radiative transfer model. The flowchart of the radiance-based method is shown in Fig. 3.5. The radiance-based method does not need in situ measurements of surface temperature and emissivity that are necessary in the temperature-based method. If an airborne radiometer is available, only the atmospheric effects between the plane and space need to be considered. This method is simpler than the temperature-based method and is also free from a complicated measurement of spectral emissivity, but the spectral response functions of the in situ radiometer and the sensor onboard satellite need to match (Tonooka et al. 2005).

3.5 Intercalibration

Intercalibration, or the so-called cross-calibration, is an operation that relates the measurements of a monitored satellite instrument to those of a well-calibrated reference satellite instrument in similar channels (Asem et al. 1987). The basic premise of intercalibration is that these two instruments should make identical measurements when they view the same target at the same time, with the same spatial and spectral response and the same viewing geometry. Since these idealized conditions never occur in reality, a series of thresholds are applied to collocate the data, which is then transformed to a comparable scale (Chander et al. 2013).

Intercalibration of satellite instruments is beneficial for two reasons (König et al. 1999). First, it can identify problems and increase the confidence in the operational calibration of individual satellites. Second, it can provide the basis for a normalized calibration, which is a prerequisite for the derivation of global products from different satellites.

The regression relationship between the radiances before and after intercalibration of the monitored instrument is given by

$$R_{\text{aic}} = a + bR_{\text{bic}} \tag{3.24}$$

where R_{aic} is the radiance after intercalibration (i.e., the recalibrated radiance of the monitored instrument), R_{bic} is the radiance before intercalibration (i.e., the measured radiance of the monitored instrument), and a and b are the intercalibration coefficients.

The objective of the intercalibration is to determine the radiance after intercalibration and then to obtain the intercalibration coefficients. There are three intercalibration methods: the ray-matching method, the radiative transfer modeling method, and the high spectral convolution method.

3.5.1 Ray-Matching Method

The ray-matching (RM) method simply uses coincident, co-angled, and co-located measurements to transfer the radiances of a well-calibrated reference instrument to those of a monitored instrument in a similar channel (Doelling et al. 2004). The flowchart of this method is shown in Fig. 3.6. For the RM method, R_{aic} in Eq. (3.24) is directly obtained from the measured radiance of the reference instrument. Although this method is straightforward, it does not account for the spectral differences between the monitored and reference instruments (Jiang et al. 2009; Jiang and Li 2009). Failure to account for small spectral differences may lead to different measurements (Minnis et al. 2002).

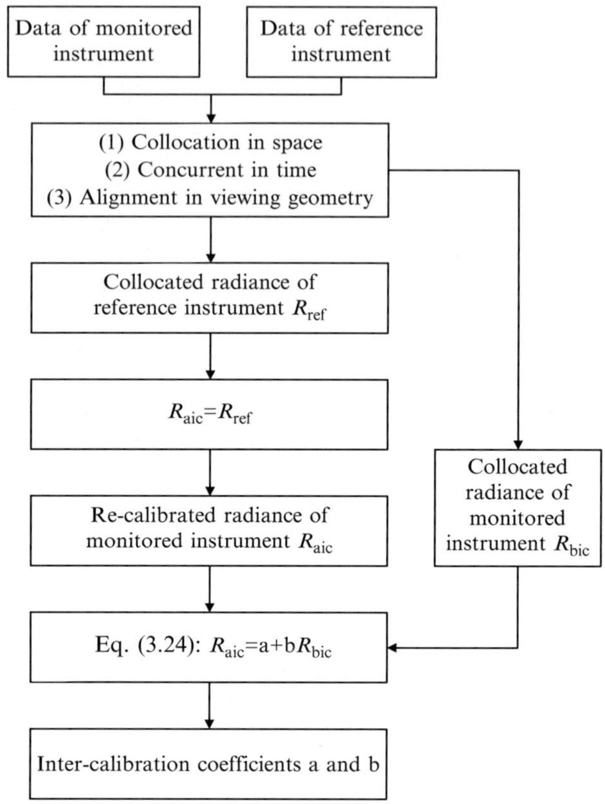

Fig. 3.6 Flowchart of the ray-matching intercalibration method

3.5.2 *Radiative Transfer Modeling Method*

The radiative transfer modeling (RTM) method is developed to take into account the spectral differences between the monitored and reference instruments based on radiative transfer calculations (Asem et al. 1987; König et al. 1999; Merchant et al. 2003). A radiative transfer model is used to simulate the radiances of the monitored and reference instruments under a variety of atmospheric and surface conditions and viewing geometry. The regression analysis is then performed between the simulated radiances of the monitored and reference instruments:

$$R_{mon} = c + dR_{ref} \tag{3.25}$$

where R_{mon} is the simulated radiance of the monitored instrument, R_{ref} is the simulated radiance of the reference instrument, and c and d are the regression coefficients.

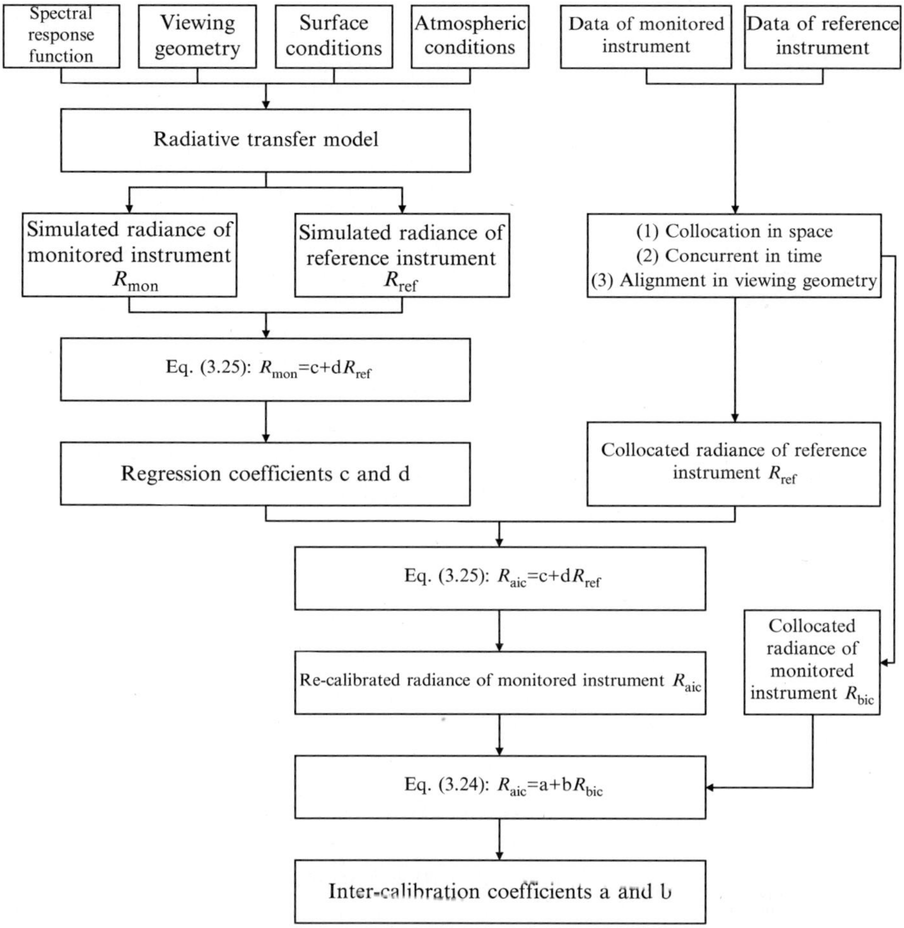

Fig. 3.7 Flowchart of the radiative transfer modeling for the intercalibration method

Once the regression coefficients are determined, R_{aic} in Eq. (3.24) can be calculated from the measured radiances of the reference instrument using Eq. (3.25). The flowchart of this method is shown in Fig. 3.7.

3.5.3 High Spectral Convolution Method

The high spectral convolution (HSC) method is developed to perform the inter-calibration of the monitored instrument using the high spectral resolution (HSR) sounders such as the AIRS/Aqua or IASI/Metop as a reference (Gunshor et al. 2004;

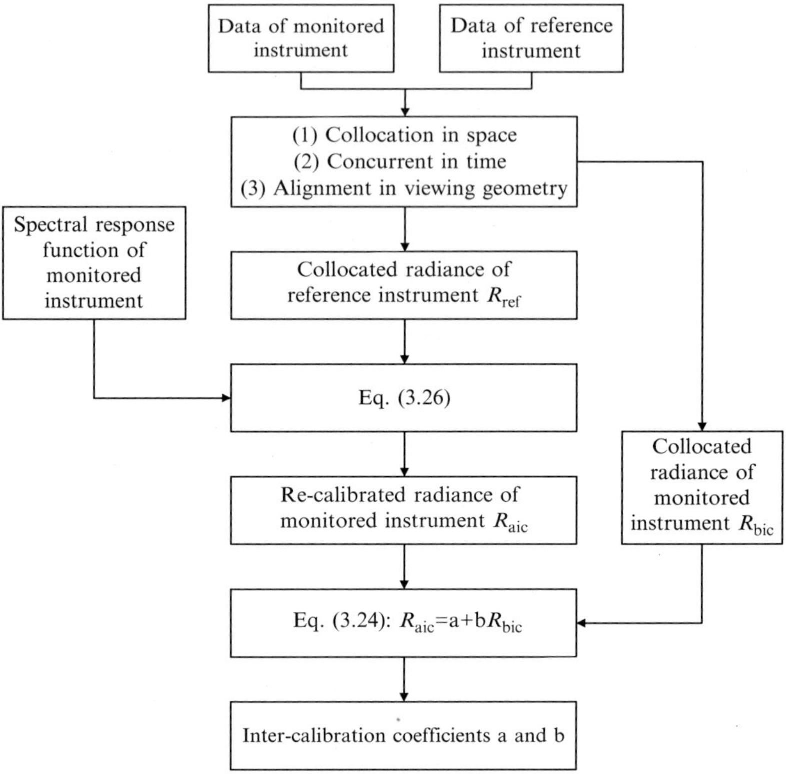

Fig. 3.8 Flowchart of the high spectral convolution intercalibration method

Tahara 2008; Jiang 2010; Shukla et al. 2012; Zhang and Gunshor 2013). Under the condition of collocation in space, concurrent in time, and alignment in viewing geometry, the radiances of the reference HSR instrument are convolved with the spectral response function of the monitored instrument (Gunshor et al. 2009):

$$R_{con} = \frac{\int_{\lambda_1}^{\lambda_2} R_{HSR}(\lambda) f(\lambda)\,d\lambda}{\int_{\lambda_1}^{\lambda_2} f(\lambda)\,d\lambda} \tag{3.26}$$

where R_{con} is the convolved radiance of the monitored instrument, R_{HSR} is the measured radiance of the reference HSR instrument, $f(\lambda)$ is the spectral response function of the monitored instrument, and λ_1 and λ_2 are the lower and upper wavelength limits of the spectral response function, respectively. For the HSC method, R_{aic} in Eq. (3.24) is obtained from the convolved radiance R_{con}. The flowchart of the HSC method is shown in Fig. 3.8.

It should be kept in mind that the HSC method is applied to the channels from the monitored instrument completely covered by the channels from the reference HSR instrument. Nevertheless, the reference HSR instrument does not cover the entire spectral range of the monitored instrument in some channels due to some spectral gaps. Two methods are proposed to compensate for the spectral gaps: the "convolution correction" method (Tobin et al. 2006) and the constrained optimization method (Tahara and Kato 2009). The HSC method has been applied by the Global Space-based Inter-Calibration System (GSICS) to perform the intercalibration of the geostationary imagers using the AIRS/Aqua or IASI/Metop as a reference (Hewison et al. 2013).

References

Asem, A., Deschamps, P. Y., & Ho, D. (1987). Calibration of METEOSAT infrared radiometer using split window channels of NOAA AVHRR. *Journal of Atmospheric and Oceanic Technology, 4*, 553–562.

Belward, A. S. (1999). International co-operation in satellite sensor calibration; the role of the CEOS working group on calibration and validation. *Advances in Space Research, 23*, 1443–1448.

Butler, J. J., & Barnes, R. A. (1998). Calibration strategy for the Earth Observing System (EOS)-AM1 platform. *IEEE Transactions on Geoscience and Remote Sensing, 36*, 1056–1061.

Chander, G., Hewison, T. J., Fox, N., Wu, X., Xiong, X., & Blackwell, W. (2013). Overview of inter-calibration of satellite instruments. *IEEE Transactions on Geoscience and Remote Sensing, 51*, 1056–1080.

Chang, T., & Xiong, X. (2011). Assessment of MODIS thermal emissive band on-orbit calibration. *IEEE Transactions on Geoscience and Remote Sensing, 49*, 2415–2425.

Doelling, D. R., Minnis, P., & Nguyen, L. (2004). Calibration comparisons between SEVIRI, MODIS and GOES data. In *Proceedings of the 2nd MSG RAO workshop*, Salzburg, Austria.

Guenther, B., Godden, G. D., Xiong, X., Knight, E. J., Qiu, S.-Y., Montgomery, H., Hopkins, M. M., Khayat, M. G., & Hao, Z. (1998). Prelaunch algorithm and data format for the Level 1 calibration products for the EOS-AM1 Moderate Resolution Imaging Spectroradiometer (MODIS). *IEEE Transactions on Geoscience and Remote Sensing, 36*, 1142–1151.

Gunshor, M. M., Schmit, T. J., & Menzel, W. P. (2004). Intercalibration of the infrared window and water vapor channels on operational geostationary environmental satellites using a single polar-orbiting satellite. *Journal of Atmospheric and Oceanic Technology, 21*, 61–68.

Gunshor, M. M., Schmit, T. J., Menzel, W. P., & Tobin, D. C. (2009). Intercalibration of broadband geostationary imagers using AIRS. *Journal of Atmospheric and Oceanic Technology, 26*, 746–758.

Hewison, T. J., Wu, X., Yu, F., Tahara, Y., & König, M. (2013). GSICS inter-calibration of infrared channels of geostationary imagers using Metop/IASI. *IEEE Transactions on Geoscience and Remote Sensing, 51*, 1056–1080.

Jiang, G.-M. (2010). Intercalibration of infrared channels of polar-orbiting IRAS/FY-3A with AIRS/Aqua data. *Optics Express, 18*, 3358–3363.

Jiang, G.-M., & Li, Z.-L. (2009). Cross-calibration of MSG1-SEVIRI infrared channels with Terra-MODIS channels. *International Journal of Remote Sensing, 30*, 753–769.

Jiang, G.-M., Yan, H., & Ma, L.-L. (2009). Intercalibration of SVISSR/FY-2C infrared channels against MODIS/Terra and AIRS/Aqua channels. *IEEE Transactions on Geoscience and Remote Sensing, 47*, 1548–1558.

König, M., Schmetz, J., & Tjemkes, S. (1999). Satellite intercalibration of IR window radiance observations. *Advances in Space Research, 23*, 1341–1348.

Liang, S. (2004). *Quantitative remote sensing of land surfaces*. Hoboken: Wiley.

Merchant, C. J., Simpson, J. J., & Harris, A. R. (2003). A cross-calibration of GMS-5 thermal channels against ATSR-2. *Remote Sensing of Environment, 84*, 268–282.

Minnis, P., Nguyen, L., Doelling, D. R., Young, D. F., Miller, W. F., & Kratz, D. P. (2002). Rapid calibration of operational and research meteorological satellite imagers, Part II: Comparison of infrared channels. *Journal of Atmospheric and Oceanic Technology, 19*, 1250–1266.

Ono, A., Sakuma, F., Arai, K., Yamaguchi, Y., Fujisada, H., Slater, P. N., Thome, K. J., Palluconi, F. D., & Kieffer, H. H. (1996). Preflight and in-flight calibration plan for ASTER. *Journal of Atmospheric and Oceanic Technology, 13*, 321–335.

Sakuma, F., & Ono, A. (1993). Radiometric calibration of the EOS ASTER instrument. *Metrologia, 30*, 231–241.

Shukla, M. V., Thapliyal, P. K., Bisht, J. H., Mankad, K. N., Pal, P. K., & Navalgund, R. R. (2012). Intersatellite calibration of Kalpana thermal infrared channel using AIRS hyperspectral observations. *IEEE Geoscience and Remote Sensing Letters, 9*, 687–689.

Slater, P. N., Biggar, S. F., Thome, K. J., Gellman, D. I., & Spyak, P. R. (1996). Vicarious radiometric calibrations of EOS sensors. *Journal of Atmospheric and Oceanic Technology, 13*, 349–359.

Smith, D. L., Delderfield, J., Drummond, D., Edwards, T., Mutlow, C. T., Read, P. D., & Toplis, G. M. (2001). Calibration of the AATSR instrument. *Advances in Space Research, 28*, 31–39.

Smith, D., Mutlow, C., Delderfield, J., Watkins, B., & Mason, G. (2012). ATSR infrared radiometric calibration and in-orbit performance. *Remote Sensing of Environment, 116*, 4–16.

Tahara, Y. (2008). New approach to inter-calibration using high spectral resolution sounder. *Meteorological Satellite Center Technical Note, 50*, 1–14.

Tahara, Y., & Kato, K. (2009). New spectral compensation method for inter-calibration using high spectral resolution sounder. *Meteorological Satellite Center Technical Note, 52*, 1–37.

Thome, K., Markham, B., Barker, J., Slater, P., & Blggar, S. (1997). Radiometric calibration of Landsat. *Photogrammetric Engineering & Remote Sensing, 63*, 853–858.

Thome, K., Arai, K., Hook, S., Kieffer, H., Lang, H., Matsunaga, T., Ono, A., Palluconi, F., Sakuma, H., Slater, P., Takashima, T., Tonooka, H., Tsuchida, S., Welch, R., & Zalewski, E. (1998). ASTER preflight and inflight calibration and the validation of level 2 products. *IEEE Transactions on Geoscience and Remote Sensing, 36*, 1161–1172.

Tobin, D. C., Revercomb, H. E., Moeller, C. C., & Pagano, T. S. (2006). Use of AIRS high spectral resolution infrared spectra to assess the calibration of MODIS on EOS Aqua. *Journal of Geophysical Research, 111*, D09S05. doi:10.1029/2005JD006095.

Tonooka, H., Palluconi, F., Hook, S., & Matsunaga, T. (2005). Vicarious calibration of ASTER thermal infrared bands. *IEEE Transactions on Geoscience and Remote Sensing, 43*, 2733–2746.

Wan, Z., Zhang, Y., Ma, X., King, M. D., Myers, J. S., & Li, X. (1999). Vicarious calibration of the Moderate-Resolution Imaging Spectroradiometer Airborne Simulator thermal-infrared channels. *Applied Optics, 38*, 6294–6306.

Xiong, X., & Barnes, W. (2006a). An overview of MODIS radiometric calibration and characterization. *Advances in Atmospheric Sciences, 23*, 69–79.

Xiong, X., & Barnes, W. (2006b). MODIS calibration and characterization. In J. J. Qu, W. Gao, M. Kafatos, R. E. Murphy, & V. V. Salomonson (Eds.), *Earth science satellite remote sensing* (Data, computational processing, and tools, Vol. 2, pp. 77–97). Beijing/New York: Tsinghua University Press/Springer.

Xiong, X., Chiang, K., Esposito, J., Guenther, B., & Barnes, W. (2003). MODIS on-orbit calibration and characterization. *Metrologia, 40*, S89–S92.

Xiong, X., Toller, G., Chiang, V., Sun, J., Esposito, J., & Barnes, W. (2005). *MODIS Level 1B algorithm theoretical basis document (Version 3)*. Greenbelt: Goddard Space Flight Center.

Xiong, X., Chiang, K.-F., Wu, A., Barnes, W., Guenther, B., & Salomonson, V. V. (2008). Multiyear on-orbit calibration and performance of Terra MODIS thermal emissive bands. *IEEE Transactions on Geoscience and Remote Sensing, 46*, 1790–1803.

Xiong, X., Wenny, B. N., Wu, A., Barnes, W., & Salomonson, V. V. (2009). Aqua MODIS thermal emissive band on-orbit calibration, characterization, and performance. *IEEE Transactions on Geoscience and Remote Sensing, 47*, 803–814.
Zhang, Y., & Gunshor, M. M. (2013). Intercalibration of FY-2C/D/E infrared channels using AIRS. *IEEE Transactions on Geoscience and Remote Sensing, 51*, 1231–1244.

Chapter 4
Retrieval of Land Surface Emissivity from Remotely Sensed Data

As an intrinsic property of natural materials, land surface emissivity (LSE) is often regarded as an indicator of material composition, especially for the silicate minerals (Sobrino et al. 2001, 2005). The importance of LSE is reflected not only in studies of soil development and erosion, bedrock mapping, and resource exploration (Gillespie et al. 1998) but also in the accurate estimates of the surface energy budgets (Jin and Liang 2006). Therefore, LSE has become one of the key input parameters in climatic, hydrological, ecological, and biogeochemical models.

This chapter inherits and updates the overview of a variety of remote sensing retrieval methods of LSE from space given by Li et al. (2013). The objective of this chapter is to give some theoretical background about LSE and to recall various retrieval methods of LSE. The definition and angular variation of emissivity, which is the basis of retrieving emissivity from space, are first provided. Subsequently, the algorithms used to retrieve LSE from empirical methods to physics-based methods are illustrated. The comparison and analysis of different methods are also given to assist the selection of methods in various circumstances. The validation method of satellite-derived LSE is finally proposed to verify the accuracy of the retrieved LSE.

4.1 Definition of LSE

The definition of spectral emissivity is given in Chap. 2. This spectral definition comes from the homogeneous isothermal surfaces. However, natural surfaces observed from space are usually heterogeneous, especially in the situation of low spatial resolution. It is difficult to find a "pure" and "isothermal" pixel in reality in this situation. Therefore, this definition should be revised for heterogeneous surfaces to meet requirements of retrieval land surface temperature (LST) from space. Several definitions are proposed to take the difficulties in relating the radiance of an ensemble of natural media at different temperatures to a blackbody distribution at the same effective temperature into account (Becker and Li 1995; Norman and Becker 1995; Li et al. 1999a, 2013).

H. Tang and Z.-L. Li, *Quantitative Remote Sensing in Thermal Infrared: Theory and Applications*, Springer Remote Sensing/Photogrammetry,
DOI 10.1007/978-3-642-42027-6_4, © Springer-Verlag Berlin Heidelberg 2014

4.1.1 r-Emissivity

According to the Kirchhoff's law, Becker and Li (1995) and Chen et al. (2004) defined the *r-emissivity* $\varepsilon_r(\theta,\varphi)$ along the viewing zenith θ and azimuth angle φ from the hemispherical–directional reflectivity (r_h), namely,

$$\varepsilon_r(\theta,\varphi) = 1 - r_\mathrm{h}(\theta,\varphi) \tag{4.1}$$

Because most of the common terrestrial surfaces follow the Kirchhoff's law (Salisbury et al. 1994), this definition sounds applicable even for heterogeneous and nonisothermal flat surfaces or rough systems.

For a flat pixel composed of N homogeneous subelements, the *r-emissivity* can be simply expressed as the area weighting of component emissivity, which has no relation with component temperature,

$$\varepsilon_r(\theta,\varphi) = \sum_{k=1}^{N} a_k \varepsilon_k(\theta,\varphi) \tag{4.2}$$

where α_k is the relative area of a subelement k where the sum of all α_k is unity. $\varepsilon_k(\theta,\varphi)$ is the emissivity for each subelement k. Becker and Li (1995) attributed this *r-emissivity* as the effective emissivity for a pixel and argued this *r-emissivity* can be considered as scale invariance. The main advantage of *r-emissivity* is its measurability from space and its scale invariability. However, this type of definition makes the definition of LST wavelength- and viewing-angle-dependent and also dependent on the distributions of surface temperature and emissivity within a pixel. More discussions about this feature can be found in Sect. 5.1.

4.1.2 e-Emissivity

e-Emissivity is defined as the ratio of the total radiation of a natural object surface to the blackbody radiation by assuming that there is an identical temperature distribution with nonisothermal pixels, namely,

$$\varepsilon_e(\theta,\varphi) = \frac{\sum_{k=1}^{N} a_k \varepsilon_k(\theta,\varphi) B_i(T_k)}{\sum_{k=1}^{N} a_k B_i(T_k)} \tag{4.3}$$

where there are N components in one pixel, and $B_i(T_k)$ is the radiance of a blackbody as a function of the temperature of subelement k. Obviously, the *e-emissivity* depends on the temperature distribution and the characteristics of the subelements.

4.1.3 Apparent Emissivity

Li et al. (1999a) pointed out that nonisothermal gray surfaces do not behave as gray of surfaces. To handle this problem, apparent emissivity was proposed to keep the Planck's function unchanged. This apparent emissivity is defined by adding an apparent emissivity increment caused by the nonisothermal surface into the *r-emissivity* (defined for flat surface) to make the definition of LST independent of viewing angle and wavelength:

$$\varepsilon_{\mathrm{app}}(\theta, \varphi) = \sum_{k=1}^{N} a_k \varepsilon_k (\theta, \varphi) + K_\lambda (T_0) \sum_{k=1}^{N} a_k \varepsilon_k (\theta, \varphi) \Delta T_k \tag{4.4}$$

where T_0 is a reference temperature that is independent of wavelength and viewing direction; ΔT_k is the temperature difference between the temperature of the subelement k and the reference temperature T_0; and $K_\lambda(T_0) = B'_\lambda(T_0)/B_\lambda(T_0)$, in which $B'_\lambda(T_0)$ is the first derivation of the Planck function B_λ at temperature T_0. Although LST (T_0) induced by the apparent emissivity is independent of the viewing angle and wavelength, the apparent emissivity itself is dependent on both the component temperature difference ΔT_k and the reference temperature T_0 and can be larger than unity if T_0 is underestimated. Furthermore, it is immeasurable from space because the reference temperature and the temperature of subelements within a pixel must be known a priori.

One should note that all definitions of emissivity are ambiguous because both the effective temperature and effective emissivity are defined only from one ensemble radiance emitted by a heterogeneous flat or rough nonisothermal surface. In fact, there are many possibilities for defining two parameters (LST and ε) from only one equation (ensemble radiance); the definition makes sense only if the defined parameters are measurable from space. Considering this criterion, *r-emissivity* defined by Eqs. (4.1) and (4.2) for both heterogeneous flat and rough nonisothermal surface is recommended for the LST and emissivity retrievals from space measurements (Li et al. 2013).

4.2 Characteristics of Emissivity

4.2.1 Angular Variation of Emissivity

Many efforts have been devoted to analyzing the directional emissivity of common terrestrial surfaces, such as soil, water, and leaves (Labed and Stoll 1991; Sobrino and Cuenca 1999; Cuenca and Sobrino 2004). Angular variation of emissivity has been observed in both the field and the laboratory (Becker et al. 1985; Labed and Stoll 1991) and results primarily from the angular variation of the pixel emissivity

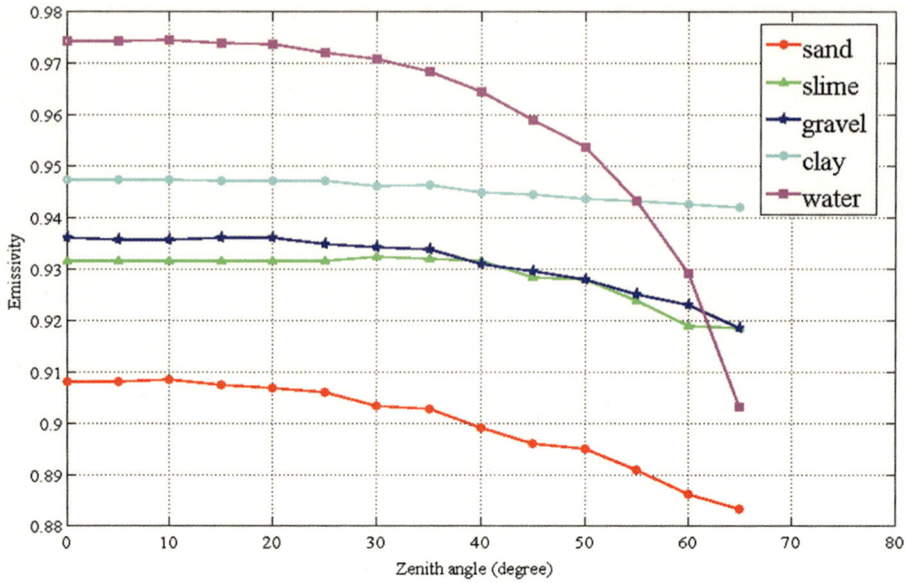

Fig. 4.1 Angular emissivity variation measured in laboratory for several natural surfaces

(inherent anisotropic emissivity of terrestrial materials) and the relative weights of different components (e.g., vegetation and background soil) with different emissivities in a three-dimensional (3D) land targets.

Commonly, the decrease of emissivity of natural surfaces is observed with increasing view zenith angle in the TIR (8–14 μm) band (Sobrino and Cuenca 1999; Cuenca and Sobrino 2004; Hori et al. 2006). Experimental results show that different land surface materials have different angular variations in emissivity; for example, arid bare soils and water generally show the highest angular dependence, as depicted in Fig. 4.1 (Snyder et al. 1998; Sobrino and Cuenca 1999; Cuenca and Sobrino 2004).

As shown in Fig. 4.1, the angular variation of emissivity may be negligible for most of the surfaces especially when the view zenith angles are low (Prata 1993). However, the terrain may exaggerate the impact of view angles on the observed emissivity due to the effect of topography (Wan 1999).

It is worth noting that the structure of surface, which may cause the change of the viewed proportions of the components, can result in a surprisingly large angular dependence of the emissivity even with Lambertian components (Snyder and Wan 1998). For example, Sobrino et al. (2005) reported that directional canopy emissivity may show different angular variations for different proportions of vegetation (Fig. 4.2).

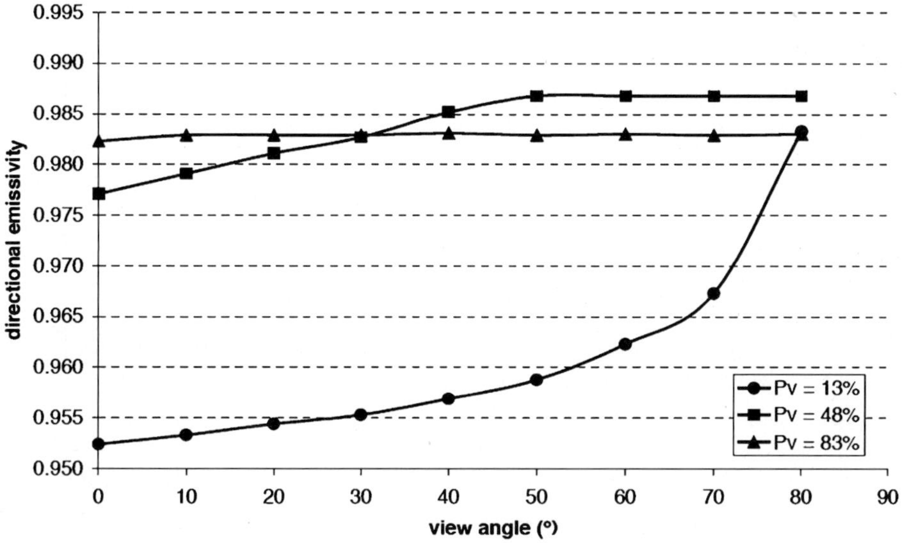

Fig. 4.2 Directional emissivity obtained with the geometric model for soil emissivity 0.94 and vegetation emissivity 0.98 at different proportions of vegetation (Pv) at nadir view (13, 48, and 83 %) (Adapted from Sobrino et al. (2005), with permission from Elsevier)

4.2.2 Spectral Variation of Emissivity

The main materials of the terrestrial ecosystem, such as rocks, soils, vegetation, water, and snow/ice, would show different spectral features. Understanding this spectral variation of emissivity in the infrared domain (3–14 μm) is important for the application of various models, for LST determination, and for the discrimination of land cover types.

Figure 4.3 shows the spectral variation of mean and standard deviation of the emissivity for a number of rocks computed from the Advanced Spaceborne Thermal Emission Reflection Radiometer (ASTER) spectral library (http://speclib.jpl.nasa.gov/) in 3–14 μm.

Three factors—the structures of the mineral molecules, the force constants between the atoms, and the long-range order of the crystal lattices—contribute to the spectral behavior of rocks (Farmer 1974; Salisbury and D'Aria 1992). The characteristics of spectral emissivity of rocks in the thermal infrared region (TIR) are mainly impacted by the aggregate Si–O stretching vibration bands (reststrahlen bands), which include strong asymmetric-stretching fundamentals between 8 and 10 μm and a weaker symmetric-stretching fundamental between 12.2 and 13 μm. All of those reststrahlen bands of quartz and carbonate can be observed even in the average spectrum. In the mid-infrared region (MIR), the shape of the

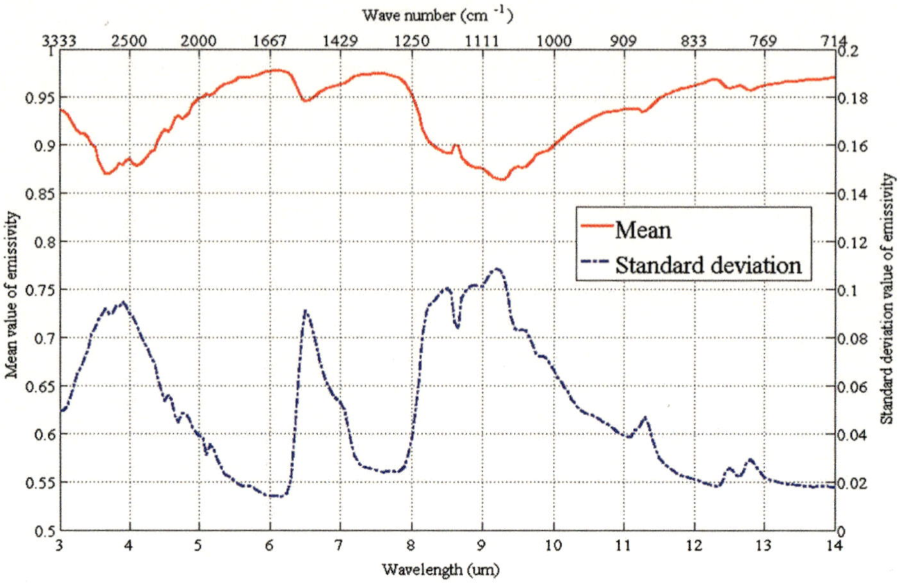

Fig. 4.3 General properties (mean and standard deviation) of the emissivity spectra for rocks in the ASTER spectral emissivity database in 3–14 μm

emissivity spectrum is mainly dominated by quartz, carbonate, tremolite, etc., and characterized by high emissivity variability with low mean and high standard deviation. A more detailed description of the spectral behavior of rocks can be found in Salisbury and D'Aria (1992).

Figure 4.4 shows the spectral variation of mean and standard deviation of the emissivity for a number of soils computed from the ASTER spectral library in 3–14 μm.

Generally, soil composition is often dominated by quartz, which is both a common mineral and resistant to weathering. Thus, there are strong reststrahlen bands between 8 and 10 μm and weaker bands between 12.2 and 13 μm and between 4.5 and 4.7 μm in soil emissivity spectra. In addition, the carbonate bands are prominent in the MIR region, especially the strongest carbonate doublet near 4.0 μm. Similar to the emissivity spectra of rock, the reststrahlen bands of quartz and carbonate can also be clearly observed even in the averaged spectrum. A more detailed description of the spectral behavior of soils can also be found in Salisbury and D'Aria (1992).

It is interesting to find the following phenomena for rocks and soils from Figs. 4.3 and 4.4: (1) The variation range is larger both over 3–5 μm and 8–10 μm regions than that in other wavelength regions. (2) For a given wavelength, the larger the variation of emissivity is, the smaller the average emissivity is. (3) The longer the wavelength is, the smaller the spectral variation of emissivity is.

Figure 4.5 shows four emissivity spectra of vegetation from the ASTER spectral library in 3–14 μm.

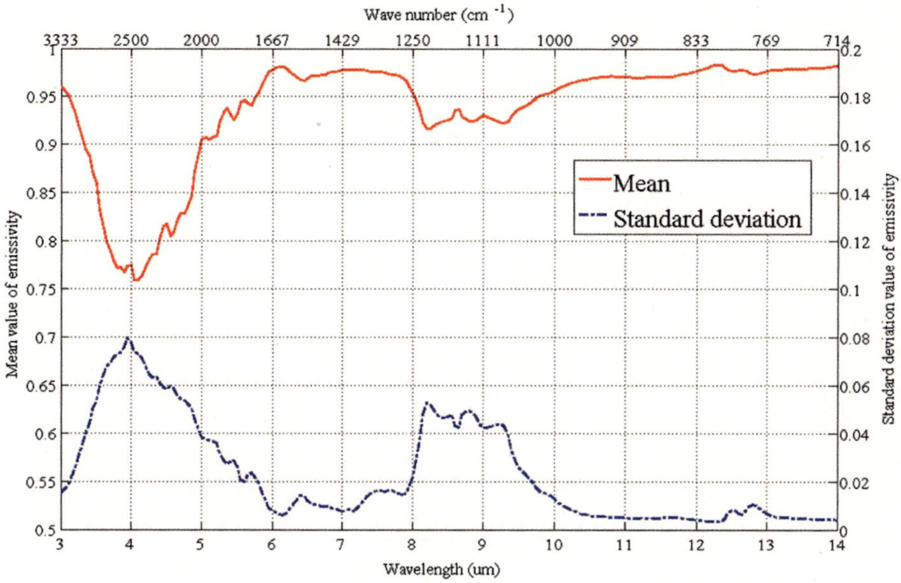

Fig. 4.4 General properties (mean and standard deviation) of the emissivity spectra for soils in the ASTER spectral emissivity database in 3–14 μm

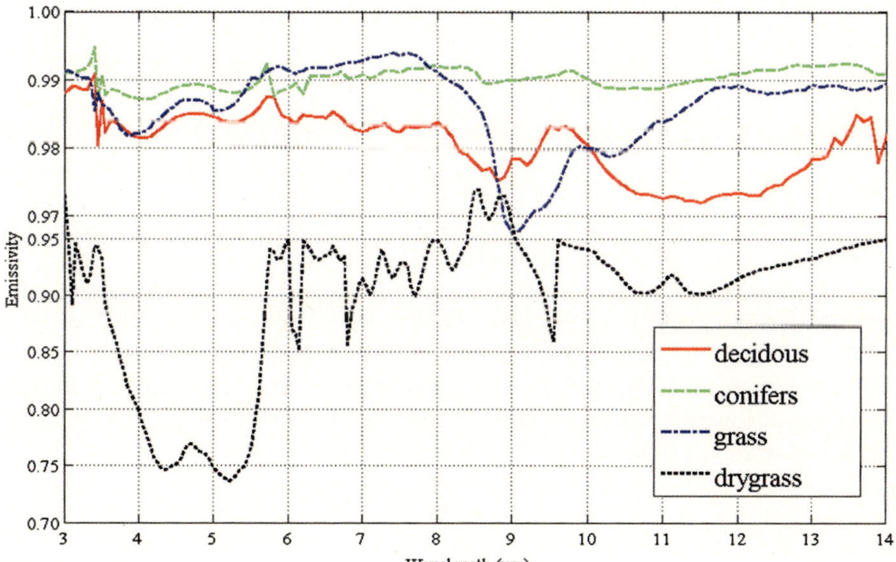

Fig. 4.5 Emissivity spectrum for four types of vegetation

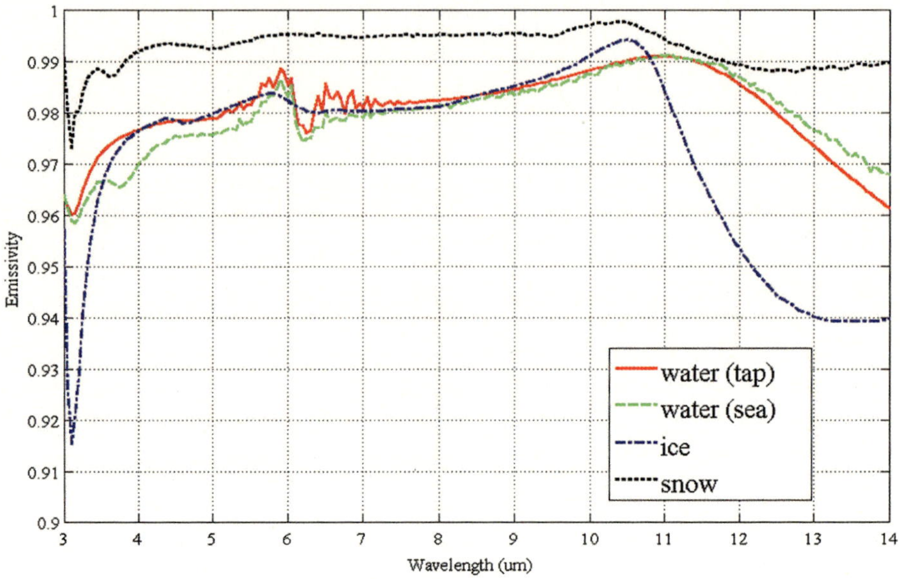

Fig. 4.6 Emissivity spectrum for water, ice, and snow

As shown above, green foliage exhibits low spectral contrast in both MIR and TIR regions, especially for conifers, which have a nearly uniform high emissivity, except for a weak reststrahlen trough near 3.43 and 3.51 μm associated with the H–C vibration bands. As for the dry grass and some senescent foliages, the emissivity is evidently reduced. Their cellulose gives the spectra double reflectance peaks between 10 and 12 μm and two minor absorption features between 8.5 and 9 μm. The emissivity peak near 3.4–3.5 μm is caused by the volume scattering. Ribeiro da Luz and Crowley (2007) found that some useful spectral information associated with leaf chemical constituents and structural aspects may be detectable from these spectral behaviors of emissivity. However, it still remains to be a major technical challenge to remotely measure the subtle emissivity features of leaves.

Figure 4.6 shows several emissivity spectra of water, snow, and ice from the ASTER spectral library in 3–14 μm.

As shown in Fig. 4.6, although water is often assumed to have an emissivity of 1.0, it departs from blackbody behavior at 11.2 μm in the TIR region. As for ice, the shape of the emissivity spectra is dominated by the volume scattering of the surface roughness in the MIR and TIR regions. Hori et al. (2006) found that the derived emissivities of snow and ice show a distinct spectral contrast at wavelengths between 10.5 and 12.5 μm, which suggests the possibility of discriminating between snow and ice from space. Different from rocks, soils, and vegetation, the emissivity spectrum of water and ice behaves as the decreasing of emissivity with increasing of wavelength.

4.3 Methodologies for Land Surface Emissivity Retrieval

Radiance measured from space can cover large spectral range, from the visible/near-infrared (VNIR), MIR, and TIR to microwave (MW). This radiance containing the combined effects of surface and atmosphere can be used to infer the LSE. The emissivity of land, unlike that of oceans, can differ significantly from unity and can vary with vegetation, surface moisture, roughness, and viewing angle (Salisbury and D'Aria 1992). Therefore, the LSEs measured in the laboratory cannot be arbitrarily used at the pixel scale. To date, various methods have been proposed to infer the LSE from space. Several use the statistical relationships between the measurements and the emissivities; others use reasonable assumptions or constraints on the basis of Planck's function and the atmospheric radiative transfer equation (RTE) to solve the undetermined problem or the ill-posed inversion process. According to the way in which the LSE is determined with LST, the emissivity retrieval methods from optical remotely sensed data can be roughly categorized into three distinct types. The first is a stepwise retrieval method that determines the LSE and the LST separately. The LSE is estimated first, and then the LST is retrieved. The second is a simultaneous retrieval method with known atmospheric information that treats both the LST and the LSE as unknowns and resolves both of them from the atmospherically corrected radiances or with approximated atmospheric profiles. The third is a further development of the simultaneous retrieval method that simultaneously retrieves the atmospheric profiles (or atmospheric quantities in the RTE) with the LST and LSE.

4.3.1 Stepwise Retrieval Method

This type of method retrieves LST using two consecutive steps. First, the LSE is (semi-)empirically determined from visible/near-infrared (VNIR) measurements or physically estimated from pairs of atmospherically corrected MIR and TIR radiances at ground level. Then, the LST is determined with this estimated LSE. Representative methods include the classification-based emissivity method (Snyder et al. 1998; Sun and Pinker 2003; Peres and DaCamara 2005), the normalized difference vegetation index (NDVI)-based emissivity method (Van de Griend and Owe 1993; Valor and Caselles 1996; Sobrino and Raissouni 2000), and the day/night temperature-independent spectral indices (TISI-based) method (Becker and Li 1990; Li and Becker 1993; Li et al. 2000).

4.3.1.1 Classification-Based Emissivity Method (CBEM)

Generally, the CBEM is based on the use of conventional land cover classification information. This method assumes that similarly classified land covers exhibit very

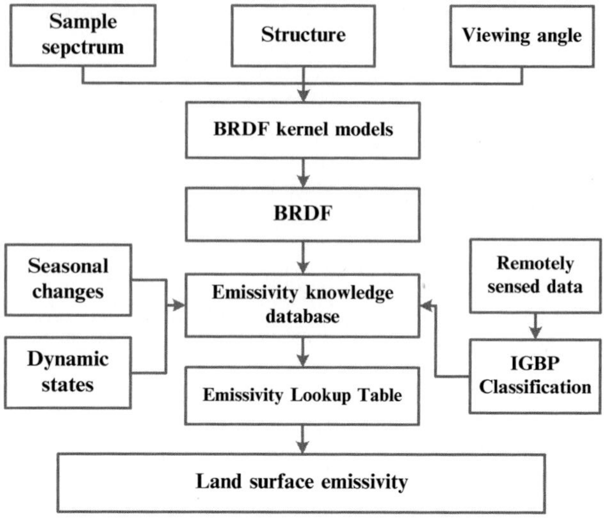

Fig. 4.7 Flowchart of LSE estimation using the classification-based emissivity method

similar LSEs. The key point of this method is to properly classify the land surface and then to assign the emissivity from classification-based look-up tables.

Snyder et al. (1998) first proposed this classification-based emissivity method by elaborately developing an emissivity knowledge database (10.8–12.3 μm), which is created by using three different BRDF kernel models with land cover-dependent spectral coefficients and structural parameters (Li and Strahler 1992; Roujean et al. 1992; Snyder and Wan 1998). The hemispherical–directional reflectance and the emissivity are then obtained via integration of BRDF over an angle range based on the Kirchhoff's law. After a series of combinatorial analysis, several emissivity classes are combined or separated from MODIS/IGBP (International Geosphere-Biosphere Programme) classes to make a good balance between the number of classes and the emissivity estimation accuracy. The analysis of the mean value and variations of emissivity for different classes shows that the emissivity for almost 70 % of the globe land surfaces can be determined with sufficient accuracy about 0.01 (Snyder et al. 1998). Once the emissivity knowledge database is developed, the LSE can be estimated directly from the MODIS/IGBP class with consideration of seasonal and dynamic states (Snyder et al. 1998). The flowchart of this type of method is depicted in Fig. 4.7.

Obviously, the mixture of several surface types within one pixel may influence the accuracy of LSE retrieval. In this case, the LSE can be estimated using a linear mixing model, as shown in Eq. (4.2). This method has already been applied to geostationary satellite data, such as METEOSAT Second Generation-1 (MSG) data (Peres and DaCamara 2005; Trigo et al. 2008) and Geostationary Operational Environmental Satellite (GOES) data (Sun and Pinker 2003).

Theoretically, the CBEM can produce accurate LSE products over the areas in which land surfaces are accurately classified and where each class has well-known emissivities (Gillespie et al. 1999). For example, the emissivities of water or of closed-canopy vegetation may be assigned reliably. In other words, the classification-based emissivity prediction is thought to be accurate for most classes, especially for high-resolution remotely sensed data (pure pixel). However, Snyder et al. (1998) pointed out several major difficulties in using the CBEM, such as the determination of surface wetness, the identification of senescent vegetation, and the uncertainty of the dynamics of snow and ice surface states. All of these difficulties and uncertainties may prevent an accurate estimation of the LSE for some of the classes. Because the existence of water can increase the LSE of any nearby material, night dew may become the primary source of error for low-emissivity classes. Furthermore, there likely exists a large variation in the emissivity within some of the classes, and the classification based on VNIR data is generally not well correlated to the LSE in the TIR region. Thus, the CBEM method may be less reliable and can produce large errors for these classes. For instance, the estimation of the LSE using the CBEM for geologic substrates is uncertain because the VNIR reflectances used to classify the land surfaces respond mainly to OH^- and Fe oxides, while the emissivities in TIR band are mainly responsive to the Si–O bond (Gillespie et al. 1999). In addition, LSE maps would have inappropriate discontinuities or appear seamed or contoured because of the discontinuities of the classification (Gillespie et al. 1999).

4.3.1.2 NDVI-Based Emissivity Method (NBEM)

This method is based on a statistical relationship between the NDVI derived from the VNIR bands and the LSE in the TIR channels. Van de Griend and Owe (1993) first reported a very high correlation between the LSE in the band covering from 8 to 14 μm and the logarithmic NDVI, that is,

$$\varepsilon = a + b \ln(\mathrm{NDVI}) \tag{4.5}$$

where *a* and *b* are the constants derived from the regression analysis. Although this method is a potentially powerful tool to estimate the LSE at a pixel scale directly from space because NDVI can be easily derived from the reflectances in VNIR bands, this type of relationship has been proven to be quite dependent on the area studied, which means the coefficients *a* and *b* obtained for one site cannot be applied to other sites (Van de Griend and Owe 1993).

Subsequently, Valor and Caselles (1996) proposed a theoretical method with NDVI to describe the "cavity effect" of emissivity and to estimate the effective LSE for a row-distributed rough system. In their method, the effective LSE can be numerically expressed as the sum of the area-weighted emissivities of each subcomponent and the cavity effect. However, this method requires a priori knowledge of some geometrical parameters of the vegetation, such as height, width, and the separation between rows, which limit the application of the method (Jimenez-Munoz et al. 2006). Considering the complexity of method proposed by Valor and Caselles (1996), Sobrino and Raissouni (2000) reduced the complexity and formulated an operational NDVI threshold method to derive the LSE from space. This method assumes that (1) the surface is only composed of soil and vegetation, (2) the emissivity of the bare soil can be linearly represented by the surface reflectivity in the red channel, and (3) the LSE changes linearly with respect to the fraction of vegetation in a pixel. Therefore, the LSE of TIR channel i can be estimated using three linear functions corresponding to conditions in which a pixel is composed of full vegetation, of full soil, or of mixed soil/vegetation content, namely,

$$\varepsilon_i = \begin{cases} a_i + b_i \rho_{\text{red}} & \text{NDVI} < \text{NDVI}_\text{s} \\ \varepsilon_{\text{v}i} P_\text{v} + \varepsilon_{\text{s}i} (1 - P_\text{v}) + \text{d}\varepsilon_i & \text{NDVI}_\text{s} < \text{NDVI} < \text{NDVI}_\text{v} \\ \varepsilon_{\text{v}i} + \text{d}\varepsilon_i & \text{NDVI} > \text{NDVI}_\text{v} \end{cases} \tag{4.6}$$

where a_i and b_i are the channel-dependent regression coefficients, ρ_{red} is the reflectivity of the red channel, and NDVI_s is the NDVI corresponding to the bare soil. $\varepsilon_{\text{v}i}$ and $\varepsilon_{\text{s}i}$ are the vegetation and soil emissivities in channel i, respectively. Both of them can be measured in the field (Rubio et al. 1997, 2003) or obtained from an emissivity database (Baldridge et al. 2009). P_v is the fraction of vegetation that can be derived either from the NDVI (Valor and Caselles 1996; Carlson and Ripley 1997; Sobrino and Raissouni 2000) or from the variable atmospherically resistant index ($\text{VARI}_{\text{green}}$) and Spectral Mixture Analysis (SMA) techniques (Sobrino et al. 2008). $\text{d}\varepsilon_i$ is the mean cavity effect and can take values of 0.02 and higher from numerical simulation (Valor and Caselles 1996), and NDVI_v is the full vegetation NDVI. NDVI_s and NDVI_v can be estimated from the histogram for the entire scene (Dash et al. 2005; Sobrino et al. 2008).

When $\text{NDVI} < \text{NDVI}_\text{s}$, the relationship between the emissivities and the red reflectivities is assumed to be linear and the coefficients can be determined from laboratory measurements of the soil spectra. However, Dash et al. (2005) argued that the correlation of this linear relationship in remotely sensed data is not strong enough. Subsequently, Sobrino et al. (2008) proposed the use of multispectral VNIR capabilities to improve this relationship. Figure 4.8 shows the relationship between the emissivities and multispectral reflectivities in MODIS 31 and 32 channels. The root-mean-square errors (RMSEs) between simulated and estimated emissivities are 0.003 for these two channels by using the emissivity spectra of soils extracted from the ASTER spectral library.

When $\text{NDVI}_\text{s} < \text{NDVI} < \text{NDVI}_\text{v}$, the mean cavity effect $\text{d}\varepsilon$ can be expressed as a linear function of P_v (Sobrino and Raissouni 2000). When $\text{NDVI} > \text{NDVI}_\text{v}$, typical constant values of $\varepsilon_{\text{v}i} = 0.985$ and $\text{d}\varepsilon_i = 0.005$ are used in TIR bands (Sobrino et al. 2008).

Because of its simplicity, this method has already been applied to various sensors with access to VNIR data (Sobrino and Raissouni 2000; Sobrino et al. 2002, 2003, 2004, 2008; Momeni and Saradjian 2007). Because only VNIR bands, which always have higher spatial resolution than TIR bands, are required, the emissivity with a finer resolution can be mapped (Sobrino et al. 2008). Moreover, an accurate atmospheric correction is not needed when estimating P_v. However, the

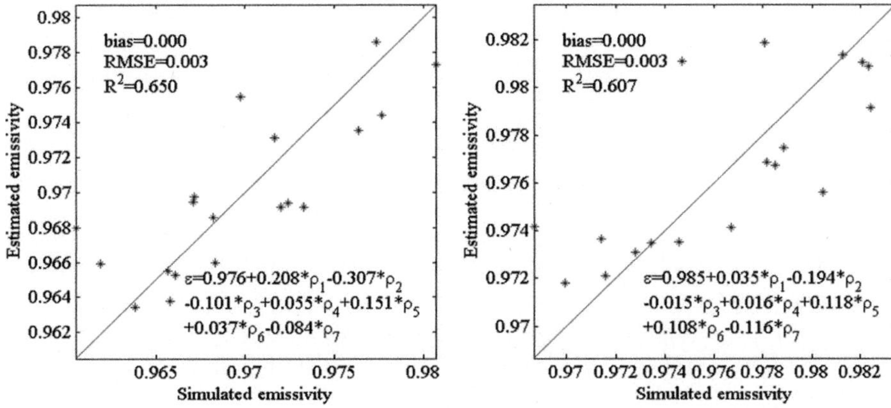

Fig. 4.8 Illustration of the relationship between the emissivities and multispectral reflectivities in MODIS 31 (*left*) and 32 (*right*) channels. ρ_i $(i = 1,\ldots,7)$ is the channel reflectivity in channel i

main problem with this method is the lack of continuity for emissivity values at $\text{NDVI} = \text{NDVI}_s$ and $\text{NDVI} = \text{NDVI}_v$ because they are calculated using different functions (Sobrino et al. 2008). From numerical analysis, Sobrino et al. (2008) pointed out that the NDVI threshold method can only provide acceptable results in the 10–12 μm interval bands because the relationship between the emissivity and reflectivity for bare soil samples does not provide satisfactory results in the 8–9.5 μm domains for some soil types. In addition, this method may hold well only for soil and vegetation mixing areas, except for senescent vegetation, and is not applicable for surfaces like water, ice, snow, and rocks (Sobrino et al. 2008). Furthermore, it requires a priori knowledge of the emissivities of soil and vegetation (Sobrino and Raissouni 2000). The determination of the soil emissivity may be the main source of error in this method (Jimenez-Munoz et al. 2006).

4.3.1.3 Day/Night Temperature-Independent Spectral Indices (TISI-Based) Method

Becker and Li (1990) and Li and Becker (1990) first proposed a day/night TISI-based method to retrieve LSE from TIR and MIR data. This method is based on the power–law approximation of Planck's law $B_i(T)$:

$$B_i(T) \approx \alpha_i T^{n_i} \tag{4.7}$$

where α_i and n_i are the channel-specific constants for reasonable variation of temperature T, and the use of the sun as an active source by taking into account the fact that in MIR channel (around 3.7 μm), the radiance emitted by the land surface itself and the reflected radiance due to sun irradiation during the day are of the same order of magnitude if the surface reflectance in this channel is about 0.1 and there is no solar reflection at all at night. Using this power–law approximation and assuming that the atmospheric corrections have been accurately performed, the TIR channel radiance R_i observed at ground level can be approximated as

$$R_i = \alpha_i T_{gi}^{n_i} = \varepsilon_i \alpha_i T_s^{n_i} + (1 - \varepsilon_i) R_{ati\downarrow} = \varepsilon_i \alpha_i T_s^{n_i} C_i \tag{4.8}$$

where T_{gi} is the brightness temperature measured at ground level in channel i, ε_i is the channel emissivity, T_s is the surface temperature, $R_{ati\downarrow}$ is the downwelling atmospheric radiance in channel i, and C_i is a correction factor that compensates for the effect of the atmospheric reflected radiance (Nerry et al. 1998).

By taking the product of N channel measurements described in Eq. (4.8) to the power d_k ($k = 1, \ldots, N$) such that

$$\sum_{k=1}^{N} d_k n_k = 0 \tag{4.9}$$

the surface temperature T_s is eliminated in the product.

Defining TISI as

$$\mathrm{TISI} = \prod_{k=1}^{N} C_k^{-d_k} \times \prod_{k=1}^{N} T_{gk}^{d_k n_k} = \prod_{k=1}^{N} (C_k \alpha_k)^{-d_k} \times \prod_{k=1}^{N} R_k^{d_k} = \prod_{k=1}^{N} \varepsilon_k^{d_k} \tag{4.10}$$

Becker and Li (1990), Li and Becker (1990), and Li et al. (1999b) demonstrated that TISI is nearly independent of surface temperature and is a pure combination of channel emissivities, that is,

$$\mathrm{TISI} \approx \prod_{k=1}^{N} \varepsilon_k^{d_k} \tag{4.11}$$

According to the general definition of TISI given by Eq. (4.10) and taking $d_i = 1$, Becker and Li (1990) and Li et al. (1999b) proposed two-channel TISI, TISI_{ij}, for two channels i and j (i is the MIR channel and j is the TIR channel) from the channel radiance R as

$$\mathrm{TISI}_{ij} = \frac{C_j^{n_i/n_j}}{C_i} \frac{\alpha_j^{n_i/n_j}}{\alpha_i} \frac{R_i}{R_j^{n_i/n_j}} \approx \frac{B_i\left(T_{gi}\right) - R_{ati\downarrow}}{B_i\left(T_{gj}\right) - R_{ati\downarrow}} \approx \frac{\varepsilon_i}{\varepsilon_j^{n_i/n_j}} \tag{4.12}$$

Assuming that the TISI_{ij} in the daytime without the contribution of solar illumination is the same as the TISI_{ij} in the nighttime, Becker and Li (1990) and Li et al. (2000) tried to extract the bidirectional reflectivity in MIR channel i by eliminating the emitted radiance during the day in this channel through comparing the TISI_{ij} in the daytime and with that in the nighttime. Choosing channel i as any channel in the MIR (3–5 μm) and channel j as any channel in TIR (10–13 μm), the bidirectional reflectivity $\rho_{bi}(\theta,\varphi,\theta_s,\varphi_s)$ can be estimated:

$$\rho_{bi}(\theta,\varphi,\theta_s,\varphi_s) = \frac{\left(\mathrm{TISI}_{ij}^{\mathrm{day}} - \mathrm{TISI}_{ij}^{\mathrm{night}}\right)\left(B_i\left(T_{gj}^{\mathrm{day}}\right) - R_{ati\downarrow}^{\mathrm{day}} - R_{sli\downarrow}\right)}{E_i \cos\theta_s \tau_i(\theta_s,\varphi_s)} \tag{4.13}$$

where $R_{sli\downarrow}$ is the channel downwelling solar diffusion radiation over the hemisphere divided by π and E_i is the solar irradiation at the top of the atmosphere in channel i.

As indicated by Eq. (4.13) and reported by Li et al. (2000), the bidirectional reflectivity in channel i can be retrieved directly from space measurements provided that all of the following four conditions are fulfilled: (1) The infrared radiometer onboard satellite has at least two channels, one within the 3–5 μm window and the other within the 10–13 μm window. (2) Multitemporal data in both MIR and TIR channels should be available, at least one during the day and the other at night. (3) The appropriate emissivity ratios (TISI) are assumed to be the same or do not change significantly between day and night. (4) The channel radiance at ground level can be obtained with good accuracy from the channel radiance at the TOA after atmospheric corrections.

To retrieve the directional emissivity, $\varepsilon_i(\theta,\varphi)$, from the bidirectional reflectivity $\rho_{bi}(\theta,\varphi,\theta_s,\varphi_s)$ extracted from Eq. (4.13), three methods have been further proposed. The first is to use an angular form factor $f_i(\theta,\varphi,\theta_s,\varphi_s)$, which was introduced by Li et al. (2000) to describe how a bidirectional reflectivity differs from that of a Lambertian reflector. With the help of this angular form factor, the directional emissivity in channel i can be estimated by

$$\varepsilon_i(\theta,\varphi) = 1 - \frac{\pi\rho_{bi}(\theta,\varphi,\theta_s,\varphi_s)}{f_i(\theta,\varphi,\theta_s,\varphi_s)} \tag{4.14}$$

in which $f_i(\theta,\varphi,\theta_s,\varphi_s)$ is inferred from that in the shortwave channels assuming the same shape of the angular form factors in these two bands.

The second is to first use the semiempirical phenomenological model of Minnaert (1941) modified by Li et al. (2000) and Petitcolin et al. (2002) to describe the angular variations of the bidirectional reflectivity by

$$\rho_{bi}(\theta,\varphi,\theta_s,\varphi_s) = \rho_0 \cos^{k-1}\theta \cos^{k-1}\theta_s\left[1 + q\sin\theta\sin\theta_s\cos(\varphi-\varphi_s)\right] \tag{4.15}$$

where ρ_0 is the reflectance for overhead sun and nadir observation, and k is a parameter varying typically between 0 and 1. The anisotropic factor q is positive if backscattering is important and negative when forward scattering is dominant. Once a series of $\rho_{bi}(\theta,\varphi,\theta_s,\varphi_s)$ are retrieved from the same type of surface at different view and illumination conditions, the parameters ρ_0, k, and q in Eq. (4.15) can be obtained. Subsequently, the directional emissivity in channel i is derived from

$$\varepsilon_i(\theta) = 1 - \frac{2\pi}{k+1}\rho_0 \cos^{k-1}\theta \tag{4.16}$$

The third is to use a kernel-driven bidirectional reflectivity model, the RossThick–LiSparse-R model, to describe the non-Lambertian reflective behavior of land surface in MIR as that in VNIR regions (Roujean et al. 1992; Wanner et al. 1995; Lucht and Roujean 2000):

$$\rho_b(\theta, \theta_i, \varphi) = k_{iso} + k_{vol} f_{vol}(\theta, \theta_i, \varphi) + k_{geo} f_{geo}(\theta, \theta_i, \varphi) \tag{4.17}$$

where k_{iso} is the isotropic scattering term, k_{vol} is the coefficient of the Roujean's volumetric kernel f_{vol}, and k_{geo} is the coefficient of the LiSparse-R geometric kernel f_{geo}. The analytical parameterizations of f_{vol} and f_{geo} can be found in Roujean et al. (1992), Lucht (1998), and Jiang and Li (2008). If a series of $\rho_{bi}(\theta,\varphi,\theta_s,\varphi_s)$ are retrieved from Eq. (4.13) with different angular configurations, the parameters k_{iso}, k_{vol}, and k_{geo} in Eq. (4.17) can be obtained. Knowing these three parameters, Jiang and Li (2008) demonstrated numerically that the directional emissivity in MIR channel could be expressed with a good approximation as

$$\begin{aligned}\varepsilon_i(\theta) = 1 - \pi k_{iso} + k_{vol}\,[0.0299 - 0.0128\exp(\theta/21.4382)] \\ + k_{geo}\left[2.0112 + 0.3410\exp\left[-2\left(\frac{\theta - 90.9545}{68.8171}\right)^2\right]\right]\end{aligned} \tag{4.18}$$

Once the directional emissivity in channel i is known, the directional emissivity in channel j is easily derived from the definition of TISI, that is,

$$\varepsilon_j(\theta, \varphi) = \left(\frac{\varepsilon_i(\theta, \varphi)}{\mathrm{TISI}_{ij}^{\mathrm{night}}}\right)^{n_j/n_i} \tag{4.19}$$

The flowchart of this method is illustrated in Fig. 4.9.

Based on the sensitivity analysis of TISI and emissivities to different error sources, Nerry et al. (1998) reported that the errors due to the approximations and the instrumental error do not exceed 1 %. Li et al. (1999b) showed that TISI may be sensitive to the uncertainties in atmospheric corrections. Nevertheless, the impacts of the uncertainties in atmospheric corrections on the emissivities are less sensitive than the temperature itself. The use of an approximate (standard) atmosphere instead of an actual atmosphere may lead to a 3 % or smaller error on LSE and a 0.5 K error on LST using the split-window method (Li and Becker 1993).

It is worth noting that the TISI-based method does not need any a priori information about the surface and can be applied to any surfaces, even those

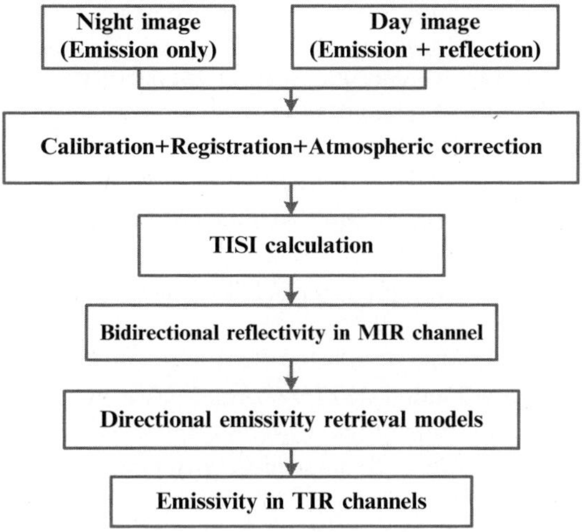

Fig. 4.9 Flowchart of LSE estimation using the TISI-based method

with strong spectral dynamics. Due to the fact that this method is based on the multitemporal data (a series of retrieved $\rho_{bi}(\theta,\varphi,\theta_s,\varphi_s)$), which requires accurate image co-registration, the retrieval errors may be large otherwise (Dash et al. 2005). Additionally, the surfaces must be observed under similar observation conditions, for example, view angle, during both day and night (Dash et al. 2005). However, Petitcolin et al. (2002) argued that TISI has mild angular dependence and remains stable over several weeks. Furthermore, the method needs both MIR and TIR data at the same time (Sobrino and Raissouni 2000). All of these problems may limit its usage in emissivity retrieval from space.

4.3.2 Simultaneous LST and LSE Retrieval Method with Known Atmospheric Information

The simultaneous retrieval methods referred to here include a group of algorithms that simultaneously retrieve the LSE and LST from the at-surface radiance. Based on some reasonable assumptions or constraints, these methods can retrieve the LSE and LST from the atmospherically corrected radiances at the ground level either by reducing the number of unknowns or by increasing the number of equations. The general strategies are to fully make use of the multitemporal or the multispectral information. The representative methods of using multitemporal information, under the assumption that the LSE is time invariant, are the two-temperature method (Watson 1992) and the physics-based day/night operational method (Wan and Li 1997). The representative methods, relying on the intrinsic spectral behavior of the LSE from multispectral information, include the reference channel method (Kahle et al. 1980), the normalization emissivity method (Gillespie 1985), the temperature emissivity separation method (Gillespie et al. 1998, 1999), the gray body emissivity method (Barducci and Pippi 1996), the iterative spectrally smooth temperature emissivity separation method (Borel 2008), and the linear emissivity constraint method (Wang et al. 2011).

4.3.2.1 Two-Temperature Method (TTM)

The idea underlying the TTM is the reduction of unknowns through multiple observations. Provided that accurate atmospheric corrections in the TIR channels have been performed and that the LSEs are time invariant, there are $2N$ measurements with $N+2$ unknowns (N channel LSEs and two LSTs) if the land surface is observed by N channels at two different times. The N LSEs and the two LSTs can therefore be simultaneously determined from the $2N$ equations if $N \geq 2$ (Watson 1992). Note that the assumption of the time-invariant LSEs implicitly requires the surface to be homogenous and have relatively stable soil moisture. The first restriction is to alleviate the LSE variation caused by pixel sizes and by viewing angles, while the second is to avoid the LSE changes with soil moisture, such as the occurrence of precipitation and dew.

The primary advantage of the TTM is that it makes no assumption about the spectral shape of the LSEs, only that they are time invariant. This method has a simple and straightforward formulation; however, the retrieval accuracy is not always guaranteed because the $2N$ equations are highly correlated and their solutions may thus be unstable and very sensitive to instrument noise and errors in the atmospheric corrections (Watson 1992; Caselles et al. 1997; Gillespie et al. 1999). Because accurate atmospheric corrections are difficult to perform without simultaneous atmospheric profile measurements, the use of approximate profiles could lead to large uncertainties in the LSE and LST retrievals. Peres and DaCamara (2004) found that increasing the number of observations and/or the temperature difference improved the retrieval accuracy, but this improvement is limited by the high correlation between TIR measurements.

In addition to the problems mentioned above, this method requires accurate geometric registration of images acquired at two different times (Watson 1992; Gillespie et al. 1999). Similar to the day/night TISI-based method, the impact of misregistration on the LSE and LST errors is small for homogeneous areas but large for heterogeneous areas (Wan 1999). A change in the satellite VZA causes a change in the LSE, consequently violating the assumption of time-invariant LSEs and decreasing the accuracy of the TTM (Li et al. 2013).

4.3.2.2 Physics-Based Day/Night Operational (D/N) Method

Inspired by the day/night TISI-based method and the TTM method, Wan and Li (1997) further developed a physics-based D/N method to simultaneously retrieve LSEs and LST from a combined use of the day/night pairs of MIR and TIR from MODIS satellite data. This method is based on the three assumptions of surface

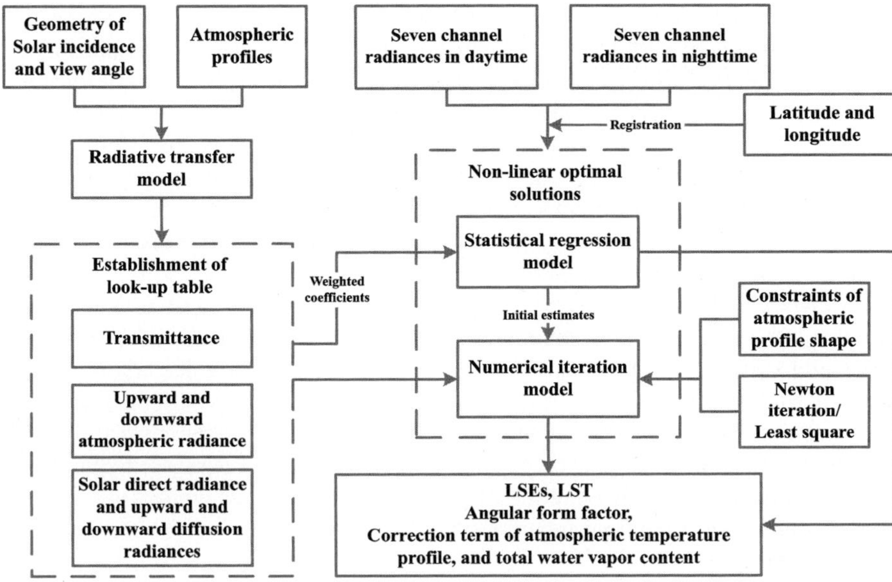

Fig. 4.10 Flowchart of LSE and LST retrieval using the D/N method

optical properties: (1) the surface emissivity does not significantly change in the day/nighttimes in several days unless rain and/or snow occurs during the short period of time, (2) the angular form factor $f_i(\theta,\varphi,\theta_s,\varphi_s)$ defined in Eq. (4.14) has very small variations (<2 %) in the wavelength range of interest in MIR, and (3) the Lambertian approximation of surface reflection for downwelling diffuse solar irradiance and atmospheric thermal irradiance does not introduce significant errors in the 3–14 μm region.

To reduce the effect of the residual error of atmospheric corrections on the retrieval, two variables, the air temperature at the surface level (T_a) and the atmospheric total water content (W), are also introduced to modify the initial atmospheric profiles in the retrieval. With two measurements (day and night) in N channels, the numbers of unknowns are $N+7$ (N channel LSEs, 2 LSTs, 2 T_a, 2 W, and 1 angular form factor in the MIR channels). Thus, to make the equations deterministic, the number of observations ($2N$) must be equal to or greater than the number of unknowns ($N+7$), which makes $N \geq 7$. Because $2N$ equations are nonlinear, a statistical regression method is used to give the initial values of $N+7$ unknowns. Next, a numerical algorithm, such as the least-squares-fit (χ^2) method, is used to find an accurate solution for $N+7$ unknowns from $2N$ measurements ($2N$ equations) (Wan and Li 1997). In the day/night algorithm, a look-up table of atmospheric parameters is also employed for high efficiency. In fact, the physics-based day/night operational (D/N) method is a development of the aforementioned TTM using two time observations. Figure 4.10 gives the detailed flowchart of the physics-based D/N method for retrieving simultaneously the LSEs and LST from MODIS data.

Compared with the TTM and TISI-based method mentioned previously, this physics-based D/N method is highlighted by several facets: (1) The contribution of solar irradiation to the radiance of the MIR channels in daytime significantly decreases the correlations among the equations and makes the solution more stable and accurate. Unlike the TISI-based method that first obtains the bidirectional reflectivity of the pixel and then calculates LSE and LST separately, the physics-based D/N method retrieves simultaneously LSEs and LST and avoids the propagation of error from stepwise retrieval. In addition, the physics-based D/N method can accurately determine LSEs and LSTs even though the LSTs are equal at the two times (daytimes and nighttimes), while the TTM with only TIR measurements requires significant difference in the temperatures. (2) The accuracy of the retrieved LSTs and LSEs is strongly improved by introducing two variables (T_a and W) to account for the uncertainties in the initial atmospheric profiles. As a result, the accuracy of the atmospheric correction is not required to be as high as that of TISI-based method and TTM. (3) The physics-based D/N method does not require 12-h interval measurements (day and night). As long as the surface emissivity does not change significantly, daytime and nighttime data collected over several days are also appropriate.

However, the physics-based D/N method still suffers from the critical problems of geometry misregistration and variations in the VZA. Wan (1999) aggregated the MODIS pixels to increase the scale from 1 to 5 km or 6 km in order to overcome the misregistration problems. Meanwhile, 16 VZA subranges are used to ensure quality of day and night VZA subranges (Wan and Li 2010). A set of new refinements were implemented to get more accurate solutions (Wan 2008). More details on the MODIS D/N method can be found in the literature (Wan 2008; Wan and Li 1997, 2010).

4.3.2.3 Reference Channel Method (RCM)

The reference channel method (RCM), developed by Kahle et al. (1980), assumes that the emissivity in one TIR channel has a constant value for all pixels. Once the atmospheric effects can be accurately corrected for and the downward atmospheric thermal radiance is known, the LST can be derived for each pixel from the measured radiance in this reference channel with the known emissivity. Finally, this LST is then used to derive emissivity values for the remaining channels:

$$\varepsilon_i = \frac{R_{gi} - R_{ati\downarrow}}{B_i\left(T_{\text{ref}}\right) - R_{ati\downarrow}} \tag{4.20}$$

where T_{ref} is the LST estimated from the reference channel, R_{gi} is the radiance observed at ground level in channel i, $R_{ati\downarrow}$ is the downward atmospheric thermal radiance, and $B_i(T_{\text{ref}})$ is the channel radiance of a blackbody at temperature T_{ref}.

Although the RCM is the simplest method for the emissivity retrieval from space, it suffers from some limitations. First, it is difficult to find a unique emissivity value that is appropriate for all surface materials in one reference channel. For example, the mean emissivity for vegetation is about 0.98, compared to 0.95 for most silicate rocks at wavelengths larger than 12 μm (Gillespie et al. 1999). An uncertainty of 1 % in the emissivity in this reference channel can result in about 0.5 K in LST and about 1–2 % of emissivity errors in other channels (Li et al. 1999b). Second, because the emissivity in the reference channel is assigned as a constant value for all pixels, there is no emissivity spatial information in this channel. Furthermore, the emissivities derived for adjacent channels are significantly affected by the constant value of emissivity in this channel and appear to be very noisy (Hook et al. 1992).

4.3.2.4 Normalization Emissivity Method (NEM)

This method was first described by Gillespie (1985) and then used by Realmuto (1990). This method assumes a constant emissivity in all N channels for a given pixel, which enables N temperatures T_{si} to be calculated for each pixel from their measured radiance, provided that the atmospheric effects are corrected for and the downward atmospheric thermal radiance is known. The maximum of those N temperatures (T_{NEM}) is considered to be the LST, namely,

$$\mathrm{LST} = T_{\mathrm{NEM}} = \max\left(T_{\mathrm{si}}\right) \text{ with } T_{\mathrm{si}} = B^{-1}\left(\frac{R_{gi} - \left(1 - \varepsilon_{\mathrm{cst}}\right) R_{ati\downarrow}}{\varepsilon_{\mathrm{cst}}}\right)$$

$$\text{and} \quad i = 1, 2 \ldots N \tag{4.21}$$

where B^{-1} is the inverse function of Planck's law, $\varepsilon_{\mathrm{cst}}$ is the assumed channel constant emissivity for a given pixel, R_{gi} is the radiance observed at ground level in channel i, and $R_{ati\downarrow}$ is the downward atmospheric thermal radiance. Finally, this estimated LST is used to derive emissivities for the other channels using Eq. (4.20) with LST instead of T_{ref}.

Mushkin et al. (2005) applied this method to the Multispectral Thermal Imager (MTI) data and assumed a different constant emissivity in the MIR and TIR bands. They found that the retrieved accuracy with the NEM is consistent with that reported for the ASTER temperature and emissivity separation method described below. Coll et al. (2001, 2003) proposed an improved NEM called the adjusted normalized-emissivity method (ANEM). In their works, the field emissivity measurements were used to adjust the initial maximum emissivity, and then the NEM was performed. However, because the ANEM requires in situ measurements, it is of limited use for exploration studies (Coll et al. 2003).

The NEM is an improvement of the RCM as the channel with the maximum emissivity can be different in the NEM for different materials. It is recommended by Li et al. (1999b) and selected by Gillespie et al. (1998) as one of the modules in the temperature and emissivity separation algorithm applicable to the ASTER data.

4.3.2.5 Temperature Emissivity Separation (TES) Method

This method was first developed by Gillespie et al. (1999) to separate the LST and the LSE from atmospherically corrected radiances. This method relies on an empirical relationship between the spectral contrast and the minimum emissivity to increase the number of equations (equivalent to reducing the number of unknowns) so that the undetermined retrieval problem becomes deterministic. The TES method comprises three mature modules: the NEM (Gillespie 1985), the spectral ratio (SR), and the maximum–minimum apparent emissivity difference method (MMD) (Matsunaga 1994).

TES first uses the NEM module to estimate the initial surface temperature and the normalized emissivities from the atmospherically corrected radiances at ground level (Gillespie et al. 1999). Subsequently, the spectral ratio module is used to calculate the ratio of the normalized emissivities ε_i to their average:

$$\beta_i = \frac{\varepsilon_i}{\frac{1}{N}\sum_{j=1}^{N}\varepsilon_j} \quad (N \text{ is the total number of TIR channels}) \tag{4.22}$$

Although the spectral ratio β cannot directly provide the actual emissivity, it has been shown to describe the shape of the emissivity spectra well even if the surface temperature is roughly estimated. Finally, on the basis of the spectral ratio, the MMD module is used to find the spectral contrast in N channels:

$$\text{MMD} = \max(\beta_i) - \min(\beta_i) \tag{4.23}$$

To recover the actual values of the emissivities, an empirical relationship between the minimum emissivity ($\varepsilon_{\min}$) in N channels and MMD is established:

$$\varepsilon_{\min} = A_0 + A_1 \times \text{MMD}^{A_2} \tag{4.24}$$

with A_0, A_1, and A_2 being the sensor dependent coefficients. Figure 4.11 shows the empirical relationship between $\varepsilon_{\min}$ and MMD, where $A_0 = 0.994$, $A_1 = -0.687$, and $A_2 = 0.737$ for ASTER sensor (Gillespie et al. 1998).

Once $\varepsilon_{\min}$ is estimated, the emissivities in other channels can be straightforward derived from the spectral ratio β_i using

$$\varepsilon_i = \beta_i \times \left(\frac{\varepsilon_{\min}}{\min(\beta_i)}\right) \quad (i = 1, \ldots, N) \tag{4.25}$$

and LST can be refined and estimated with the estimated LSEs. Figure 4.12 shows the flowchart of this TES method.

The main advantage of the TES is that it combines attractive features of three precursors and uses an empirical relationship between the range of emissivities and the minimum emissivity in the N channels to retrieve the LST and LSEs. Consequently, it can be applied to any kind of natural surface without considering

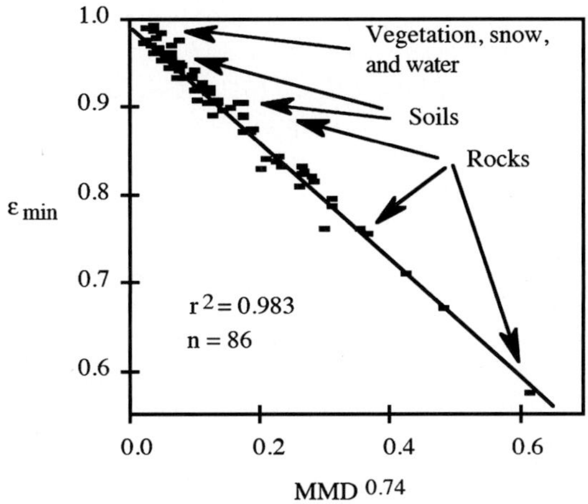

Fig. 4.11 The empirical relationship between ε_{min} and MMD found for the ASTER sensor using the laboratory and field emissivity spectra (Adapted from Gillespie et al. (1998), with permission from IEEE)

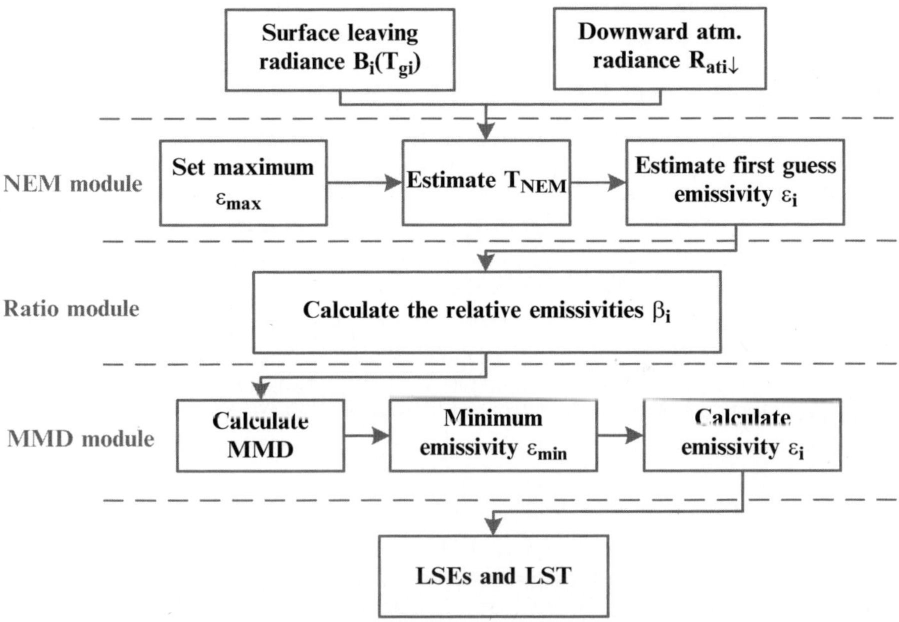

Fig. 4.12 Flowchart of TES method for retrieving LSEs and LST

spectral variations in the emissivity, especially for surfaces with high spectral contrast emissivities such as rocks and soils (Gillespie et al. 1998; Sobrino et al. 2008). Numerical simulation and some field validations have demonstrated that the TES can retrieve the LST to within about ±1.5 K and the LSEs to within about ±0.015 when the atmospheric effects are accurately corrected (Gillespie et al. 1998, 1999; Abrams 2000). More detailed information about TES algorithm can be found in Gillespie et al. (1998, 1999).

The main assumption of TES is that the relationship between minimum emissivity and spectral contrast, that is, Eq. (4.24), holds true for the entire gamut of surface materials (Gillespie et al. 1999). Gillespie et al. (1999) have tested this assumption and reported this assumption appears to be valid for most surfaces. However, Payan and Royer (2004) found that the metals do not obey this empirical relationship. However, the nature surface is mainly composed of water, soils, vegetation, and snow rather than metals. Therefore, the violation of the empirical relationship by the metals does not degrade the retrieved accuracy any more when dealing with the Earth observation data using the TES method. Because of its simplicity and the needlessness of a priori knowledge on the surface, the TES method has already attracted more and more attentions (Schmugge et al. 2002; Dash et al. 2005). At the same time, some changes in the original TES method are proposed to improve the retrieval accuracies, including integrating broadband emissivity in the iterative algorithm for high-contrast emissivity surfaces (Payan and Royer 2004), adjusting radiances with respect to the channel 13 (10.6 μm) of ASTER for low-contrast surfaces (Coll et al. 2007), removal of the iterative correction for downwelling irradiance and the threshold test for spectral contrast (Gustafson et al. 2006), and a water adjustment and the compensation for partially vegetated surfaces by a fractional vegetation cover adjustment (Hulley and Hook 2009a).

However, some reports have indicated that the TES method exhibited significant errors in the LST and LSEs of surfaces with low spectral contrast emissivity (e.g., water, snow, and vegetation) and under hot and wet atmospheric conditions (Gillespie et al. 1999, 2011; Sawabe et al. 2003; Coll et al. 2007; Hulley and Hook 2009b, 2011). Sabol et al. (2009) pointed out that the low-emissivity contrast and high-emissivity contrast have been treated differently in original version of TES. Consequently, the retrieved LSEs are too low and the LST is too high in the original version for the materials (e.g., soils, vegetation, and water/snow) that are plotted above the regression line in the scatter plot of LSE_{min} and MMD. That is why some studies have reported that inaccurate atmospheric corrections may produce LST errors of 2–4 K for bare soil (Dash et al. 2002). For warm and wet atmospheric conditions, the cause of significant errors is different. The uncertainties in the atmospheric corrections will result in a large apparent emissivity contrast. This effect is more serious over gray body surfaces (Hulley and Hook 2011). To minimize atmospheric correction errors, Gillespie et al. (2011) improved the TES method by using a water vapor scaling (WVS) approach proposed by Tonooka (2005).

As shown by numerical simulations, the uncertainties on the LST and LSE retrievals increase when the number of channels is reduced, making the TES method inapplicable to most operational sensors (Sobrino et al. 2008). Moreover, sensor calibration errors and noise in the TIR channels also cause uncertainties in the retrieved LST and LSEs (Jimenez-Munoz et al. 2006; Sobrino et al. 2008; Gillespie et al. 2011). In addition, TES scales low- and high-contrast surfaces differently, which leads to step discontinuities at the edges of gray body units such as water, forests, and crops (Sobrino et al. 2007). To overcome these problems, Sabol et al. (2009) recently replaced the power relationship of LSE_{min} and MMD in the original TES method with a linear expression and applied the new relationship available for all materials to alleviate such discontinuities. This revision was reported to reduce slightly the accuracy for both rock surfaces and gray bodies but can improve the accuracy for near-gray body surfaces.

4.3.2.6 Gray Body Emissivity (GBE) Method

This method assumes that the LSE has a flat spectrum, that is, the LSE is independent of the wavelength, for wavelengths larger than 10 μm to reduce the number of unknowns and stabilize the retrieval algorithm (Barducci and Pippi 1996). On the basis of the assumption that

$$\frac{d\varepsilon}{d\lambda} = 0 \quad \text{or} \quad \varepsilon_i = \varepsilon_j \tag{4.26}$$

where ε_i and ε_j denote the emissivity at the *i*th and *j*th channel, the number of unknowns is reduced and the undetermined retrieval problem becomes deterministic, too.

The main advantage of the GBE is that no additional assumption about the shape of the emissivity spectrum is required, except the assumption that it is flat in some wavelength interval. In theory, the LSEs and LST can be simultaneously retrieved as long as at least two channels have the same LSE (not necessarily the gray body) in the wavelength interval of interest. However, the limitations of the method are evident. The application of the GBE method to space-based measurements requires accurate atmospheric corrections in the TIR channels and at least two channels with the same LSE. Similar to the TTM, this method is very sensitive to instrument noise and errors in the atmospheric corrections because the TIR measurements are highly correlated. Moreover, requiring spectrally flat LSEs often hampers the use of the GBE in multispectral TIR data unless at least two channels with the same LSE can be identified. This problem can be more or less overcome with hyperspectral TIR data because it is easier to find at least two channels with the same LSE in hyperspectral data than in multispectral data, and hundreds or even thousands of channels can further improve the retrieval accuracy. Therefore, the GBE method is thought to be more applicable to hyperspectral TIR data.

4.3.2.7 Iterative Spectrally Smooth Temperature Emissivity Separation (ISSTES) Method

Hyperspectral TIR data provide much more detailed spectral information about the atmosphere and land surface. Borel (1997, 1998, 2008) reported that a typical emissivity spectrum is rather smooth compared with the spectral features introduced by the atmosphere. Figure 4.13 shows the differential spectra curves between one typical emissivity spectra curve and one spectra curve of downward atmospheric radiance, as well as several estimated emissivity spectra curves under different

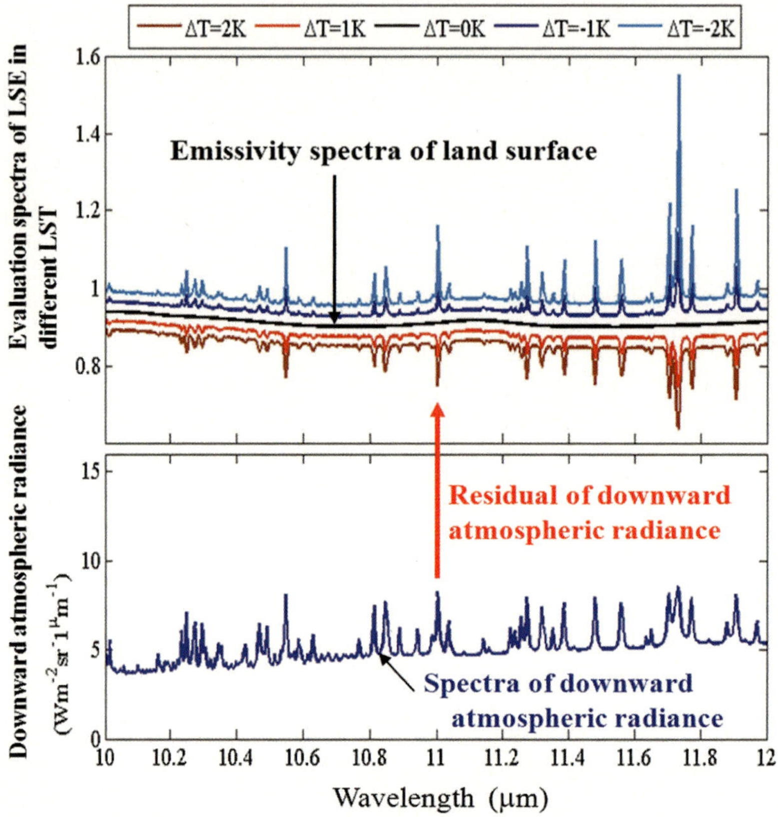

Fig. 4.13 Illustration of the impact of estimated LST on the retrieved LSE with atmospheric spectral features. (*Top*) The emissivity spectra curves under different LSTs. ΔT is the temperature difference between the estimated and actual LST. (*Bottom*) The selected spectra of downward atmospheric radiance

estimated LSTs. According to Eq. (4.20), if the LST is not accurately estimated, the corresponding LSE spectrum will exhibit the atmospheric spectral features, that is, there are some sharp convexities or concavities caused by the atmospheric absorption lines on the estimated emissivity spectrum. The overestimated LST will cause underestimated LSEs, and vice versa. The best estimates of the LSEs and LST should be obtained when the spectral smoothness of the retrieved LSE is maximized.

Based on this property, the ISSTES method has been developed to iteratively retrieve the LSEs and LST from hyperspectral TIR data. For three neighboring channels, $i-1$, i, and $i+1$, in hyperspectral TIR band, Borel (1997, 1998) defined a spectral smoothness (SM) as

$$\mathrm{SM} = \sum \left(\varepsilon_i - \frac{\varepsilon_{i-1} + \varepsilon_i + \varepsilon_{i+1}}{3} \right)^2. \tag{4.27}$$

The optimal estimate of LST corresponding to the minimum of SM can be iteratively found. Once this optimal LST is determined, the emissivity spectrum can be easily derived from the hyperspectral TIR data using Eq. (4.20).

Different formulations including the first and second derivative criteria for the spectral smoothness criterion have been proposed (Borel 1997, 1998; Kanani et al. 2007; Cheng et al. 2010; OuYang et al. 2010). However, they all statistically lead to the same performance for the hyperspectral TIR data, implying that the performance is not sensitive to the choice of smoothness function.

Ingram and Muse (2001) analyzed the method's sensitivity to smoothness assumptions and measurement noise and found that the retrieval accuracy caused by the smoothness assumptions is negligible for typical materials but is significantly dependent on the signal-to-noise ratio (SNR), that is, a high accuracy can be obtained with a high SNR. Similar to most methods presented above, the atmospheric correction needs to be accurately performed, and its impact on the retrieval results is the greatest among all influences. The retrieval accuracy is also sensitive to shifts in the central wavelengths and bandwidths of the TIR channels (Borel 2008). In addition, Wang et al. (2011) reported that the occurrence of singular points may lead to difficulties in finding an acceptable solution when the LST is close to the effective temperature of the downward atmospheric radiance.

4.3.2.8 Linear Emissivity Constraint Temperature Emissivity Separation Method (LECTES)

Inspired by the GBE initially proposed by Barducci and Pippi (1996), Wang et al. (2011) proposed a new TES method to retrieve simultaneously both LST and LSEs from atmospherically corrected hyperspectral TIR data. This method assumes that the emissivity spectrum can be divided into M segments in which the emissivity in each segment varies linearly with the wavenumber or wavelength. Thus, the emissivity spectrum can be reconstructed using a piecewise linear function with slopes a_k and intercept b_k ($k = 1, \ldots, M$) as shown in Fig. 4.14.

In mathematics, these piecewise linear relationships expressed in wavenumber form can be written as

$$\varepsilon = \begin{cases} a_1 v + b_1 & v_1 < v \le v_2 \\ a_2 v + b_2 & v_2 < v \le v_3 \\ \ldots\ldots & \ldots\ldots \\ a_M v + b_M & v_M < v \le v_{M+1} \end{cases} \tag{4.28}$$

where v is the wavenumber, and M is the number of linear variation segments in emissivity spectrum. It is evident that the LSEs and LST can be simultaneously obtained, provided that the number of equations N (corresponding to the N channel measurements) is equal to or greater than the number of unknowns ($2M + 1$

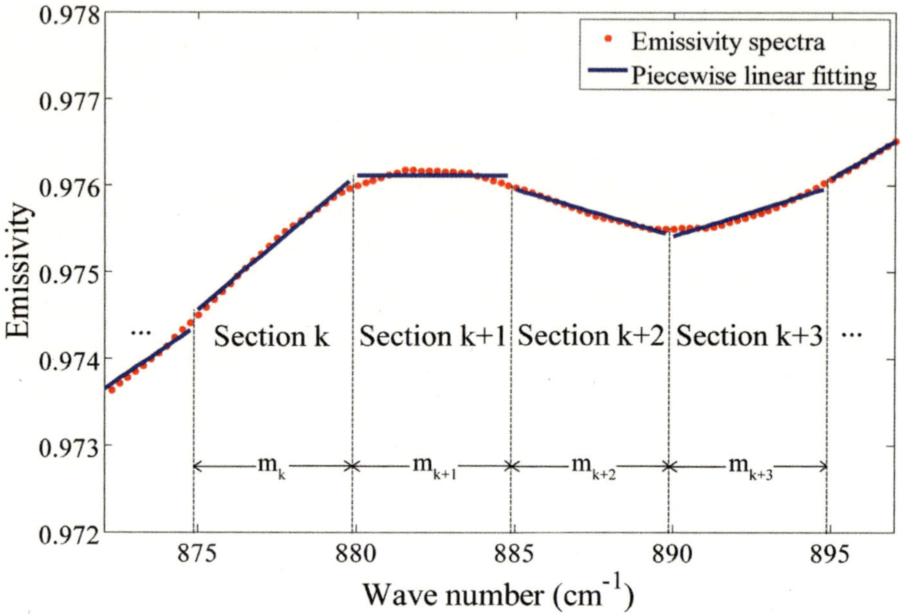

Fig. 4.14 Sketch map of the piecewise linear emissivity spectra fitting. The spectra in this figure are a type of soil, the abscissa is the wavenumber, and the ordinate is the emissivity. The *red dot line* is the actual emissivity spectrum, and the *blue lines* are the fitting spectra (Adapted from Wang et al. (2011), with permission from IEEE)

unknowns corresponding to 1 LST, $M\ a_k$, and $M\ b_k$). The requirement of $N \geq 2M + 1$ is easily fulfilled for hyperspectral TIR data because a huge number of narrow channels are available in a hyperspectral TIR sensor.

Wang et al. (2011) carried out a series of sensitivity analyses and concluded that the errors introduced by the assumption of linear emissivity can be neglected if the segment length is well chosen. A segment length of about 10 cm^{-1} is suggested. Compared with the ISSTES method, this method produces less frequent singular points and is more resistant to both white noise and uncertainty in the downward atmospheric radiance. Because atmospheric spectral features are more significant under wet and warm atmospheric conditions, the LECTES method performs better in wet and warm atmospheres than in dry and cold atmospheres. Similar to the ISSTES method, this method is only suitable for hyperspectral TIR data and requires accurate atmospheric corrections. However, because the number of unknowns can be greatly reduced with a piecewise linear function, this method exhibits great potential as a technique for simultaneously retrieving the LST, the LSE, and atmospheric profiles.

4.3.3 *Simultaneous LSEs, LST, and Atmospheric Profile (Atmospheric Quantities) Retrieval Method*

Although the simultaneous LSEs and LST retrieval methods reviewed in Sect. 4.3.2 can accurately obtain the LSEs and LST if the atmospheric effects are well corrected for, accurate atmospheric profiles are usually unavailable synchronously with TIR measurements, and thus the accuracy of the retrieved LSEs and LST can be degraded. An ideal solution is to simultaneously retrieve the LSEs, LST, and atmospheric parameters (e.g., atmospheric profiles) (Ma et al. 2002). Because the narrow bandwidth offered by hyperspectral TIR sensors with thousands of channels can improve the vertical resolution and allow atmospheric profiles and surface parameters (LSEs and LST) to be obtained more accurately (Chahine et al. 2001), several methods have been proposed to retrieve simultaneously the surface and atmospheric parameters. The representatives of these methods are the artificial neural network (ANN) method (Wang et al. 2010, 2013) and the two-step physical retrieval method (Ma et al. 2000, 2002).

4.3.3.1 Artificial Neural Network (ANN) Method

Because an ANN can robustly perform highly complex, nonlinear, parallel computations, ANNs have become increasingly utilized in simultaneous retrievals of LSEs, LST, and atmospheric profiles (Aires et al. 2002b; Blackwell 2005; Wang et al. 2010, 2013). ANNs resemble the brain in two aspects: they acquire knowledge through a learning process and store the acquired knowledge using interneuron connection strengths (Mas and Flores 2008). Therefore, ANNs represent massively parallel distributed processors that can acquire experiential knowledge and make that knowledge available for use.

The main advantages of ANN methods over conventional retrieval methods are their ability to learn complex patterns, generalization to noisy environments, and incorporation of both knowledge and physical constraints (Mas and Flores 2008). Because of ANNs' powerful nonlinear retrieval abilities, a number of attempts have been made to develop neural networks to retrieve both the surface and atmospheric biophysical variables without exact knowledge of the complex physics mechanisms. For example, Mao et al. (2008) used an ANN to estimate the LST and LSE, while Aires et al. (2002b) and Blackwell (2005) used an ANN to retrieve atmospheric profiles. To reduce the effect of coupling between the surface and atmosphere on the retrieval accuracy, Aires et al. (2002a) proposed using an ANN to retrieve both the atmospheric and surface temperatures, and Wang et al. (2010, 2013) attempted to establish a neural network to simultaneously retrieve the LSEs, LST, and atmospheric profiles from hyperspectral TIR data. The preliminary results demonstrated that ANNs can be used to simultaneously retrieve the LST, LSEs, and atmospheric profiles from hyperspectral TIR data with acceptable accuracy for some applications.

However, because ANNs perform like black boxes and can produce corresponding outputs from any given inputs, the retrieval process cannot be well controlled, and it is difficult to interpret the weights assigned to each input and improve the output due to the complex nature of the network. In addition, the implementation of an ANN depends largely on its architecture and the training data (Mas and Flores 2008). It is difficult to determine the architectures and learning schemes for an ANN, which are directly related to its ability to learn and generalize. Although one or two hidden layers are recognized to be enough for most problems (Sontag 1992; Aires et al. 2002b; Mas and Flores 2008), a number of experiments are still required to determine what architecture-related parameters will improve the accuracy, such as the number of input and hidden nodes, the initial weight range, the activation functions, the learning rate and momentum, and the stopping criterion. To date, no ANN architecture is universally accepted for a particular problem. The characteristics of the training data, such as the size and the representativeness, are also of considerable importance. The use of too few or unrepresentative training samples results in a network that cannot accurately retrieve the outputs, while the use of too many training samples requires more time for learning. Because physical understanding is not required, ANN methods may be regarded as empirical methods. However, their results can be used to provide initial guesses for further improvements in the physical retrieval methods (Motteler et al. 1995). More detailed information about the application of ANNs can be found in the work of Mas and Flores (2008).

4.3.3.2 Two-Step Physical Retrieval Method (TSRM)

Because the measured radiance at the TOA is a function of the surface and atmospheric parameters, the surface and atmospheric variables can be theoretically obtained by selecting appropriate channels even from multispectral data (Menzel et al. 2006). Ma et al. (2000) made an initial attempt at simultaneously retrieving the LST and atmospheric profiles by assuming that the LSEs is invariant within the MIR channels and also invariant within the TIR channels and by ignoring the solar contribution in MIR channels. However, these rough assumptions may degrade the retrieval accuracies in the troposphere. Along this line of reasoning, Ma et al. (2002) further considered the solar contribution and proposed an extended two-step physical retrieval method that simultaneously extracts the LSEs, the LST, and the atmospheric profiles from MODIS data.

The main idea underlying the TSRM inherits that of atmospheric profile retrieval. The first step is to tangent-linearize the atmospheric RTE with respect to the LSEs, LST, and atmospheric temperature–humidity profiles, namely,

$$\delta T_{\mathrm{B}} = W_{\mathrm{T_s}} \delta T_{\mathrm{s}} + W_{\varepsilon} \delta \varepsilon + \int_0^{p_{\mathrm{s}}} W_{\mathrm{T}} \delta T \,\mathrm{d}p + \int_0^{p_{\mathrm{s}}} W_{q_{\mathrm{w}}} \delta \ln q_{\mathrm{w}} \mathrm{d}p \tag{4.29}$$

where the perturbation δ is with respect to an a priori mean condition. T_B is a brightness temperature vector. W_{Ts}, W_{ε}, W_T, and W_{qw} are the weighting functions of land surface temperature (T_s), surface emissivity (ε), atmospheric temperature (T), and water vapor (q_w), respectively.

Given initial guesses for those atmospheric and surface variables, a set of equations based on the tangent-linearized RTE can be derived using the remotely sensed measurements (Smith 1972; Li et al. 1994; Ma et al. 1999). At the same time, the principle component analysis (PCA) technique and the Tikhonov regularization method are employed to reduce the number of unknowns and stabilize the ill-posed problem (Smith and Woolf 1976; Ma et al. 2000), which makes the solution of these equations stable and deterministic. According to the statistical analysis in the work of Ma et al. (2000, 2002), only five temperature and three water vapor eigenvectors can explain all of the information of 40 atmospheric temperature and water vapor levels, respectively. In the second step, the Newtonian iteration algorithm is utilized with the regularized solution as the initial guess to obtain the final maximum likelihood solution of the LSEs, LST, and atmospheric temperature–humidity profiles. Here, the Tikhonov regularization is used to stabilize the ill-posed problem and to obtain a meaningful solution, while the Newton iterative algorithm is used to further improve the solution. Thus, the name of the technique is "two steps."

There are at least three assumptions involved in this method: (1) the RTE can be tangent-linearized around an initial guess, (2) a constant angular form factor is used for the solar beam in the MIR region to simplify the RTE, and (3) the PCA can be used to reduce the number of unknowns without significant loss of accuracy. These assumptions ensure the existence of stable and accurate solutions without a priori atmospheric corrections, as opposed to other conventional methods. However, Ma et al. (2002) found that the solutions are highly dependent on the initial guess. Therefore, one possible improvement to this type of method is to improve the initial guess. As pointed out earlier, the results of an ANN can be used as the initial guesses in the physical retrieval method. It is worth noting that the physical nature of the algorithm requires an adequate number of channels in each specific window, and its complex nature may lead to a low computational efficiency. These two properties hamper the use of this method in practice.

4.4 Comparison and Validation of Satellite-Derived LSEs

4.4.1 Comparison and Analysis of Different Methods

The methodologies for estimating LSEs from space were briefly reviewed above. With their advantages and limitations, these methods have different accuracies and are applicable for various sensors and applications. It is important to evaluate them on an identical standard and to give theoretical advice on their applications.

Because the various methods have been proposed and developed under different circumstances and for various applications, only a few studies directly compare several methods. For instance, Li et al. (1999b) have evaluated six methods, including TISI, RCM, and NEM, for extracting the relative spectral emissivity from TIR data. They showed that (1) all of the methods are sensitive to the atmospheric uncertainties, (2) the systematic error has little effect on the relative emissivity retrieval, (3) instrumental noise from 0.1 to 0.3 K can lead to an error in the relative emissivity ranging from 0.002 to 0.005, and (4) the TISI and NEM methods are recommended because they are slightly superior to others.

Because the TISI-based and TES methods are theoretically robust and do not require a priori information about the emissivity or surface type (Dash et al. 2005; Sobrino et al. 2008), whereas the NBEM is operationally simple and can give a satisfactory estimate of the LSEs for soil and vegetation mixed areas to some extent (Sobrino et al. 2002; Dash et al. 2005), TISI-based, TES, and NBEM are often taken as reference methods for comparison to others. Sobrino et al. (2002) compared several retrieval methods and found that NBEM and NEM give the same absolute emissivity values with differences between 1 and 0.2 %, while NBEM and TES give the greatest differences of around 2 %. Jacob et al. (2004) compared MODIS TISI-based and TES retrievals over Africa and Jornada and found that the retrieval results agree well, with root-mean-square deviations ranging from 0.006 to 0.016. Dash et al. (2005) compared the performance of the TISI-based method and NBEM and found that the dynamic range of LSE is compressed in the NBEM. The difference between the LSE obtained by the NBEM and the TISI-based methods ranges between −0.038 and 0.032, but the peak of the histogram of LSE difference for vegetated areas is 0.002, which further confirms that NBEM is more suitable for vegetated areas. Momeni and Saradjian (2007) evaluated the emissivities of MODIS bands 31 and 32 retrieved from the NBEM and D/N methods and found that they agree relatively well with each other. Wang and Liang (2009) evaluated the ASTER and MODIS emissivity products at six Surface Radiation Budget Network (SURFRAD) sites and concluded that Collection 5 broadband emissivity is 0.01 larger than that of MODIS Collection 4 products and ASTER emissivity. Obviously, these comparisons contradict each other to some extent and highlight the importance of the LSE validation.

Other methods, for a variety of reasons, have not been subject to a comparative analysis. However, their assumptions, advantages, and drawbacks are comprehensively described in the previous section, which gives full insight into the understanding of the emissivity retrieval methods. To give a full insight into the understanding of those methods, the assumptions, advantages, and limitations of each of these methods are recapitulated in Table 4.1.

Table 4.1 Comparisons of a variety of commonly applied LSE retrieval methods

Method		Assumptions	Advantages	Disadvantages	References
Stepwise retrieval methods	CBEM	The same class have the same emissivity	Simplicity; accurate with the classes having well-known emissivities; TIR bands and accurate atmospheric correction are not required; emissivity can be obtained at the higher spatial resolution	Require the a prior knowledge of emissivity database for classes and the corresponding classification map; depend on classification accuracy and seasonal and dynamic states; less accurate for coarse resolution and less reliable for the classes with contrast emissivities; discontinuities	Snyder et al. (1998) and Gillespie et al. (1999)
	NBEM	Surface is composed of soil and vegetation and the variation of emissivity is linearly related to the fraction of vegetation in a pixel	Simplicity; the cavity effects are taken into account; suitable for various instruments; TIR bands and accurate atmospheric correction are not required; emissivity can be obtained at the higher spatial resolution	Require a priori knowledge of the emissivities of soil and vegetation, the vegetation fraction, NDVI thresholds for soil and vegetation, as well as geometric structure of vegetation; less accurate beyond the 10–12 μm spectral interval; inapplicable over other surfaces such as water, ice, snow, and rocks and failing for the senescent vegetation; discontinuities	Van de Griend and Owe (1993), Valor and Caselles (1996), and Sobrino and Raissouni (2000)
	TISI	Emissivity ratios are the same or do not change significantly between two times, i.e., day and night	Nearly independent of LST; approximate atmospheric corrections are sufficient; physical basis and suitable for various surfaces	Require approximate atmospheric corrections; require at least two channel measurements, one in MIR and another in TIR atmospheric windows, acquired at day- and nighttimes; require accurate geometry registration and similar viewing angles at both day- and nighttimes	Li and Becker (1993), Li t al. (2000), and Jiang et al. (2006)

(continued)

Table 4.1 (continued)

Method		Assumptions	Advantages	Disadvantages	References
Simultaneous LSE and LST retrieval methods with known atmospheric information	TTM	Emissivity is invariant with time	More suitable for geostationary satellites data	Multi-temporal TIR data, similar viewing angles, accurate atmospheric corrections and geometry registration, and large temperature difference between two times are required; solution more sensitive to instrument noises and errors in atmospheric corrections	Watson (1992)
	D/N	LSEs do not change significantly in several days; angular form factor has very small variation in MIR channels	Does not require a priori accurate atmospheric profiles; solutions are more stable and accurate by introducing MIR channels and by modifying the atmospheric profiles; physical basis and suitable for various surfaces	Require multi-temporal data in several channels in the MIR and TIR atmospheric windows, accurate geometric registration, and similar viewing angles; approximate shapes of the atmospheric profiles need to be given a priori; retrieval process is complicated and initial guess values are required	Wan (1999) and Wan and Li (1997)
	RCM	Emissivity at an assigned channel has a constant value for all pixels	Simplicity	Require accurate atmospheric corrections in TIR channels; inappropriate to assign a unique value for all the surface materials in a specific channel; accuracy of emissivity and temperature depends largely on the assigned constant value	Kahle et al. (1980)
	NEM	Assuming that the maximum emissivity among all channels is the same for all pixels	Simplicity; channel with the maximum emissivity is not specified and may be different between pixels; having better performance than RCM	Require accurate atmospheric corrections in TIR channels; accuracy of emissivity and temperature depends largely on the assigned maximum emissivity value	Gillespie (1985) and Realmuto (1990)

TES	The relationship between minimum emissivity and spectral contrast holds true for the entire gamut of surface materials	Refine the values of the maximum emissivity used in the NEM; do not need any assumptions on emissivity; retrieve simultaneously LST and LSEs for any kinds of surfaces	Require accurate atmospheric corrections in TIR channels and at least three TIR bands in atmospheric windows; accuracy depends on atmospheric compensation and the empirical relationship between the minimum emissivity and MMD; uncertainty is more serious for gray bodies; step discontinuities at the edges of a gray body may be caused due to the inaccurate inherent scaling behaviors	Gillespie et al. (1998, 1999) and Sobrino et al. (2008)
GBE	There exists a flat region in emissivity spectrum	No other assumptions on the shape of emissivity spectrum except for a flat region in emissivity spectrum; do not need to assign channel emissivity	Require accurate atmospheric corrections in TIR channels and at least two channels having the same emissivity; assumption is not often satisfied for most situations; difficult to find two channels having the same emissivity in multispectral TIR sensors; solution is more sensitive to instrument noise and errors in the atmospheric corrections	Barducci and Pippi (1996)
ISSTES	Surface emissivity spectrum is rather smoother than the spectral features introduced by the atmosphere	Performance is insensitive to the choice of a smoothness function; high accuracy can be obtained with high SNR	Require accurate atmospheric corrections; only suitable for the hyperspectral IR data; sensitive to the spectral shift and varying of FWHM; occurrence of singular points leads to difficulties in finding an acceptable solution	Borel (1997, 1998, 2008)
LECTES	Emissivity spectrum can be divided into *M* segments and emissivity in each segment changes linearly with wavelength	Reduce the occurrence of singular points and is resistant to white noise; perform well under wet and warm atmospheric conditions	Require accurate atmospheric corrections and a priori knowledge of downward atmospheric radiances; only suitable for hyperspectral TIR data; sensitive to shifts in the central wavelengths of the TIR channels	Wang et al. (2011)

(continued)

Table 4.1 (continued)

Method		Assumptions	Advantages	Disadvantages	References
Simultaneous LSE, LST, and atmospheric profile retrieval methods	ANN	No special assumption	Ability to learn from complex patterns; generalizable to noisy environments; incorporate knowledge and various physical constraints	Highly dependent on the architecture of the ANN and the training data; difficult to determine the appropriate architectures and learning schemes and representative training data sets; retrieval process cannot be well controlled; difficult to interpret the weights assigned to each input and improve the output	Aires et al. (2001, 2002a) and Wang et al. (2013)
	TSRM	A specular surface reflection and a constant angular form factor are used to simplify the RTE; PCA can be used to reduce the number of unknowns without loss of too much accuracy	Do not need a priori atmospheric corrections; PCA and Tikhonov regularization can be used to make the solution more stable and accurate	Complexity; low computational efficiency limits the application; require adequate numbers of channels and an initial guess of LSEs, LST, and atmospheric temperature–humidity; the solution is more dependent on the initial guess	Ma et al. (2000, 2002) and Li et al. (2007)

4.4.2 Validation of Satellite-Derived LSE

4.4.2.1 Ground Measurements of LSE

Generally, the direct ground measurements of LSEs are still challenging because of the couple of LSEs and LST in the measured radiance. Even when the downward thermal radiance is known, the LSEs and LST cannot be obtained directly from only one equation due to the ill-posed problem described earlier. Compared with the LSE retrieval methods from space, the ground measurements of LSEs are not influenced by the absorption of the atmosphere and by the upward atmospheric contribution.

Obviously, the equations of ground measurements will become resolvable by either reducing the numbers of unknown or increasing the numbers of equation. Wan et al. (1994) reported that most ground measurements of emissivity depend on the following basic assumptions: (1) the surface temperature does not change in short time during measurements or the correlation between the surface temperature and variations in the external radiation source is negligible; (2) the surface emissivity does not change during measurements yet; and (3) the surface is a Lambertian surface or a specular reflecting surface unless a complete set of bidirectional reflectance is also measured.

Based on these assumptions, one feasible solution to perform ground measurement of emissivity is to change the environmental radiance to get an additional equation. To date, there have been at least three representative methods to get the required additional equation. The first method is to conduct the radiance measurements under sunshine and shadow conditions (Wan et al. 1994; Wan and Li 2008, 2010). The sample is measured at both MIR and TIR bands, which are similar to seven bands of MODIS (bands 20, 22, 23, 29, 31–33), under sunshine and shadow conditions, respectively. The shadow condition is produced by blocking the solar beam directly. The diffuse gold plate is also measured in the same procedure as those for sample under sunshine and shadow conditions. Because the emissivity and the temperature of the diffuse gold plate are known in advance, the emissivity, the anisotropic factor in MIR channels, and the temperatures at two conditions for the sample and other parameters involved in the RTE can be resolved from those 28 equations (each measurement step would introduce 7 equations). This method is similar to but simpler than the physical D/N method presented earlier (Wan and Li 2008).

The second method is to perform measurements with and without cloth chamber under the clear sky (Zhang et al. 2004; Tian et al. 2008). This method is similar to the sunshine and shadow method; however, the method becomes simpler. The active source, sun, is not used and simultaneous measurements at both MIR and TIR channels are not required and only one TIR channel is enough. Both the sample and the diffuse gold plate are measured at hot environment (inside the cloth chamber) and cool environment (outside the cloth chamber, namely, sky environment). The environmental radiances are assumed invariant when measuring the sample and the gold plate. In addition, the change of hot and cool environments is performed within several seconds; the temperature variation of the sample can be neglected. There are four unknowns for one TIR band (the emissivity and the temperature of sample, hot and cool environmental radiance) when the emissivity and the temperature of the diffuse gold plate are known. Therefore, the emissivity of sample can be obtained from the four measurement equations.

The third method is to use the illumination of CO_2 laser beam in one or several TIR bands under clear sky (Zhang 1988; Nerry et al. 1991). The introduction of active source CO_2 laser will help to determine the directional reflectance in TIR bands by comparing the two measurements with and without the illumination of CO_2 laser beam. Then the surface emissivity can be calculated by the Kirchhoff's law.

4.4.2.2 Validation and Analysis of Different Methods

Undoubtedly, the quality and accuracy of the LSEs estimated using different methods can only be assessed by validation, which is a work of comparing the products to be validated with similar products derived from other independent sources. The feedback from the validation is invaluable to the improvement of quality of the generated products (Liang et al. 2002; Wang et al. 2007).

There are two distinct methods to validate the products (Li et al. 2013). The first, known as the direct method, directly compares the ground-based measurements with satellite-derived products. The second, known as the indirect method, indirectly validates the nonvalidated product with the different satellite-derived products, models simulations, or other information and applications.

The direct validation of emissivity, which is straightforward, can be accomplished by comparing the retrieval emissivity with that measured in the laboratory or in situ. However, this direct comparison is very challenging because of the heterogeneity of land surface and the difficulty in the definition of the emissivities themselves (Petitcolin and Vermote 2002). The emissivities measured in the laboratory or in situ are often not representative of the effective emissivity on a pixel scale due to the spatial variability of LSE. This problem can only be prevented over highly homogeneous and relatively flat surfaces, for example, lakes, dense vegetation, or a uniform large area of silt playa (Wan et al. 2002). The key requirements for a good direct validation site are a size larger than several pixels; a homogeneous surface in terms of material, emissivity, and temperature; easy accessibility for the deployment of instruments; and minor interference between the validation activities and the normal life activities (Wan et al. 2002). To reduce the uncertainty of measurements, the average of these measured emissivities is regarded as the "true" emissivity in a homogeneous site and can then be compared with the satellite-derived emissivity. Using this validation method, several studies are devoted to describing the quality and accuracy of different emissivity retrieval methods for various sensors.

Wan et al. (2004) compared the LSEs retrieved from MODIS D/N method with those measured from laboratory. Figure 4.15a shows the retrieved LSEs over the Sahara desert and those measured from sand samples collected in beach in the

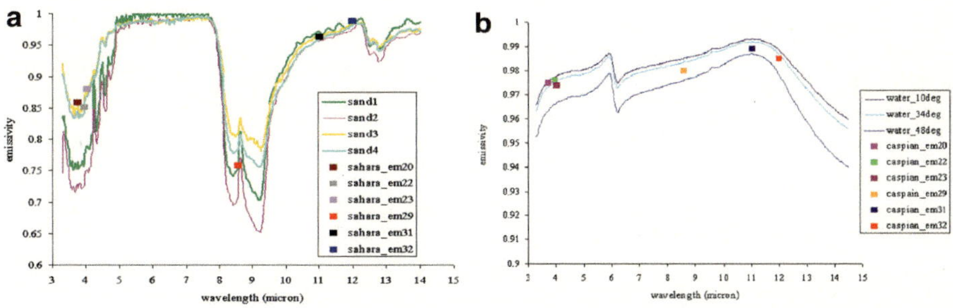

Fig. 4.15 Comparison between retrieved and measured LSEs. (**a**) Sand samples. (**b**) The sea water sample. *Lines* are measured LSE spectra in laboratory, while *squares* are the MODIS-retrieved LSEs at six channels from MIR to TIR (Adapted from Wan et al. (2004), with permission from Taylor & Francis)

laboratory, while Fig. 4.15b shows the retrieved LSEs in Caspian Sea and the theoretical emissivity values of the sea water. It is shown that the retrieved LSEs are highly correlated with the measured LSEs.

Hulley et al. (2009) validated the LSE products in the North American ASTER Land Surface Emissivity Database (NAALSED) over bare surfaces with laboratory measurements of sand samples collected at nine pseudo-invariant sand dune sites (Fig. 4.16). Results show that the retrieved LSEs by ASTER TES are usually in qualitative agreement with field or laboratory measurements.

It is reported that the RMSEs between the retrieved and measured LSEs are about 0.01–0.03 depending on the type of land surface and the retrieved method used (Coll et al. 2001; Schmugge et al. 2002; Sobrino et al. 2002; Jimenez-Munoz et al. 2006; Wang et al. 2007; Hulley et al. 2009, 2010; Hulley and Hook 2009a; Sabol et al. 2009; Wan and Li 2010). However, there are still some contradictive conclusions about the retrieved accuracies of LSEs by some certain methods (Li et al. 2013). This shows the importance and necessity of validating the LSEs in the future.

The main difficulty of direct validation of emissivity is to find homogeneous areas. In practice, the surface is always heterogeneous on the pixel scale. Indirect validation, which refers to the comparisons between different satellite-derived products, model simulations, or other information, is therefore an alternative and can reflect the performance and uncertainty for the emissivity-retrieved methods. An operational and effective validation approach is to scale the fine-resolution measurements up to the low-resolution measurement (Liang et al. 2002; Morisette et al. 2002; Wan et al. 2002; Wan 2008). Once the products at the finer resolution are validated, the accuracy of the products at the lower resolution can be assessed through product aggregation at the finer resolution. This upscale-validation method may be effective for emissivity validation; however, only a few attempts have been made to date (Hulley and Hook 2009b). Several researchers proposed using other indirect methods to validate LSE retrieved from space, such as checking the consistency of LST (Petitcolin and Vermote 2002) and LSEs (Hulley et al. 2010), comparing with the simulated data (Dash et al. 2003, 2005). Momeni and Saradjian

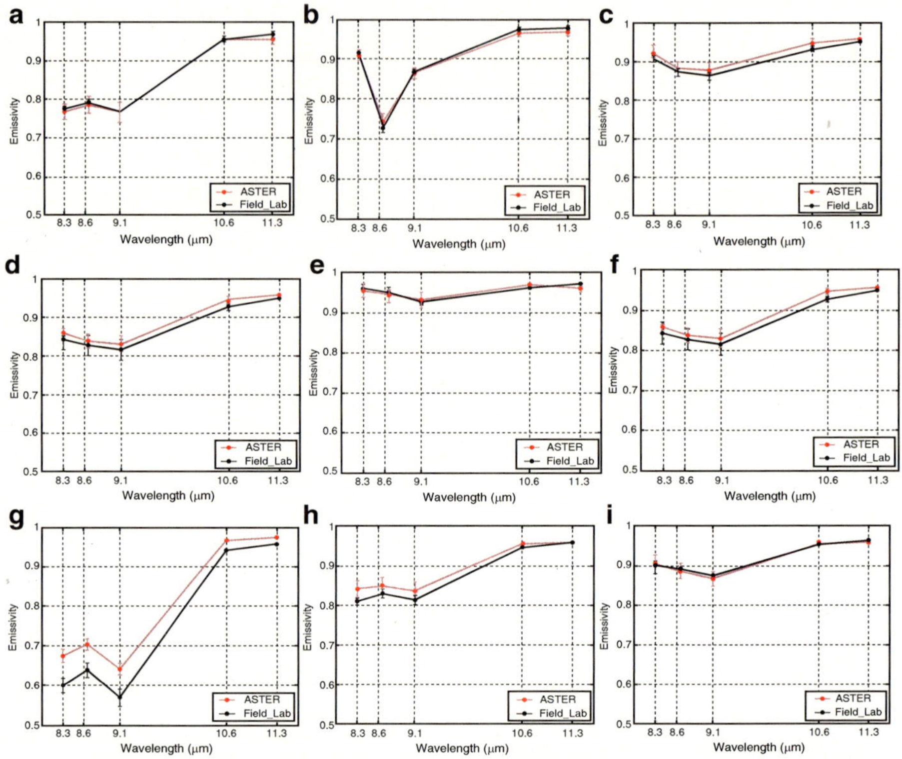

Fig. 4.16 LSE spectra comparisons between ASTER TES products and laboratory measurements for (**a**) Algodones Dunes, California; (**b**) White Sands, New Mexico; (**c**) Great Sands, Colorado; (**d**) Kelso Dunes, California; (**e**) Moses Lake, Washington; (**f**) Killpecker, Wyoming; (**g**) Coral Pink Sands, Utah; (**h**) Little Sahara, Utah; and (**i**) Stovepipe Wells, California. *Red lines* are ASTER TES products, and *black lines* are laboratory measurements (Adapted from Hulley et al. (2009), with permission from Elsevier)

(2007) suggested that the validation of LST retrieved by the D/N method is also an indirect way to assess the accuracy of the estimated LSEs because they are simultaneously derived from the same $2N$ equations. However, this suggestion needs to be further investigated and verified.

References

Abrams, M. (2000). The Advanced Spaceborne Thermal Emission and Reflection Radiometer (ASTER): Data products for the high spatial resolution imager on NASA's Terra platform. *International Journal of Remote Sensing, 21*(5), 847–859.

Aires, F., Prigent, C., Rossow, W. B., & Rothstein, M. (2001). A new neural network approach including first guess for retrieval of atmospheric water vapor, cloud liquid water path, surface temperature, and emissivities over land from satellite microwave observations. *Journal of Geophysical Research, 106*, 14887–14907.

Aires, F., Chédin, A., Scott, N. A., & Rossow, W. B. (2002a). A regularized neural net approach for retrieval of atmospheric and surface temperatures with the IASI instrument. *Journal of Applied Meteorology, 41*, 144–159.

Aires, F., Rossow, W. B., Scott, N. A., & Chédin, A. (2002b). Remote sensing from the infrared atmospheric sounding interferometer instrument, 2. Simultaneous retrieval of temperature, water vapor, and ozone atmospheric profiles. *Journal of Geophysical Research, 107*, 4620.

Baldridge, A. M., Hook, S. J., Grove, C. I., & Rivera, G. (2009). The ASTER spectral library version 2.0. *Remote Sensing of Environment, 113*(4), 711–715.

Barducci, A., & Pippi, I. (1996). Temperature and emissivity retrieval from remotely sensed images using the "grey body emissivity" method. *IEEE Transactions on Geoscience and Remote Sensing, 34*, 681–695.

Becker, F., & Li, Z.-L. (1990). Temperature-independent spectral indices in thermal infrared bands. *Remote Sensing of Environment, 32*, 17–33.

Becker, F., & Li, Z.-L. (1995). Surface temperature and emissivity at various scales: Definition, measurement and related problems. *Remote Sensing Reviews, 12*, 225–253.

Becker, F., Ramanantsizehena, P., & Stoll, M. (1985). Angular variation of the bidirectional reflectance of bare soils in the thermal infrared band. *Applied Optics, 24*, 365–375.

Blackwell, W. J. (2005). A neural-network technique for the retrieval of atmospheric temperature and moisture profiles from high spectral resolution sounding data. *IEEE Transactions on Geoscience and Remote Sensing, 43*, 2535–2546.

Borel, C. C. (1997). Iterative retrieval of surface emissivity and temperature for a hyperspectral sensor. In *First JPL workshop on remote sensing of land surface emissivity* (pp. 1–5). Pasadena: Jet Propulsion Laboratory.

Borel, C. C. (1998). Surface emissivity and temperature retrieval for a hyperspectral sensor. In *Proceedings of the 1998 IEEE International Geoscience and Remote Sensing Symposium* (pp. 546–549). Seattle.

Borel, C. C. (2008). Error analysis for a temperature and emissivity retrieval algorithm for hyperspectral imaging data. *International Journal of Remote Sensing, 29*, 5029–5045.

Carlson, T. N., & Ripley, D. A. (1997). On the relation between NDVI, fractional vegetation cover, and leaf area index. *Remote Sensing of Environment, 62*, 241–252.

Caselles, V., Valor, E., Coll, C., & Rubio, E. (1997). Thermal band selection for the PRISM instrument, 1. Analysis of emissivity-temperature separation algorithms. *Journal of Geophysical Research-Atmospheres, 102*, 11145–11164.

Chahine, M. T., Aumann, H., Goldberg, M., McMillin, L., Rosenkranz, P., Staelin, D., Strow, L., & Susskind, J. (2001). *AIRS-team retrieval for core products and geophysical parameters*. Algorithm Theoretical Basis Document JPL D-17006: AIRS Level 2 ATBD Version 2.2.

Chen, L. F., Li, Z.-L., Liu, Q. H., Chen, S., Tang, Y., & Zhong, B. (2004). Definition of component effective emissivity for heterogeneous and non-isothermal surfaces and its approximate calculation. *International Journal of Remote Sensing, 25*, 231–244.

Cheng, J., Liang, S., Wang, J., & Li, X. (2010). A stepwise refining algorithm of temperature and emissivity separation for hyperspectral thermal infrared data. *IEEE Transactions on Geoscience and Remote Sensing, 48*(3), 1588–1597.

Coll, C., Caselles, V., Rubio, E., Sospedra, F., & Valor, E. (2001). Temperature and emissivity separation from calibrated data of the Digital Airborne Imaging Spectrometer. *Remote Sensing of Environment, 76*(2), 250–259.

Coll, C., Valor, E., Caselles, V., & Niclos, R. (2003). Adjusted normalized emissivity method for surface temperature and emissivity retrieval from optical and thermal infrared remote sensing data. *Journal of Geophysical Research-Atmospheres, 108*, 4739.

Coll, C., Caselles, V., Valor, E., Niclòs, R., Sánchez, J. M., Galve, J. M., & Mira, M. (2007). Temperature and emissivity separation from ASTER data for low spectral contrast surfaces. *Remote Sensing of Environment, 110*, 162–175.

Cuenca, J., & Sobrino, J. A. (2004). Experimental measurements for studying angular and spectral variation of thermal infrared emissivity. *Applied Optics, 43*, 4598–4602.

Dash, P., Göttsche, F. M., Olesen, F. S., & Fischer, H. (2002). Land surface temperature and emissivity estimation from passive sensor data: Theory and practice-current trends. *International Journal of Remote Sensing, 23*(13), 2563–2594.

Dash, P., Gottsche, F. M., & Olesen, F. S. (2003). Emissivity and temperature estimation from MSG SEVIRI data; Method validation with simulated and NOAA-14 AVHRR data. *Advances in Space Research, 32*, 2241–2246.

Dash, P., Gottsche, F. M., Olesen, F. S., & Fischer, H. (2005). Separating surface emissivity and temperature using two-channel spectral indices and emissivity composites and comparison with a vegetation fraction method. *Remote Sensing of Environment, 96*, 1–17.

Farmer, V. C. (1974). *The infrared spectra of minerals*. London: Mineralogical Society.

Gillespie, A. R. (1985). Lithologic mapping of silicate rocks using TIMS. In *Proceedings of the thermal infrared multispectral scanner data user's workshop* (pp. 29–44). Pasadena: Jet Propulsion Laboratory, California Institute of Technology.

Gillespie, A., Rokugawa, S., Matsunaga, T., Cothern, J. S., Hook, S., & Kahle, A. B. (1998). A temperature and emissivity separation algorithm for Advanced Spaceborne Thermal Emission and Reflection Radiometer (ASTER) images. *IEEE Transactions on Geoscience and Remote Sensing, 36*, 1113–1126.

Gillespie, A. R., Rokugawa, S., Hook, S. J., Matsunaga, T., & Kahle, A. B. (1999). Temperature/emissivity separation algorithm theoretical basis document, Version 2.4. *NASA/GSFC*, pp. 1–64.

Gillespie, A. R., Abbott, E. A., Gilson, L., Hulley, G., Jiménez-Muñoz, J. C., & Sobrino, J. A. (2011). Residual errors in ASTER temperature and emissivity standard products AST08 and AST05. *Remote Sensing of Environment, 115*(12), 3681–3694.

Gustafson, W. T., Gillespie, A. R., & Yamada, G. J. (2006, September 25–29). Revisions to the ASTER temperature/emissivity separation algorithm. In *2nd international symposium on Recent Advances in Quantitative Remote Sensing* (pp. 770–775). Torrent, Valencia, Spain.

Hook, S. J., Gabell, A. R., Green, A. A., & Kealy, P. S. (1992). A comparison of techniques for extracting emissivity information from thermal infrared data for geologic studies. *Remote Sensing of Environment, 42*, 123–135.

Hori, M., Aoki, T., Tanikawa, T., Motoyoshi, H., Hachikubo, A., Sugiura, K., Yasunari, T., Eide, H., Storvold, R., Nakajima, Y., & Takahashi, F. (2006). In-situ measured spectral directional emissivity of snow and ice in the 8–14 μm atmospheric window. *Remote Sensing of Environment, 100*(4), 486–502.

Hulley, G. C., & Hook, S. J. (2009a). The North American ASTER land surface emissivity database (NAALSED) version 2.0. *Remote Sensing of Environment, 113*(9), 1967–1975.

Hulley, G. C., & Hook, S. J. (2009b). Intercomparison of versions 4, 4.1 and 5 of the MODIS Land Surface Temperature and Emissivity products and validation with laboratory measurements of sand samples from the Namib desert, Namibia. *Remote Sensing of Environment, 113*, 1313–1318.

Hulley, G. C., & Hook, S. J. (2011). Generating consistent land surface temperature and emissivity products between ASTER and MODIS data for earth science research. *IEEE Transactions on Geoscience and Remote Sensing, 49*(4), 1304–1315.

Hulley, G. C., Hook, S. J., & Baldridge, A. M. (2009). Validation of the North American ASTER Land Surface Emissivity Database (NAALSED) version 2.0 using pseudo-invariant sand dune sites. *Remote Sensing of Environment, 113*, 2224–2233.

Hulley, G. C., Hook, S. J., & Baldridge, A. M. (2010). Investigating the effects of soil moisture on thermal infrared land surface temperature and emissivity using satellite retrievals and laboratory measurements. *Remote Sensing of Environment, 114*(7), 1480–1493.

Ingram, P. M., & Muse, A. H. (2001). Sensitivity of iterative spectrally smooth temperature/emissivity separation to algorithmic assumptions and measurement noise. *IEEE Transactions on Geoscience and Remote Sensing, 39*, 2158–2167.

Jacob, F., Petitcolin, F., Schmugge, T., Vermote, E., French, A., & Ogawa, K. (2004). Comparison of land surface emissivity and radiometric temperature derived from MODIS and ASTER sensors. *Remote Sensing of Environment, 90*, 137–152.

Jiang, G. M., & Li, Z.-L. (2008). Intercomparison of two BRDF models in the estimation of the directional emissivity in MIR channel from MSG1-SEVIRI data. *Optics Express, 16*, 19310–19321.

Jiang, G. M., Li, Z.-L., & Nerry, F. (2006). Land surface emissivity retrieval from combined mid-infrared and thermal infrared data of MSG-SEVIRI. *Remote Sensing of Environment, 105*, 326–340.

Jimenez-Munoz, J. C., Sobrino, J. A., Gillespie, A., Sabol, D., & Gustafson, W. T. (2006). Improved land surface emissivities over agricultural areas using ASTER NDVI. *Remote Sensing of Environment, 103*, 474–487.

Jin, M., & Liang, S. (2006). An improved land surface emissivity parameter for land surface models using global remote sensing observations. *Journal of Climate, 19*(12), 2867–2881.

Kahle, A. B., Madura, D. P., & Soha, J. M. (1980). Middle infrared multispectral aircraft scanner data: Analysis for geological applications. *Applied Optics, 19*, 2279–2290.

Kanani, K., Poutier, L., Nerry, F., & Stoll, M. (2007). Directional effects consideration to improve out-doors emissivity retrieval in the 3–13 μm domain. *Optics Express, 15*, 12464–12482.

Labed, J., & Stoll, M. P. (1991). Angular variation of land surface spectral emissivity in the thermal infrared: Laboratory investigations on bare soils. *International Journal of Remote Sensing, 12*, 2299–2310.

Li, Z.-L., & Becker, F. (1990). Properties and comparison of temperature-independent thermal infrared spectral indices with NDVI for HAPEX data. *Remote Sensing of Environment, 33*, 165–182.

Li, Z.-L., & Becker, F. (1993). Feasibility of land surface-temperature and emissivity determination from AVHRR data. *Remote Sensing of Environment, 43*, 67–85.

Li, X. W., & Strahler, A. H. (1992). Geometric-optical bidirectional reflectance modeling of the discrete crown vegetation canopy: Effect of crown shape and mutual shadowing. *IEEE Transactions on Geoscience and Remote Sensing, 30*, 276–292.

Li, J., Zhou, F., & Zeng, Q. (1994). Simultaneous non-linear retrieval of atmospheric temperature and absorbing constituent profiles from satellite infrared sounder radiances. *Advances in Atmospheric Sciences, 11*, 128–138.

Li, X. W., Strahler, A. H., & Friedl, M. A. (1999a). A conceptual model for effective directional emissivity from nonisothermal surfaces. *IEEE Transactions on Geoscience and Remote Sensing, 37*, 2508–2517.

Li, Z.-L., Becker, F., Stoll, M. P., & Wan, Z. M. (1999b). Evaluation of six methods for extracting relative emissivity spectra from thermal infrared images. *Remote Sensing of Environment, 69*, 197–214.

Li, Z.-L., Petitcolin, F., & Zhang, R. H. (2000). A physically based algorithm for land surface emissivity retrieval from combined mid-infrared and thermal infrared data. *Science in China Series E-Technological Sciences, 43*, 23–33.

Li, J., Li, J., Weisz, E., & Zhou, D. K. (2007). Physical retrieval of surface emissivity spectrum from hyperspectral infrared radiances. *Geophysical Research Letters, 34*, L16812.

Li, Z.-L., Wu, H., Wang, N., Qiu, S., Sobrino, J. A., Wan, Z., Tang, B., & Yan, G. (2013). Land surface emissivity retrieval from satellite data. *International Journal of Remote Sensing, 34*(9–10), 3084–3127.

Liang, S. L., Fang, H. L., Chen, M. Z., Shuey, C. J., Walthall, C., Daughtry, C., Morisette, J., Schaaf, C., & Strahler, A. (2002). Validating MODIS land surface reflectance and albedo products: Methods and preliminary results. *Remote Sensing of Environment, 83*, 149–162.

Lucht, W. (1998). Expected retrieval accuracies of bidirectional reflectance and albedo from EOS-MODIS and MISR angular sampling. *Journal of Geophysical Research, 103*, 8763–8778.

Lucht, W., & Roujean, J. (2000). Considerations in the parametric modeling of BRDF and albedo from multiangular satellite sensor observations. *Remote Sensing Reviews, 18*, 343–379.
Ma, X. L., Schmit, T. J., & Smith, W. L. (1999). A nonlinear physical retrieval algorithm-its application to the GOES-8/9 sounder. *Journal of Applied Meteorology, 38*, 501–513.
Ma, X. L., Wan, Z. M., Moeller, C. C., Menzel, W. P., Gumley, L. E., & Zhang, Y. L. (2000). Retrieval of geophysical parameters from Moderate Resolution Imaging Spectroradiometer thermal infrared data: Evaluation of a two-step physical algorithm. *Applied Optics, 39*, 3537–3550.
Ma, X. L., Wan, Z. M., Moeller, C. C., Menzel, W. P., & Gumley, L. E. (2002). Simultaneous retrieval of atmospheric profiles, land-surface temperature, and surface emissivity from Moderate-Resolution Imaging Spectroradiometer thermal infrared data: Extension of a two-step physical algorithm. *Applied Optics, 41*, 909–924.
Mao, K., Shi, J., Tang, H., Li, Z.-L., Wang, X., & Chen, K. S. (2008). A neural network technique for separating land surface emissivity and temperature from ASTER imagery. *IEEE Transactions on Geoscience and Remote Sensing, 46*, 200–208.
Mas, J. F., & Flores, J. J. (2008). The application of artificial neural networks to the analysis of remotely sensed data. *International Journal of Remote Sensing, 29*, 617–663.
Matsunaga, T. (1994). A temperature-emissivity separation method using an empirical relationship between the mean, the maximum, and the minimum of the thermal infrared emissivity spectrum. *Journal of the Remote Sensing Society of Japan, 14*(2), 230–241.
Menzel, W. P., Seemann, S. W., Li, J., & Gumley, L. E. (2006). *MODIS atmospheric profile retrieval algorithm theoretical basis document*, Version 6. *NASA/GSFC*, pp. 1–40.
Minnaert, M. (1941). The reciprocity principle of linear photometry. *Astrophysical Journal, 93*, 403–410.
Momeni, M., & Saradjian, M. R. (2007). Evaluating NDVI-based emissivities of MODIS bands 31 and 32 using emissivities derived by Day/Night LST algorithm. *Remote Sensing of Environment, 106*, 190–198.
Morisette, J. T., Privette, J. L., & Justice, C. O. (2002). A framework for the validation of MODIS Land products. *Remote Sensing of Environment, 83*, 77–96.
Motteler, H. E., Strow, L. L., McMillin, L., & Gualtieri, J. A. (1995). Comparison of neural networks and regression-based methods for temperature retrievals. *Applied Optics, 34*, 5390–5397.
Mushkin, A., Balick, L. K., & Gillespie, A. R. (2005). Extending surface temperature and emissivity retrieval to the mid-infrared (3–5 μm) using the Multispectral Thermal Imager (MTI). *Remote Sensing of Environment, 98*, 141–151.
Nerry, F., Stoll, M. P., & Kologo, N. (1991). Scattering of a CO2 laser beam at 10.6 um by bare soils: Experimental study of the polarized bidirectional scattering coefficient; model and comparison with directional emissivity measurements. *Applied Optics, 30*, 3984–3994.
Nerry, F., Petitcolin, F., & Stoll, M. P. (1998). Bidirectional reflectivity in AVHRR Channel 3: Application to a region in Northern Africa. *Remote Sensing of Environment, 66*, 298–316.
Norman, J. M., & Becker, F. (1995). Terminology in thermal infrared remote sensing of natural surfaces. *Agricultural and Forest Meteorology, 77*, 153–166.
OuYang, X., Wang, N., Wu, H., & Li, Z.-L. (2010). Errors analysis on temperature and emissivity determination from hyperspectral thermal infrared data. *Optics Express, 18*, 544–550.
Payan, V., & Royer, A. (2004). Analysis of temperature emissivity separation (TES) algorithm applicability and sensitivity. *International Journal of Remote Sensing, 25*(1), 15–37.
Peres, L. F., & DaCamara, C. C. (2004). Land surface temperature and emissivity estimation based on the two-temperature method: Sensitivity analysis using simulated MSG/SEVIRI data. *Remote Sensing of Environment, 91*, 377–389.
Peres, L. F., & DaCamara, C. C. (2005). Emissivity maps to retrieve land-surface temperature from MSG/SEVIRI. *IEEE Transactions on Geoscience and Remote Sensing, 43*, 1834–1844.
Petitcolin, F., & Vermote, E. (2002). Land surface reflectance, emissivity and temperature from MODIS middle and thermal infrared data. *Remote Sensing of Environment, 83*, 112–134.

Petitcolin, F., Nerry, F., & Stoll, M. P. (2002). Mapping temperature independent spectral indice of emissivity and directional emissivity in AVHRR channels 4 and 5. *International Journal of Remote Sensing, 23*, 3473–3491.

Prata, A. J. (1993). Land surface temperatures derived from the Advanced Very High-Resolution Radiometer and the Along-Track Scanning Radiometer: 1. Theory. *Journal of Geophysical Research-Atmospheres, 98*, 16689–16702.

Realmuto, V. J. (1990). Separating the effects of temperature and emissivity: Emissivity spectrum normalization. In *Proceedings of the second TIMS workshop* (pp. 23–27). Pasadena: Jet Propulsion Laboratory.

Ribeiro da Luz, B., & Crowley, J. K. (2007). Spectral reflectance and emissivity features of broad leaf plants: Prospects for remote sensing in the thermal infrared (8.0–14.0 μm). *Remote Sensing of Environment, 109*(4), 393–405.

Roujean, J. L., Leroy, M., & Deschamps, P. Y. (1992). A bidirectional reflectance model of the Earth's surface for the correction of remote sensing data. *Journal of Geophysical Research, 97*, 20455–20468.

Rubio, E., Caselles, V., & Badenas, C. (1997). Emissivity measurements of several soils and vegetation types in the 8–14, μm wave band: Analysis of two field methods. *Remote Sensing of Environment, 59*(3), 490–521.

Rubio, E., Caselles, V., Coll, C., Valor, E., & Sospedra, F. (2003). Thermal–infrared emissivities of natural surfaces: Improvements on the experimental set-up and new measurements. *International Journal of Remote Sensing, 24*(24), 5379–5390.

Sabol, D. E., Jr., Gillespie, A. R., Abbott, E., & Yamada, G. (2009). Field validation of the ASTER temperature–emissivity separation algorithm. *Remote Sensing of Environment, 113*(11), 2328–2344.

Salisbury, J. W., & D'Aria, D. M. (1992). Emissivity of terrestrial materials in the 8–14 mm atmospheric window. *Remote Sensing of Environment, 42*, 83–106.

Salisbury, J. W., Wald, A., & D'Aria, D. M. (1994). Thermal-infrared remote-sensing and Kirchhoff law: 1. Laboratory measurements. *Journal of Geophysical Research-Solid Earth, 99*, 11897–11911.

Sawabe, Y., Matsunaga, T., Rokugawa, S., & Hoyano, A. (2003). Temperature and emissivity separation for multi-band radiometer and validation ASTER TES algorithm. *Journal of Remote Sensing Society of Japan, 23*, 364–375.

Schmugge, T., French, A., Ritchie, J. C., Rango, A., & Pelgrum, H. (2002). Temperature and emissivity separation from multispectral thermal infrared observations. *Remote Sensing of Environment, 79*(2), 189–198.

Smith, W. L. (1972). Satellite techniques for observing the temperature structure of the atmosphere. *Bulletin of the American Meteorological Society, 53*, 1074–1082.

Smith, W. L., & Woolf, H. M. (1976). The use of eigenvectors of statistical covariance matrices for interpreting satellite sounding radiometer observations. *Journal of Atmospheric Sciences, 33*, 1127–1140.

Snyder, W. C., & Wan, Z. M. (1998). BRDF models to predict spectral reflectance and emissivity in the thermal infrared. *IEEE Transactions on Geoscience and Remote Sensing, 36*, 214–225.

Snyder, W. C., Wan, Z., Zhang, Y., & Feng, Y. Z. (1998). Classification-based emissivity for land surface temperature measurement from space. *International Journal of Remote Sensing, 19*, 2753–2774.

Sobrino, J. A., & Cuenca, J. (1999). Angular variation of thermal infrared emissivity for some natural surfaces from experimental measurements. *Applied Optics, 38*, 3931–3936.

Sobrino, J. A., & Raissouni, N. (2000). Toward remote sensing methods for land cover dynamic monitoring: Application to Morocco. *International Journal of Remote Sensing, 21*, 353–366.

Sobrino, J. A., Raissouni, N., & Li, Z.-L. (2001). A comparative study of land surface emissivity retrieval from NOAA data. *Remote Sensing of Environment, 75*, 256–266.

Sobrino, J. A., Jimenez-Munoz, J. C., Labed-Nachbrand, J., & Nerry, F. (2002). Surface emissivity retrieval from Digital Airborne Imaging Spectrometer data. *Journal of Geophysical Research-Atmospheres, 107*, 4729.

Sobrino, J. A., Kharraz, J. E., & Li, Z.-L. (2003). Surface temperature and water vapour retrieval from MODIS data. *International Journal of Remote Sensing, 24*, 5161–5182.
Sobrino, J. A., Jimenez-Munoz, J. C., & Paolini, L. (2004). Land surface temperature retrieval from LANDSAT TM 5. *Remote Sensing of Environment, 90*, 434–440.
Sobrino, J. A., Jimenez-Munoz, J. C., & Verhoef, W. (2005). Canopy directional emissivity: Comparison between models. *Remote Sensing of Environment, 99*, 304–314.
Sobrino, J. A., Jimenez-Munoz, J. C., Balick, L., Gillespie, A. R., Sabol, D. A., & Gustafson, W. T. (2007). Accuracy of ASTER level-2 thermal infrared standard products of an agricultural area in Spain. *Remote Sensing of Environment, 106*, 146–153.
Sobrino, J. A., Jimenez-Munoz, J. C., Soria, G., Romaguera, M., Guanter, L., Moreno, J., Plaza, A., & Martincz, P. (2008). Land surface emissivity retrieval from different VNIR and TIR sensors. *IEEE Transactions on Geoscience and Remote Sensing, 46*, 316–327.
Sontag, E. D. (1992). Feedback stabilization using two-hidden-layer nets. *IEEE Transactions on Neural Networks, 3*, 981–990.
Sun, D. L., & Pinker, R. T. (2003). Estimation of land surface temperature from a Geostationary Operational Environmental Satellite (GOES-8). *Journal of Geophysical Research-Atmospheres, 108*, 4326.
Tian, J., Zhang, R., Su, H., Sun, X., Chen, S., & Xia, J. (2008). An automatic instrument to study the spatial scaling behavior of emissivity. *Sensors, 8*(2), 800–816.
Tonooka, H. (2005). Accurate atmospheric correction of ASTER thermal infrared imagery using the WVS method. *IEEE Transactions on Geoscience and Remote Sensing, 43*, 2778–2792.
Trigo, I. F., Peres, L. F., DaCamara, C. C., & Freitas, S. C. (2008). Thermal land surface emissivity retrieved from SEVIRI/Meteosat. *IEEE Transactions on Geoscience and Remote Sensing, 46*(2), 307–315.
Valor, E., & Caselles, V. (1996). Mapping land surface emissivity from NDVI: Application to European, African, and South American areas. *Remote Sensing of Environment, 57*, 167–184.
Van de Griend, A. A., & Owe, M. (1993). On the relationship between thermal emissivity and the normalized difference vegetation index for natural surfaces. *International Journal of Remote Sensing, 14*, 1119–1131.
Wan, Z. M. (1999). *MODIS land surface temperature algorithm theoretical basis document*, Version 3.3. *NASA/GSFC*, pp. 1–77.
Wan, Z. M. (2008). New refinements and validation of the MODIS land-surface temperature/emissivity products. *Remote Sensing of Environment, 112*, 59–74.
Wan, Z. M., Ng, D., & Dozier, J. (1994). Spectral emissivity measurements of land-surface materials and related radiative transfer simulations. *Advances in Space Research, 14*(3), 91–94.
Wan, Z. M., & Li, Z.-L. (1997). A physics-based algorithm for retrieving land-surface emissivity and temperature from EOS/MODIS data. *IEEE Transactions on Geoscience and Remote Sensing, 35*, 980–996.
Wan, Z., Zhang, Y., Zhang, Q., & Li, Z.-L. (2002). Validation of the land-surface temperature products retrieved from TerraModerate Resolution Imaging Spectroradiometer data. *Remote Sensing of Environment, 83*, 163–180.
Wan, Z. M., & Li, Z.-L. (2008). Radiance-based validation of the V5 MODIS land-surface temperature product. *International Journal of Remote Sensing, 29*, 5373–5395.
Wan, Z. M., & Li, Z.-L. (2010). MODIS land surface temperature and emissivity products. In B. Ramachandran, C. O. Justice, & M. J. Abrams (Eds.), *Land remote sensing and global environmental changes* (pp. 563–577). London: Springer.
Wan, Z. M., Zhang, Y., Zhang, Q., & Li, Z.-L. (2004). Quality assessment and validation of the MODIS global land surface temperature. *International Journal of Remote Sensing, 25*, 261–274.
Wang, K., & Liang, S. L. (2009). Evaluation of ASTER and MODIS land surface temperature and emissivity products using long-term surface longwave radiation observations at SURFRAD sites. *Remote Sensing of Environment, 113*, 1556–1565.

Wang, K., Wan, Z., Wang, P., Sparrow, M., Liu, J., & Haginoya, S. (2007). Evaluation and improvement of the MODIS land surface temperature/emissivity products using ground-based measurements at a semi-desert site on the western Tibetan Plateau. *International Journal of Remote Sensing, 28*, 2549–2565.

Wang, N., Tang, B. H., Li, C., & Li, Z.-L. (2010). A generalized neural network for simultaneous retrieval of atmospheric profiles and surface temperature from hyperspectral thermal infrared data. In *IEEE International Geoscience and Remote Sensing Symposium* (pp. 1055–1058). Honolulu: IEEE.

Wang, N., Wu, H., Nerry, F., Li, C. R., & Li, Z.-L. (2011). Temperature and emissivity retrievals from hyperspectral thermal infrared data using linear spectral emissivity constraint. *IEEE Transactions on Geoscience and Remote Sensing, 49*, 1291–1303.

Wang, N., Li, Z.-L., Tang, B. H., Zeng, F., & Li, C. (2013). Retrieval of atmospheric and land surface parameters from satellite-based thermal infrared hyperspectral data using a neural network technique. *International Journal of Remote Sensing, 34*(9–10), 3485–3502.

Wanner, W., Li, X., & Strahler, A. (1995). On the derivation of kernels for kernel-driven models of bidirectional reflectance. *Journal of Geophysical Research, 100*, 21077.

Watson, K. (1992). Two-temperature method for measuring emissivity. *Remote Sensing of Environment, 42*, 117–121.

Zhang, R. H. (1988). A proposed approach to determine the infrared emissivities of terrestrial surfaces from airborne or spaceborne platforms. *International Journal of Remote Sensing, 9*(3), 591–595.

Zhang, R. H., Li, Z.-L., Tang, X. Z., Sun, X. M., Su, H. B., Zhu, C., & Zhu, Z. L. (2004). Study of emissivity scaling and relativity of homogeneity of surface temperature. *International Journal of Remote Sensing, 25*(1), 245–259.

Chapter 5
Land Surface Temperature Retrieval from Thermal Infrared Data

5.1 Background

Land surface temperature (LST) is not only a good indicator of both the energy equilibrium of the earth's surface and greenhouse effects but also one of the key variables controlling fundamental biospheric and geospheric interactions between the earth's surface and its atmosphere (Sellers et al. 1988; Prata et al. 1995; Brunsell and Gillies 2003; Zhang et al. 2008). As the direct driving force in the exchange of longwave radiation and turbulent heat fluxes in the surface–atmosphere interface, it can play either a direct role such as when estimating longwave fluxes or an indirect role such as when estimating latent and sensible heat fluxes (Mannstein 1987; Sellers et al. 1988). Moreover, many other applications, such as evapotranspiration modeling (Serafini 1987; Bussieres et al. 1990; Su 2002; Kalma et al. 2008), soil moisture estimating (Price 1990), urban climatic, hydrological, vegetation monitoring, environmental studies (Bastiaanssen et al. 1998; Kogan 2001; Su 2002; Arnfield 2003; Voogt and Oke 2003; Weng et al. 2004; Weng 2009, Hansen et al. 2010), and biogeochemical studying (Schmugge and André 1991; Running et al. 1994), rely on the knowledge of LST by providing information on the temporal and spatial variations of the surface equilibrium state (Kerr et al. 2000), and it has been recognized as one of the high-priority parameters of the International Geosphere and Biosphere Program (IGBP) (Townshend et al. 1994). Because of the strong heterogeneity of land surface characteristics in vegetation, topography, and soil (Liu et al. 2006; Neteler 2010), the LST changes rapidly in space as well as in time (Vauclin et al. 1982; Prata et al. 1995). An adequate characterization of the LST distribution and its temporal evolution, therefore, requires measurements with detailed spatial and temporal sampling. These measurements cannot be obtained using ground-based systems due to the complexity and inaccuracy of conventional LST measurements and the errors in spatial scaling resulting from inadequate sampling and inaccurate scaling methods. With the development of remote sensing from space, satellite data offer the only possibility for measuring the LST at the largest scales with sufficient high temporal resolution and with complete spatially

H. Tang and Z.-L. Li, *Quantitative Remote Sensing in Thermal Infrared: Theory and Applications*, Springer Remote Sensing/Photogrammetry,
DOI 10.1007/978-3-642-42027-6_5, © Springer-Verlag Berlin Heidelberg 2014

averaged rather than point values. Consequently, the retrieval of LST from satellite measurements is of increasing attraction with respect to current earth observations.

To better understand the earth system at the global scale, LST must be retrieved at an accuracy of about 1 K (Wan and Dozier 1996). However, direct estimation of the LST accurately from the radiation emitted in the thermal infrared (TIR) spectral region is a very difficult task because the radiances measured by the radiometers onboard satellites depend not only on surface parameters (temperature and emissivity) but also on atmospheric effects (Li and Becker 1993; Ottlé and Stoll 1993; Prata et al. 1995). Therefore, besides radiometric calibration and cloud screening, the determination of LSTs from space-based TIR measurements requires both emissivity and atmospheric corrections (Vidal 1991; Li and Becker 1993). The history of retrieving LST from satellite TIR data can date back to the 1970s (McMillin 1975). Up to now, many studies have been carried out, and different approaches have been proposed to derive LSTs from satellite TIR data by using a variety of methods to deal with the effects of emissivity and atmosphere (Price 1983, 1984; Susskind et al. 1984; Becker and Li 1990a; Kerr et al. 1992; Wan and Dozier 1996; Pozo Vazquez et al. 1997; Wan and Li 1997; Gillespie et al. 1998; Qin et al. 2001; Tonooka 2001; Jiménez-Muñoz and Sobrino 2003; Tang et al. 2008; Li et al. 2013a). Consequently, there have been quite a large number of publications on LST retrieval algorithms and methods.

This chapter inherits and updates the overview given by Li et al. (2013a) of satellite- derived LST methods from TIR data and the work of Becker and Li (1995) on the LST definition. Section 5.2 gives the LST definition at satellite pixel scale. Section 5.3 highlights the difficulties and problems in the retrieval of LST from space measurements. Various representative LST retrieval methods and models are detailed and categorized in Sects. 5.4 and 5.5. Comparison and analysis of different methods are depicted in Sect. 5.6. Finally, Sect. 5.7 presents several LST retrieval validation methods.

5.2 Definition of Land Surface Temperature

The definition of the surface temperature may depend strongly on the type of application and the method of measurement. How to physically interpret the results of the LST measurement remains a crucial question.

To determine surface heat flux at large scale from space measurement, it is necessary to introduce a bulk temperature for the whole ground pixel area, and this temperature should be measurable from space and should be related to the corresponding fluxes. The strong vertical and horizontal heterogeneity of land surfaces makes it very difficult to define an overall surface temperature for a given pixel at a given scale, for instance, from in situ to aircraft or various satellite measurements. So far, there have been many definitions for such bulk temperature, and each of them is based upon different assumptions regarding the integration process used to build up such a temperature from the very different temperatures within the pixel (Sutherland and Bartholic 1977; Becker 1981; Price 1982; Caselles et al. 1988; Li and Becker 1993; Becker and Li 1995).

Becker and Li (1995) proposed a definition of radiometric temperature $\langle T_{sr}\rangle$ to take into account the heterogeneous and nonisothermal surfaces for a given pixel at different scales. In the definition, the following three criteria should be satisfied: (1) The defined temperature is directly measurable from space by a radiometer. (2) The defined temperature must be coherent when the surface changes from heterogeneity to homogeneity. (3) The definition of temperature is scale invariant in space.

As for a large heterogeneous pixel, for example, observed by a satellite, and assuming that it contains n different homogeneous subelements, each element k is characterized by a surface parameter such as temperature T_{sk} or emissivity $\varepsilon_{\lambda k}$. This parameter is defined from the measured radiance $R_{\lambda k}$ by a model M, which relates $R_{\lambda k}$ to T_{sk} (or $\varepsilon_{\lambda k}$) by the relation

$$T_{sk} = M\left(R_{\lambda k}\right) \tag{5.1}$$

For the whole pixel, at the satellite's scale, the overall radiance $\langle R\rangle_\lambda$ may be described as a function of the n measured radiances $R_{\lambda k}$ ($k = 1, \ldots, n$) at local scale by

$$\langle R\rangle_\lambda = g\left(R_{\lambda 1}, R_{\lambda 2}, \ldots R_{\lambda n}\right) \tag{5.2}$$

while the corresponding parameter $\langle T_{sr}\rangle$ (or $\langle\varepsilon\rangle_\lambda$) is related to T_{sk} (or $\varepsilon_{\lambda k}$) ($k = 1, n$) by

$$\langle T_{sr}\rangle = f\left(T_{s1}, T_{s2}, \ldots T_{sn}\right) \tag{5.3}$$

At large scale, there exists a model $\Re$ (which could be inferred from the model M) relating $\langle R\rangle_\lambda$ to $\langle T_{sr}\rangle$ by

$$\langle T_{sr}\rangle = \Re\left(\langle R\rangle_\lambda\right) \tag{5.4}$$

Table 5.1 gives the scheme of this change of scale for the parameters and the models described above. Generally, neither the functions f and g nor the models M and $\Re$ are identical except in the case where M is the linear function and f, g are the average functions.

Let $\mathrm{d}\omega_k$ be a small solid angle from which a homogeneous area within the pixel with a monochromatic radiance $R_{\lambda k}$ is observed at ground level by a radiometer having a field of view Ω $\left(\Omega = \sum_{k=1}^{n} \mathrm{d}\omega_k\right)$. The radiance received by this radiometer can be written as

Table 5.1 Illustration of the parameter relationships at different scales

	Descriptive parameters	Models	Measured parameters
Local scale (all parameters are well defined)	$T_{s1}, T_{s2}, \ldots T_{sn}$ or $\varepsilon_{\lambda 1}, \varepsilon_{\lambda 2}, \ldots \varepsilon_{\lambda n}$	$T_{sk} = M\,(R_{\lambda k})$ $(k = 1, n)$	$R_{\lambda 1}, R_{\lambda 2}, \ldots, R_{\lambda n}$
Way to change scale	$\langle T_{sr}\rangle = f(T_{s1}, T_{s2}, \ldots T_{sn})$ or $\langle\varepsilon\rangle_\lambda = f(\varepsilon_{\lambda 1}, \varepsilon_{\lambda 2}, \ldots \varepsilon_{\lambda n})$	$M \to \Re$	$\langle R\rangle_\lambda = g(R_{\lambda 1}, R_{\lambda 2}, \ldots R_{\lambda n})$
Large scale (parameters to be redefined)	$\langle T_{sr}\rangle$ or $\langle\varepsilon\rangle_\lambda$	$\langle T_{sr}\rangle = \Re(\langle R\rangle_\lambda)$	$\langle R\rangle_\lambda$

$$\langle R\rangle_\lambda = \sum_{k=1}^{n} R_{\lambda k} S_k \tag{5.5}$$

where S_k is the normalized fractional area of the homogeneous subpixel k, that is, $S_k = \mathrm{d}\omega_k/\Omega$, and n is the number of homogeneous areas within the pixel. Equation (5.5) is, by the law of radiometry, the composition function of g in Eq. (5.2).

The definition of an equivalent $\langle T_{sr}\rangle$ from Eq. (5.5) is not a simple problem due to the nonlinearities of the equations and because of two unknown variables $\langle\varepsilon\rangle_\lambda$ and $\langle T_{sr}\rangle$ to be defined with this unique relation.

5.2.1 Definition of Temperature for a Flat Surface

For a flat surface k and for a given wavelength λ, the measured radiance $R_{\lambda k}$ at ground level can be expressed as

$$R_{\lambda k} = \varepsilon_{\lambda k} B_\lambda\,(T_{sk}) + (1 - \varepsilon_{\lambda k})\,R_{at\_\lambda\downarrow} \tag{5.6}$$

where $\varepsilon_{\lambda k}$ is the surface spectral emissivity of the element k, and $B_\lambda(T_{sk})$ is the Planck function with surface temperature T_{sk} for element k. Equation (5.6) is the inverse of the model M.

Inserting Eq. (5.6) into Eq. (5.5), one gets

$$\langle R\rangle_\lambda = \sum_{k=1}^{n} \varepsilon_{\lambda k} B_\lambda\,(T_{sk})\,S_k + \left(1 - \sum_{k=1}^{n} \varepsilon_{\lambda k} S_k\right) R_{at\_\lambda\downarrow}$$

To preserve the form of Eq. (5.6) for a heterogeneous surface, that is to wish $M = \Re$, one can define $\langle T_{sr}\rangle$ as

$$\langle R\rangle_\lambda = \langle\varepsilon\rangle_\lambda B_\lambda\,(\langle T_{sr}\rangle) + (1 - \langle\varepsilon\rangle_\lambda)\,R_{at\_\lambda\downarrow} \tag{5.7}$$

which yields

$$\langle \varepsilon \rangle_\lambda = \sum_{k=1}^{N} \varepsilon_{\lambda k} S_k, \tag{5.8}$$

$$\langle T_{sr} \rangle_\lambda = B_\lambda^{-1} \left[\frac{\sum_{k=1}^{N} \varepsilon_{\lambda k} B_\lambda (T_{sk}) S_k}{\langle \varepsilon \rangle_\lambda} \right] \tag{5.9}$$

Equation (5.8) is the composition function f for emissivity, while Eq. (5.9) is that for temperature. Note that the Eq. (5.8) has the same form as Eq. (5.5), which means in that case $f = g$. This is the particular case where the relationship between $\varepsilon_{\lambda k}$ (emissivity) and $R_{\lambda k}$ (radiance) is linear as shown in Eq. (5.6), which is not the case for the definition of $\langle T_{sr} \rangle_\lambda$; thus in that case $f \neq g$.

Had we required that $f = g$ for both emissivity and temperature, this would have led to another definition of the surface temperature, namely,

$$\langle T_{sm} \rangle = \sum_{k=1}^{N} T_{sk} S_k \tag{5.10}$$

In the TIR bands, this definition does not preserve at all scales, and therefore this temperature would not be equal to that measured from space by a TIR radiometer because it is not possible to write the radiance $\langle R \rangle_\lambda$ in the form of Eq. (5.7), which shows that $M \neq \mathfrak{R}$.

5.2.2 Definition of Temperature for a Rough Surface

For a rough surface, Eqs. (5.6) and (5.8) are no longer valid and new expressions have to be derived (Sutherland and Bartholic 1977; Becker 1981; Caselles et al. 1988). We still impose Eq. (5.7), that is, $M = \mathfrak{R}$, in the definition and define $\langle \varepsilon \rangle_\lambda$ from the Kirchhoff's law as

$$\langle \varepsilon \rangle_\lambda = 1 - \langle \rho \rangle_\lambda \tag{5.11}$$

where $\langle \rho \rangle_\lambda$ is the hemispheric reflectivity and is measurable directly.

On the basis of Eq. (5.7), $\langle T_{sr} \rangle$ can be defined for a rough surface as

$$\langle T_{sr} \rangle = B_\lambda^{-1} \left[\frac{\langle R \rangle_\lambda - \langle \rho \rangle_\lambda R_{at\_\lambda\downarrow}}{\langle \varepsilon \rangle_\lambda} \right] \tag{5.12}$$

It is worth noting that Eqs. (5.11) and (5.12) are the general definition forms of land surface emissivity and temperature at pixel scale, respectively. Equations (5.8) and (5.9) can be regarded as the special case of definition of Eqs. (5.11) and (5.12) for a flat surface. All those quantities in Eqs. (5.11) and (5.12) are directly measurable from space at the same scale. The definitions of $\langle T_{sr}\rangle$ and $\langle\varepsilon\rangle_\lambda$ are scale invariant because in an embedded scale system, the definition at a scale (n) of $\langle T_{sr}\rangle$ and $\langle\varepsilon\rangle_\lambda$ can be expressed in terms of scale (n − 1) or (n − 2) or of any smaller scale as demonstrated in Becker and Li (1995).

It should be pointed out that this radiometric temperature $\langle T_{sr}\rangle$ is also called skin temperature since it corresponds to the radiation emitted from a depth of the order of the penetration depth, that is, of the order of the wavelength (λ).

By means of an exponential approximation of the Planck's function in the vicinity of a temperature T_0 (Slater 1980), that is,

$$B_\lambda(T) \cong m_\lambda T^{n_\lambda} \tag{5.13}$$

with

$$n_\lambda(T_0) = \frac{C_2}{\lambda T_0}\left(1 + \frac{1}{\exp(C_2/\lambda T_0) - 1}\right) \tag{5.14a}$$

$$m_\lambda(T_0) = \frac{B_\lambda(T_0)}{T_0^{n_\lambda(T_0)}} \tag{5.14b}$$

Equation (5.9) or Eq. (5.12) can be rewritten as

$$\langle T_{sr}\rangle = \langle T_{sm}\rangle\left[\sum_{k=1}^{n}\left(1 + \frac{\varepsilon_{\lambda k} - \langle\varepsilon\rangle_\lambda}{\langle\varepsilon\rangle_\lambda}\right)\left(1 + \frac{T_{sk} - \langle T_{sm}\rangle}{\langle T_{sm}\rangle}\right)^{n_\lambda} S_k\right]^{1/n_\lambda}.$$

For a reference temperature $T_0 = 300$ K, $n_{3.7\ \mu\text{m}} \approx 13$, $n_{11\ \mu\text{m}} \approx 4.5$, and $n_{mv} = 1$ for wavelength at 3.7 μm, 11 μm, and microwave, respectively.

If the second order of Taylor's expansion is used, the relationship between $\langle T_{sr}\rangle$ and $\langle T_{sm}\rangle$ is simplified to

$$\langle T_{sr}\rangle = \langle T_{sm}\rangle\left[1 + \sigma_{T\varepsilon} + \frac{1}{2}(n_\lambda - 1)\sigma_T^2\right] \tag{5.15}$$

with

$$\sigma_{T\varepsilon} = \sum_{k=1}^{N}\left[S_k\left(\frac{T_{sk} - \langle T_{sm}\rangle}{\langle T_{sm}\rangle}\right)\left(\frac{\varepsilon_{\lambda k} - \langle\varepsilon\rangle_\lambda}{\langle\varepsilon\rangle_\lambda}\right)\right] \tag{5.16a}$$

$$\sigma_T^2 = \sum_{k=1}^{N}\left[S_k\left(\frac{T_{sk} - \langle T_{sm}\rangle}{\langle T_{sm}\rangle}\right)^2\right] \tag{5.16b}$$

It is worth noting that Eq. (5.16a) is the covariance of emissivity and temperature, and Eq. (5.16b) is the variance of temperature. If the surface is approximately homogeneous, that is, the variation of $\varepsilon_{\lambda k}$ is very small or the $\varepsilon_{\lambda k}$ is the same for all subpixels k, the covariance $\sigma_{T\varepsilon}$ in Eq. (5.15) is vanished, which leads to

$$\langle T_{sr}\rangle = \langle T_{sm}\rangle \left[1 + \frac{1}{2}(n_\lambda - 1)\,\sigma_T^2\right] \tag{5.17}$$

in this case, it is possible to get the average temperature $\langle T_{sm}\rangle$ and the variance σ_T^2 of LST within a pixel at a given scale from at least two channel measurements. This is the particular advantage of temperature definition, which can be used to estimate the temperature variation σ_T^2 within a pixel using the measurements made at different wavelengths. It is also shown that if the variations of LSE and LST are small within a pixel, $\langle T_{sr}\rangle$ is equivalent to the average of LST within this pixel.

It should be pointed out that though the radiometric temperature $\langle T_{sr}\rangle$ depends on the wavelength as shown in Eqs. (5.12) and (5.15), the variation of $\langle T_{sr}\rangle$ with wavelength in 8–14 μm is small and can be regarded as a constant for practical use (Becker and Li 1995).

For simplicity, in the following, we use T_s or LST and ε or LSE to replace $\langle T_{sr}\rangle$ and $\langle\varepsilon\rangle_\lambda$, respectively.

5.3 Difficulties and Problems in the Retrieval of LST from Space Measurements

On the basis of radiative transfer theory described in Sect. 2.2, for a cloud-free atmosphere in thermodynamic equilibrium, an infrared sensor onboard a satellite viewing the earth's surface measures the radiation from the earth and its atmosphere along the line of sight. As illustrated in Figs. 5.1 and 2.6, the radiance $B_i(T_i)$ measured at the top of the atmosphere (TOA) in channel i can be written as (Li et al. 2000)

$$B_i(T_i) = B_i\left(T_{gi}\right)\,\tau_i + R_{at_i\uparrow} + R_{sl_i\uparrow} \tag{5.18}$$

where $B_i(T)$ is the radiance in channel i corresponding to the emission of a black-body at temperature T. T_i and T_{gi} are the brightness temperatures at the TOA and at ground level, respectively. τ_i is the total atmospheric transmittance along the target to sensor path in channel i. $R_{at_i\uparrow}$ is the upward atmospheric thermal radiance, and $R_{sl_i\uparrow}$ is the upward solar diffusion radiance resulting from atmospheric scattering of the solar radiance. $B_i(T_{gi})$ is the channel radiance observed at ground level, which can be written as (Li et al. 2000)

$$B_i\left(T_{gi}\right) = \varepsilon_i B_i(T_s) + (1-\varepsilon_i)\left(R_{at_i\downarrow} + R_{sl_i\downarrow}\right) + \rho_{b_i} R_{gi}^s \tag{5.19}$$

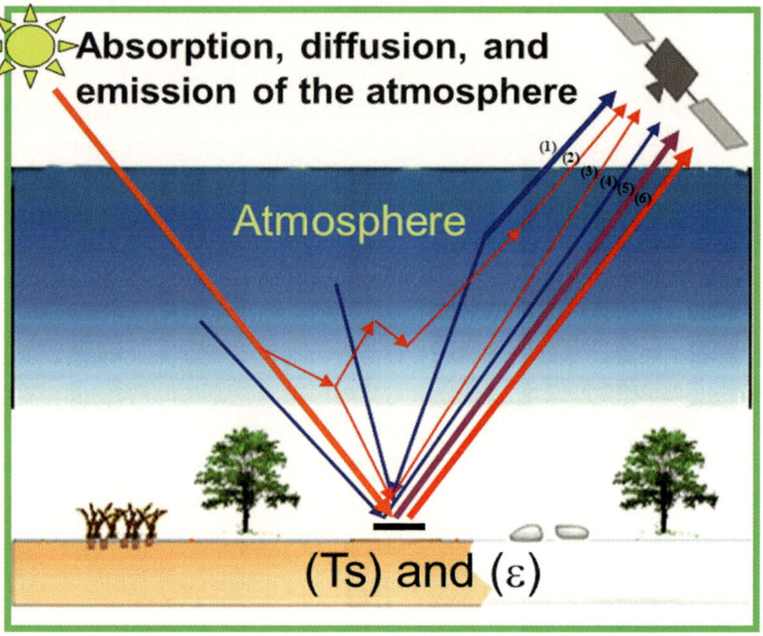

Fig. 5.1 Illustration of radiative transfer equation in infrared regions. (1) Represents the upward atmospheric thermal radiance, (2) the upward solar diffusion radiance, (3) the contribution of the downward solar diffusion radiance reflected by the surface to the measured radiance, (4) the contribution of the downward atmospheric thermal radiance reflected by the surface to the measured radiance, (5) the contribution of the radiance emitted directly by the surface to the measured radiance, and (6) the contribution of the direct solar radiation reflected by the surface to the measured radiance

in which ε_i and T_s are the effective surface emissivity and surface temperature, respectively, in channel i. $R_{at_i\downarrow}$ is the downward atmospheric thermal radiance and $R_{sl_i\downarrow}$ is the downward solar diffusion radiance. ρ_{b_i} is the bidirectional reflectivity of the surface; R^s_{gi} is the solar irradiance at the ground level in channel i. For simplicity, the solar geometry angles and viewing geometry angles are omitted in Eqs. (5.18) and (5.19).

It is noted that all variables/parameters in Eqs. (5.18) and (5.19) are channel-effective values, which are defined in Eq. (2.49). Equations (5.18) and (5.19) are actually approximations to the theoretical radiative transfer equation (RTE) in which monochromatic quantities are replaced with channel-effective values, but these approximations or simplifications require several important preconditions, which are given in Sect. 2.2.3.

From Eqs. (5.18) to (5.19), one can see that the LST is directly linked to the satellite-based TIR data through the RTE. The retrieval of the LST from remotely sensed TIR data, even if at the ground level, is very difficult because the downward atmospheric radiance and the surface-emitted radiance are coupled together through the LSE. Consequently, accurately determining the LST from space requires not only atmospheric correction but also the LSE correction.

Because the contribution of solar radiation at the TOA is negligible in the 8–14 μm window during both day and night and in the 3–5 μm window at night, the solar-related items in Eqs. (5.18) and (5.19) can be neglected without loss of accuracy, the TOA channel radiance $B_i(T_i)$ in TIR can be written as

$$B_i(T_i) = \varepsilon_i B_i(T_s)\,\tau_i + (1-\varepsilon_i)\,R_{at_i\downarrow}\tau_i + R_{at_i\uparrow} \tag{5.20}$$

According to Eq. (2.44), if an atmospheric equivalent temperature $T_{a\uparrow}$ is used, the upward atmospheric radiance $R_{at_i\uparrow}$ can be rewritten as

$$R_{at_i\uparrow} = (1-\tau_i)\,B_i\left(T_{a\uparrow}\right) \tag{5.21}$$

Taking the first order of the Taylor's expansion of the Planck function in the vicinity of temperature T_0, that is,

$$B_i(T) = B_i(T_0) + \frac{\partial B_i}{\partial T}(T - T_0) \tag{5.22}$$

Equation (5.20) can be rewritten as

$$T_s = T_i + \frac{1-\tau_i}{\tau_i}\left(T_i - T_{a\uparrow}\right) + \frac{1-\varepsilon_i}{\varepsilon_i}\left[\frac{\left(B_i(T_i) - R_{at_i\downarrow}\right)}{\partial B/\partial T} + \frac{1-\tau_i}{\tau_i}\left(T_i - T_{a\uparrow}\right)\right] \tag{5.23}$$

5.3.1 Atmospheric Effect

As shown in Eq. (5.23), the second term on the right-hand side of this equation represents the impact of atmosphere on the LST retrieval, ΔT_{atm}, that is,

$$\Delta T_{\text{atm}} = \frac{1-\tau_i}{\tau_i}\left(T_i - T_{a\uparrow}\right) \tag{5.24}$$

Equation (5.24) implies that the smaller the τ_i or $T_{a\uparrow}$ is, the larger the atmospheric effect is. Because T_i is generally larger than $T_{a\uparrow}$, the atmosphere acts as an absorber. In case that T_i is lower than $T_{a\uparrow}$, which happens often in the inversion of atmospheric temperature profile, the atmosphere behaves as an emitter. In other words, for the dry and cold atmosphere, $\tau \approx 1$ or for the inversion of atmospheric temperature profile, $T_{a\uparrow} \approx T_i$, ΔT_{atm} is small; otherwise, for the wet and warm atmosphere, ΔT_{atm} is large. As an example, Fig. 5.2 shows the effect of different atmospheres on the LST retrieval. One can see that the warm and wet tropical atmospheres can affect the LST retrieval as large as 11 K.

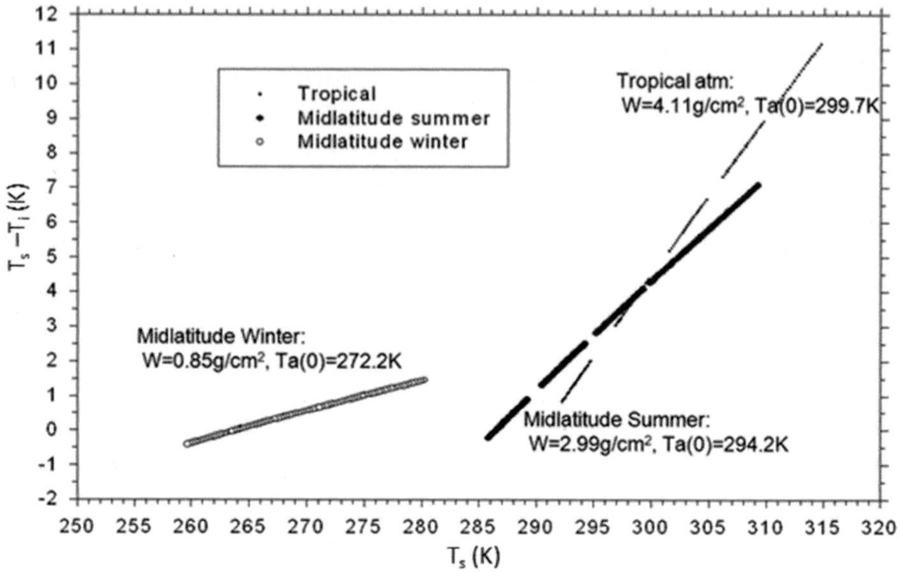

Fig. 5.2 Effects of different standard atmospheres on the retrieval of LST (W is the total column atmospheric water vapor content, and $T_a(0)$ is the near surface air temperature of the atmospheric profile)

5.3.2 *Emissivity Effect*

The third term on the right-hand side of Eq. (5.23) represents the impact of LSE on the LST retrieval, ΔT_ε, that is,

$$\Delta T_\varepsilon = \frac{1-\varepsilon_i}{\varepsilon_i}\left[\frac{(B_i(T_i) - R_{ati}\downarrow)}{\partial B / \partial T} + \frac{1-\tau_i}{\tau_i}(T_i - T_{a\uparrow})\right] \tag{5.25}$$

Defining $\beta_i = \frac{R_{ati}\downarrow}{B_i(T_i)}$ and taking the exponential approximation of the Planck's function given in Eq. (5.13), one can get

$$\Delta T_\varepsilon = \frac{1-\varepsilon_i}{\varepsilon_i}\left[\frac{T_i}{n_i}(1-\beta_i) + \Delta T_{atm}\right] \tag{5.26}$$

From Eq. (5.26), it is worth noting that

1. For a given temperature, when the total column water vapor in the atmosphere increases, β_i increases, which results in ΔT_ε decreases. If no atmosphere, $\beta_i = 0$, ΔT_ε reaches its maximum $\left(\frac{1-\varepsilon_i}{\varepsilon_i}\frac{T_i}{n_i}\right)$; thus, the drier the atmosphere is, the larger the impact of LSE on LST retrieval is.

2. For a given atmosphere, the higher the T_s is, the smaller (larger) the β_i (ΔT_{atm}) is, and the larger the ΔT_ε is, otherwise the smaller the ΔT_ε is.
3. Because ΔT_ε is inversely proportional to n_i, thus for a given ε_i, the longer the wavelength of channel is, the larger the ΔT_ε is.
4. ΔT_ε generally varies from $20\frac{1-\varepsilon_i}{\varepsilon_i}$ to $60\frac{1-\varepsilon_i}{\varepsilon_i}$. The wetter the atmosphere is, the closer the value of ΔT_ε is to 20. Contrarily, the drier the atmosphere is, the closer the value of ΔT_ε is to 60.

5.3.3 Difficulties and Problems

From the above analysis, we can see that the effects of atmosphere and LSE on the LST retrieval are significant and must be corrected for. However, applying these corrections is not a simple task, and some key difficulties and problems involved in the retrieval of the LST must be overcome and resolved. These key difficulties and problems are the following:

1. The retrieval of the LST from space is mathematically underdetermined and unsolvable (Kealy and Hook 1993). The RTE described in Eqs. (5.18) and (5.19) shows that if the radiance is measured in N channels, there will always be $N + 1$ unknowns, corresponding to N emissivities in each channel and an unknown LST for N equations. Such an ill-posed problem makes the solution of the RTE sets underdetermined at ground level even if the atmospheric quantities involved in Eqs. (5.18) and (5.19) are accurately estimated. To make the retrieval of LST deterministic, the LSEs must be known, or the LST and LSEs have to be simultaneously solved with the aid of some assumptions or constraints on the LSEs (Gillespie et al. 1996; Dash et al. 2002).
2. Measurements in the TIR region are highly correlated, implying that instrumental noise and errors in the atmospheric corrections exert strong influences on the accuracy of the LST retrieval and the solutions of the equation are unstable. Figure 5.3 illustrates the correlation of the two measurements at 11 μm and 12 μm in the LST retrieval. As shown in this figure, if there are no measurement errors in these two channels, actual LST can be retrieved. Otherwise, if there is only 0.1 K measurement error at 12 μm, the retrieval error in LST can reach to 1.2 K. In addition, this correlation represents a problem even if the LST is made solvable either by reducing the number of unknowns or by increasing the number of equations through reasonable assumptions or constraints on the LSEs (Gillespie et al. 1996; Li et al. 2013a, b). These highly correlated measurements make LST retrieval unstable and have hampered the methodological development of LST retrieval.
3. It is difficult to decouple the LST, the LSEs, and the downward atmospheric radiance in the measured radiances. As seen from Eq. (5.19) and Fig. 5.1, the downward atmospheric radiance and the surface-emitted radiance are coupled together through LSEs. The nonunity LSE of a natural surface reduces

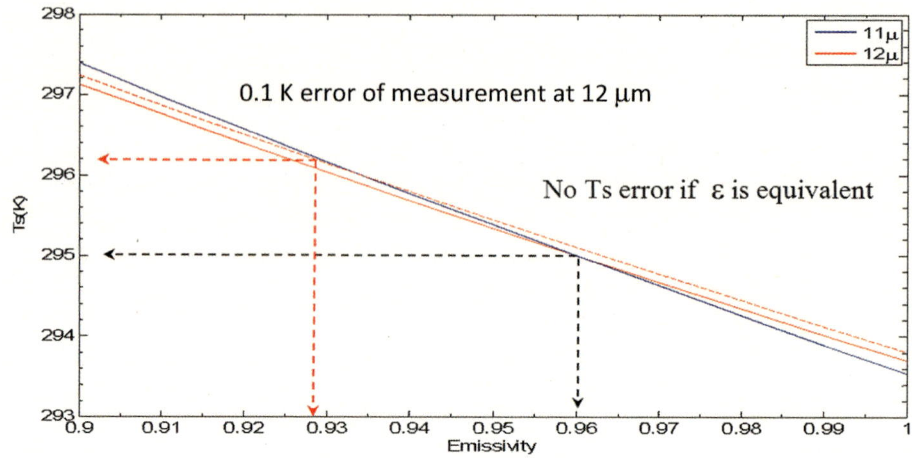

Fig. 5.3 Illustration of the LST retrieval error resulted from two TIR channels with/without channel measurement error at 12 μm

the surface-emitted radiance while increasing the reflection of the downward atmospheric radiance back to the atmosphere, which compensates partly for the reduction in the surface-emitted radiance. This process can reduce or increase the total surface-leaving radiance depending on the atmospheric and surface conditions. It is therefore impossible to separate, on a physical basis, the contributions of the LST from the contributions of the LSEs and the atmosphere in the observed radiance in passive infrared thermography radiometry.

4. The atmospheric corrections are difficult to implement. The presence of the atmosphere between the surface and the sensors disturbs the radiances measured by a radiometer at the TOA (see Figs. 5.1 and 2.6). These radiances result primarily from emission–reflection at the surface modulated by the effects of the absorption, diffusion, and emission of the atmosphere (see Figs. 5.1 and 2.6). The atmospheric corrections thus consist of correcting the radiance measured by the sensors for the effects of atmospheric absorption, emission, and emission–reflection. However, the great variability of the vertical profiles of atmospheric water vapor and temperature makes these effects highly variable (Perry and Moran 1994). As shown in Eq. (5.24) and Fig. 5.2, the effect of dry and cold atmospheres on ΔT_{atm} is small and can be ignored sometimes, but for warm and wet atmospheres, ΔT_{atm} may reach to 20 K, which indicate that one has to consider the vertical profiles of atmospheric water vapor and temperature in the LST retrieval algorithm.
5. It is difficult to remove the effects of solar in the MIR measurements. As described in (2), the TIR measurements are highly correlated which makes LST retrieval unstable even if the solution of the RTE sets becomes deterministic through some assumptions and constraints on the LSEs. Because the direct solar irradiation reflected by the surface is on the same order of magnitude as the

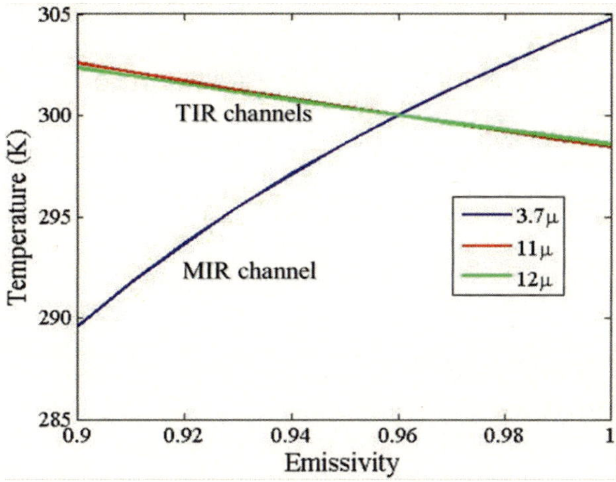

Fig. 5.4 Illustration of the decorrelation of MIR channel and TIR channels in the LST retrieval

radiance directly emitted by the surface in the middle infrared (MIR) if the surface albedo is about 0.1, the introduction of the MIR channels in LST retrieval significantly reduces the correlation of the RTE sets and greatly improves the accuracy of the estimated LST (Li et al. 2013a). As an example, Fig. 5.4 shows the decorrelation of MIR channel and TIR channels in LST and LSE retrievals. One can see that the introduction of the MIR channel in LST and LSE retrievals significantly reduces the correlation of the RTE sets and can make the solution of the RTE more stable and accurate. Additionally, MIR channels are less sensitive to the water vapor in the atmosphere compared with TIR channels, and the LST retrieval from the MIR is only half, even one third, as sensitive to errors in emissivity as that from the TIR (Mushkin et al. 2005). Therefore, LST retrieval with the MIR instead of the TIR sounds more appropriate. However, solar effects are difficult to eliminate in the MIR during the daytime because the separation of solar irradiation from the total energy measured in the MIR requires not only the accurate atmospheric information but also the knowledge of the bidirectional reflectivity of the surface. As all of those information and parameters are always unknown and their accuracy is degraded by several factors (Adams et al. 1989; Mushkin et al. 2005), their uncertainties can lead to a large error on the LST from the MIR measurements (Gillespie 2005). As an old saying goes, every coin has two sides. The introduction of the MIR channels may benefit the retrieval of the LST, but would also introduce other uncertain terms.

6. The surface temperature used in Eq. (5.19) is called the radiometric temperature (or the skin temperature) that corresponds to the radiation emitted by a surface with depths less than the penetration depth of a given wavelength (Becker and Li 1995; Norman and Becker 1995). The penetration depth is usually within a few millimeters in the TIR region (Wan 1999). This radiometric temperature physically differs from other definition of temperatures, such as the thermodynamic temperature defined for a medium in thermal equilibrium

and measured by a thermometer. For homogeneous and isothermal surfaces, the radiometric and thermodynamic temperatures are reported to be equivalent. As the thermodynamic temperature is actually hard to measure in reality, even for water, the radiometric temperature is often the only practical measure for the homogeneous and isothermal surface. However, most surfaces are not in equilibrium, and for heterogeneous and nonisothermal surfaces, these two temperatures are different. Considering that the spatial resolution of the current onboard systems varies from 10^{-2} to 10 km^2, there may be several surface types with different temperatures within one pixel, which complicates the physical understanding of the LST values retrieved from space and the relation of the radiometric temperature at large scales to other temperatures used in different applications. To date, no consensus has been reached on the definition of the LST for heterogeneous and nonisothermal surfaces, but the definition by Becker and Li (1995) and in Eqs. (5.9) and (5.12), which depends on the distributions of the LST and the LSE within a pixel, is measurable from space and is thus recommended for LST retrieval from space.

7. It is very difficult to validate the LST derived from satellite pixel scale with kilometer resolution. This is true due to the following facts: (1) the difficulty of conducting in situ LST measurements and (2) the difficulty of obtaining representative LST data at the scale of a single pixel. Generally, temperatures over the land surfaces vary strongly in space and time (Prata et al. 1995), and it is not unusual for the LST to vary by more than 10 K over just a few centimeters of distance or by more than 1 K in less than one minute. The most important problem is how to relate the satellite-derived LSTs to those measured at ground level, especially at large scale. The strong spatial heterogeneity and temporal variation of the LST limit ground-based validation only to several relatively homogeneous surfaces, such as lakes, deserts, and densely vegetated land. Furthermore, how to perform a representative measurement of the LST of a complex heterogeneous surface is also a critical question. Scaling methods should be developed to relate the measurements at different scales and help validate the retrieved LST (Liu et al. 2006; Wu and Li 2009; Li et al. 2013a).

Although there exist many difficulties and problems described above for retrieving LST from space measurements, over the past several decades, LST estimation from satellite TIR measurements has significantly progressed. Many algorithms have been proposed to treat the characteristics of various sensors onboard different satellites and utilizing different assumptions and approximations for the RTE and LSEs. These algorithms can be roughly grouped into three categories: single-channel methods, multichannel methods, and multiangle methods, provided that the LSEs are known a priori. If the LSEs are not known, then the algorithms can be categorized into three types: stepwise retrieval method, simultaneous retrieval of LSEs and LST with known atmospheric information, and simultaneous retrieval with unknown atmospheric information.

5.4 LST Retrieval with Known LSEs

5.4.1 *Single-Channel Method*

The single TIR channel method, also called the single-channel method, uses the radiance measured by the satellite sensor in a single channel chosen within an atmospheric window and corrects the radiance for residual atmospheric absorption and emission using atmospheric transmittance/radiance code that requires input data of the atmospheric profiles. The LST is then retrieved from the radiance measured in this channel by inverting the RTE (Price 1983; Susskind et al. 1984; Chédin et al. 1985; Ottlé and Vidal-Madjar 1992; Li et al. 2004a):

$$T_s = B^{-1}\left[\frac{B_i(T_i) - R_{at_i\uparrow} - (1-\varepsilon_i)\,R_{at_i\downarrow}\tau_i}{\varepsilon_i\tau_i}\right] \tag{5.27}$$

in which B^{-1} is the inverse of the Planck function. Once the channel emissivity ε_i and the atmospheric quantities τ_i, $R_{at_i\uparrow}$, and $R_{at_i\downarrow}$ are known in advance, one can easily obtain the LST by inverting the Planck's law with Eq. (5.27). It should be noted that accurate determination of the LST using this method requires high-quality atmospheric transmittance/radiance code to estimate the atmospheric quantities, good knowledge of the channel LSE, an accurate atmospheric profile, and a correct consideration of the topographic effects (Sobrino et al. 2004b). Figure 5.5 gives the flowchart of the single TIR channel method for retrieving LST from satellite data.

Generally, the accuracy of atmospheric transmittance/radiance code is primarily limited by the radiative transfer model (RTM) used in the code and by the uncertainties in atmospheric molecular absorption coefficients and aerosol absorption/scattering coefficients (Wan 1999). The most popular atmospheric RTMs, such

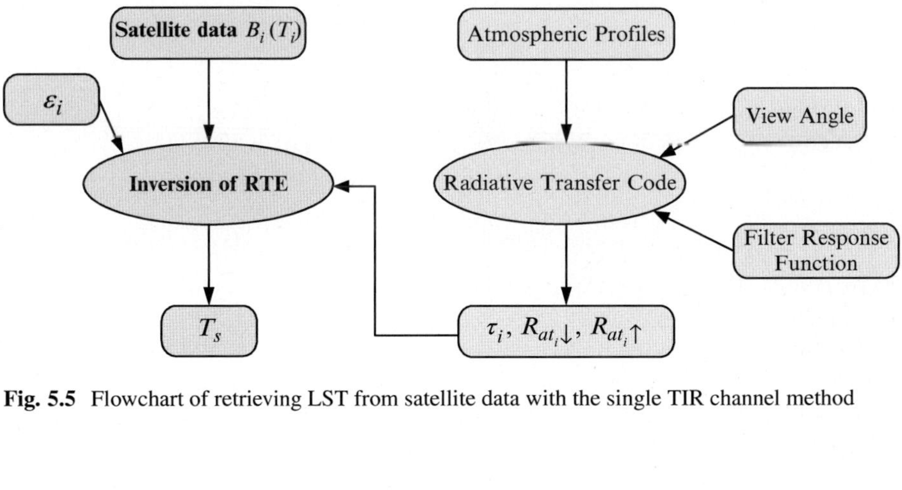

Fig. 5.5 Flowchart of retrieving LST from satellite data with the single TIR channel method

as the series of MODTRAN (Berk et al. 2003) and 4A/OP (Chaumat et al. 2009), have been widely used to perform atmospheric corrections and/or to simulate satellite TIR data. A few studies have shown that the uncertainty of the different RTMs ranges from 0.5 % to 2 % within known atmospheric windows, such as 3.4–4.1 μm and 8–13 μm, leading to uncertainties in the retrieved LST of 0.4 to 1.5 K (Wan 1999). Studies have also demonstrated that an error of 1 % in the LSE can give rise to an error in the LST ranging from 0.3 K for a hot and humid atmosphere to 0.7 K for a cold and dry atmosphere (Dash et al. 2002). As the single channel is usually chosen around 10 μm where the LSE for most land surfaces can be known within a few percent, the uncertainty in LSE may lead to an error of 1 K to 2 K in LST if the single-channel method is used. It is worth noting that the incomplete characterization of atmospheric profiles used in compensation for atmospheric absorption and path radiance is a big problem even if the RTM itself is completely error-free (Gillespie et al. 2011).

During the past decade, in order to reduce the dependence on radiosounding data, several single-channel algorithms have been proposed to estimate the LST from satellite data provided that the LSE is known. Qin et al. (2001) proposed a method to estimate the LST specifically from Landsat-5 (Thematic Mapper channel 6, TM 6) data using only the near-surface air temperature (T_0) and the total column water vapor content (W) instead of atmospheric profiles as input, with the aid of the empirical linear relationships between τ_i and W, as well as between the mean atmospheric temperature ($T_{a\uparrow}$) and T_0. The proposed single-channel algorithm for obtaining LST from TM 6 is written by

$$T_s = \frac{1}{C}\left\{a\left(1 - C - D\right) + \left[b\left(1 - C - D\right) + C + D\right]T_i - DT_{a\uparrow}\right\} \tag{5.28}$$

with $C = \varepsilon_i \tau_i$, $D = (1 - \tau_i)[1 + (1 - \varepsilon_i)\tau_i]$, a and b are the sensor-dependent constants; for TM 6, $a = -67.355351$, $b = 0.458606$. For TM 6 and atmosphere in mid-latitude summer, $T_{a\uparrow}$ is given by

$$T_{a\uparrow} = 16.01 + 0.926T_0 \tag{5.29}$$

and for the range of W from 0.4 to 1.6 g/cm^2, τ_i is estimated by

$$\tau_i = 0.9743 - 0.0801W \quad \text{(for high } T_0\text{)} \tag{5.30a}$$

$$\tau_i = 0.9820 - 0.0961W \quad \text{(for high } T_0\text{)} \tag{5.30b}$$

More details about this algorithm and its sensitivity can be found in the work of Qin et al. (2001).

Jiménez-Muñoz and Sobrino (2003) and Jiménez-Muñoz et al. (2009) developed a generalized single-channel algorithm for retrieving LST from any TIR channel with a full width at half maximum (FWHM) of ~1 μm, provided that LSE and W are known:

$$T_s = \gamma \left[\varepsilon_i^{-1} \left(\psi_1 B_i(T_i) + \psi_2 \right) + \psi_3 \right] + \delta \tag{5.31}$$

with

$$\gamma = \left\{ \frac{C_2 B_i(T_i)}{T_i^2} \left[\frac{\lambda^4}{C_1} B_i(T_i) + \lambda^{-1} \right] \right\}^{-1} \tag{5.32a}$$

$$\delta = -\gamma B\left(T_i\right) + T_i \tag{5.32b}$$

in which $B_i(T_i)$ is in W/(m$^2 \cdot$ Sr $\cdot$ μm), T_i is in K, λ is the effective central wavelength (11.457 μm for band TM 6), $C_1 = 1.19104 \times 10^8$ W $\cdot$ (μm)^4/(Sr $\cdot$ m^2), and C_2 =14387.7 μm $\cdot$ K. ψ_1, ψ_2, and ψ_3 are the atmospheric quantities, which can be obtained as a function of W from the following empirical equations. For TM 6, they can be written as

$$\psi_1 = 0.14714W^2 - 0.15583W + 1.1234 \tag{5.33a}$$

$$\psi_2 = -1.1836W^2 - 0.37607W - 0.52894 \tag{5.33b}$$

$$\psi_3 = -0.04554W^2 + 1.8719W - 0.39071 \tag{5.33c}$$

This generalized single-channel algorithm requires the minimum input data and can be applied to different thermal sensors using the same equation. Cristóbal et al. (2009) found the inclusion of the near-surface air temperature together with the water vapor content in the single-channel method improves LST retrieval, especially with intermediate and high atmospheric water vapor content. Sobrino et al. (2004b), Sobrino and Jiménez-Muñoz (2005), and Jiménez-Muñoz and Sobrino (2010) analyzed and compared the aforementioned algorithms and pointed out that all of the single-channel algorithms that use empirical relationships provide poor results at high atmospheric water vapor contents because the relationships included in the algorithms are unstable at high water vapor concentrations.

It should be kept in mind that the single-channel methods involve a simple inversion of the RTE, provided that the LSE and the atmospheric profiles are known in advance. These methods can provide theoretically accurate LST retrieval, but are currently still far from practical use, because it requires that the vertical and horizontal distribution of temperature and water vapor in the atmosphere be accurately known at the same time of sensor overpass. This requirement is generally not possible to meet under most realistic conditions. Currently, the possible sources of atmospheric profiles are usually obtained either from ground-based atmospheric radiosoundings, satellite vertical sounders, or from the output of meteorological forecasting models.

As far as ground-based radiosoundings are concerned, they indeed do not fit our needs. Due to the high variability of the atmospheric water vapor in space and in time, the use of ground-based radiosoundings far from the collocated area of interest and/or far from the synchronous time of satellite overpass may result in large errors up to 10 K on surface temperature retrieval (Cooper and Asrar 1989). Let alone the radiosounding measurement errors that are reported on the order of 0.5 K for temperature and 10 % for water vapor (Ma et al. 1999), there are two uncertainties in radiosounding measurements: (1) The actual sounding location and place where the balloon is set free may be several kilometers apart (up to 60 km). (2) The atmospheric status will change when the balloon rises, which means the atmospheric profiles at different heights are measured at different time. Those uncertainties about atmosphere will further propagate into the retrieved LST. In addition, as radiosoundings are not currently available with sufficient spatial density or in coincidence with the time of the satellite overpass (Coll et al. 2005), therefore, the use of ground-based radiosoundings can be only used occasionally for validation purposes at some special sites.

As for the atmospheric profiles derived from the measurements of satellite vertical sounders (Susskind et al. 1984; Chédin et al. 1985), they can, in theory, be used to retrieve the LST from concurrent TIR data in the atmospheric window using the single-channel method. Unfortunately, the accuracy of the retrieved atmospheric profiles near the surface is insufficient for the single-channel method (Susskind et al. 1984; Chédin et al. 1985; Ottlé and Stoll 1993). Large errors in the LST can be produced if inaccurate atmospheric profiles are used.

Nowadays, some reanalysis of the outputs of meteorological forecasting models are another alternative to radiosoundings and are available from different sources such as the National Centers for Environmental Prediction (NCEP) operational global analysis data (http://www.cdc.noaa.gov/data/reanalysis/) and the European Centre for Medium-Range Weather Forecast (ECMWF) reanalysis (ERA-40) operational model data (http://www.ecmwf.int/products/data/archive/descriptions/e4/index.html) except for radiosoundings. Coll et al. (2012a) and Jiménez-Muñoz et al. (2010) reported either atmospheric profile products or reanalysis may yield reasonable results to meet the required accuracy for many cases. However, it is time-consuming to process those massive profile data sets by running them through atmospheric RTMs (e.g., MODTRAN). In addition, atmospheric profile data are either discrete or too coarse in space and time (available only at UTC times 0, 6, 12, and 18 h and at a spatial resolution of $0.25° \times 0.25°$ latitude/longitude). It is therefore necessary to interpolate the atmospheric quantities in terms of the viewing zenith angle, the terrain altitude, and the space and time (Schroedter et al. 2003; Jiang et al. 2006; Tang and Li 2008). It obviously affects more or less the accuracy of the estimated LST, especially for the cases with high atmospheric water vapor content.

Consequently, based on existing systems and ancillary data, to obtain LST with enough accuracy on a global scale, one must rely on alternative methods, which, even though less accurate theoretically, have the advantage of being suitable for global applications and can be run "operationally," without sophisticated ancillary data. The so-called split-window techniques (SWT), which use at least two thermal infrared channels, have been proposed and developed for the LST retrieval for terrestrial surface.

5.4.2 Multichannel Method

As discussed above, the single-channel method requires an accurate RTM and accurate knowledge of the atmospheric profiles over the study area at the satellite overpass. These requirements are difficult or even impossible to satisfy in most practical situations. To obtain the LST from satellite TIR data with sufficient accuracy on a global or regional scale, other methods must be developed. When there are two adjacent infrared channels at 11 and 12 μm or more, corresponding to different atmospheric transmissions, available in a sensor, it is possible to use the differential absorption of these channel observations to estimate the atmospheric contribution to the signal. This method is the so-called split-window (SW) technique. The basis of this method is that the atmospheric attenuation suffered by the surface-emitted radiance is proportional to the difference between the at-sensor radiances measured simultaneously in two adjacent thermal infrared channels (McMillin 1975). This method was firstly proposed by McMillin (1975) to estimate the sea surface temperature (SST) from satellite measurements without any information about the atmospheric profiles. Since then, a variety of SW techniques have been developed and modified to successfully retrieve the SST (Deschamps and Phulpin 1980; Llewellyn-Jones et al. 1984; McClain et al. 1985; Barton et al. 1989; Sobrino et al. 1993; França and Carvalho 2004; Niclòs et al. 2007).

Encouraged by the success of the SW method for estimating the SST from space measurements, many efforts and studies have been made since the late 1980s to extend the SW method to retrieve the LST. With modifications to treat the spatial–temporal and spectral variations of the LSE, the large difference between the LST and the air temperature, the total column water vapor (W) in the atmosphere, and the viewing zenith angle (VZA), a variety of SW algorithms for LST retrieval have been proposed. These algorithms assume that the LSEs in both SW channels are known a priori (Price 1984; Becker 1987, Becker and Li 1990b; Sobrino et al. 1991, 1994, 1996; Coll et al. 1994; Prata 1994a, b; Ulivieri et al. 1994; Wan and Dozier 1996; Tang et al. 2008; Atitar and Sobrino 2009). The following will introduce the different SW algorithms found in the literature.

5.4.2.1 Methodological Development of Split-Window Algorithm

On the basis of radiative transfer theory for TIR region, for a cloud-free atmosphere under local thermodynamic equilibrium, the radiances measured by two adjacent TIR channels at the satellite consist of only the emitted radiances from both the surface and the atmosphere; Eqs. (5.20) and (5.21) can be combined and written simply for channels i and j, provided that both channels are in the TIR spectrum and very close (usually one within 10–11 μm and another within 11–12 μm):

$$B_i(T_i) = \varepsilon_i B_i(T_s)\,\tau_i + (1-\varepsilon_i)\,R_{at_i\downarrow}\tau_i + (1-\tau_i)\,B_i\left(T_{a\uparrow}\right) \tag{5.34}$$

$$B_j\left(T_j\right) = \varepsilon_j B_j(T_s)\,\tau_j + \left(1-\varepsilon_j\right) R_{at_j\downarrow}\tau_j + \left(1-\tau_j\right) B_j\left(T_{a\uparrow}\right) \tag{5.35}$$

There are two different ways to develop the SW method for retrieving LST. One is the radiance-based method, which is deduced from the derivation of Planck function to wavelength, and another is the temperature-based method deduced from the derivation of Planck function to temperature.

Radiance-Based SW Method

Since the central wavelengths (λ_i, λ_j) of the two adjacent channels (i, j) are not very different, taking the first-order Taylor series expansion of the radiance in channel j around the central wavelength (λ_i) of channel i, $B_j(T_j)$ measured in channel j can be rewritten as

$$B_j(T_j) = B_i(T_j) + (\lambda_j - \lambda_i)\left(\frac{\partial B}{\partial \lambda}\right)_{\lambda_i} \tag{5.36}$$

Here, $B_i(T_j)$ represents the corresponding radiance in channel i with the brightness temperature T_j measured in channel j. Applying Eq. (5.36) to Eq. (5.35), the latter becomes

$$\begin{aligned} B_i(T_j) &= \varepsilon_j \tau_j B_i(T_s) + (1-\tau_j) B_i(T_{a\uparrow}) + (1-\tau_j) R_{at_j\downarrow}\tau_j \\ &+ \varepsilon_j \tau_j (\lambda_j - \lambda_i)\left(\frac{\partial B_i(T_s)}{\partial \lambda} - \frac{1}{\varepsilon_j}\frac{\partial B_i(T_{a\uparrow})}{\partial \lambda}\right) \\ &+ (\lambda_j - \lambda_i)\left(\frac{\partial B_i(T_{a\uparrow})}{\partial \lambda} - \frac{\partial B_i(T_j)}{\partial \lambda}\right) \end{aligned} \tag{5.37}$$

Provided that the temperatures T_i, T_j, T_s, and $T_{a\uparrow}$ are close to each other in magnitude, the last two terms in the right-hand side of Eq. (5.37) involving the difference in Planck derivatives are small. Neglecting these two terms and eliminating $B_i(T_{a\uparrow})$ from Eqs. (5.34) and (5.37), the radiance $B_i(T_s)$ can be derived as (Prata 1993)

$$\begin{aligned} B_i(T_s) = \frac{1}{\varepsilon_i}\left(\frac{1}{1+\eta\tau_j\Delta\varepsilon/\varepsilon_i}\right) B_i(T_i) + \frac{\eta}{\varepsilon_j}\left(\frac{1}{1+(1+\eta)\tau_i\Delta\varepsilon/\varepsilon_j}\right) \\ \times \left(B_i(T_i) - B_i(T_j)\right) + \alpha \end{aligned} \tag{5.38}$$

with

$$\eta = \frac{1-\tau_i}{\tau_i - \tau_j} \tag{5.39}$$

$$\Delta\varepsilon = \varepsilon_i - \varepsilon_j \tag{5.40}$$

$$\alpha = \frac{(1-\tau_i)\,\tau_j\,(1-\varepsilon_j)\,R_{at_j\downarrow} - (1-\tau_j)\,\tau_i\,(1-\varepsilon_i)\,R_{at_i\downarrow}}{\varepsilon_i\tau_i\,(1-\tau_j) - \varepsilon_j\tau_j\,(1-\tau_i)} \tag{5.41}$$

Equation (5.38) is a general formulation of radiance-based SW and gives the basic framework for testing the validity of the SW method over land surface. It demonstrates explicitly how the atmosphere and the emissivity affect the satellite measurement of surface temperature. If the atmospheric effects can be specified and the emissivity can be characterized at the scale of the field of view, the LST can be estimated with this equation.

If the land surface behaves as a blackbody ($\varepsilon_i = \varepsilon_j = 1$), Eq. (5.38) is then simplified as

$$B_i(T_s) = B_i(T_i) + \eta\left(B_i(T_i) - B_i(T_j)\right) \tag{5.42}$$

This is the split-window equation developed by Prabhakara et al. (1974) and McMillin (1975) for SST retrieval.

If the land surface behaves as a gray body ($\varepsilon_i = \varepsilon_j$), that is, there is no spectral dependence. In this case, $\varepsilon_i = \varepsilon_j = \varepsilon$, $\Delta\varepsilon = 0$, and Eq. (5.38) becomes

$$B_i(T_s) = \frac{1}{\varepsilon}B_i(T_i) + \frac{\eta}{\varepsilon}\left(B_i(T_i) - B_i(T_j)\right) + \frac{(1-\varepsilon)}{\varepsilon}\left(\eta\tau_j R_{at_j\downarrow} - (1+\eta)\,\tau_i R_{at_i\downarrow}\right) \tag{5.43}$$

This type of radiance-based SW is only applicable to land surfaces which have little or no spectral emissivity dependence between 10 and 13 μm (e.g., irrigated crops).

Temperature-Based SW Method

Note that the brightness temperature is commonly used by scientists in TIR than radiance since brightness temperatures are more easily validated and can be compared without reference to a wavelength. To obtain a temperature-based SW algorithm involving brightness temperatures measured at TOA, similar to the development of the radiance-based SW, taking the first-order Taylor series expansion of the radiance around a temperature T_0 as given in Eq. (5.22) and $B_i(T_i)/\left(\frac{\partial B}{\partial T}\right)_{T_i} \approx \frac{T_i}{n_i}$ and eliminating $T_{a\uparrow}$ from Eqs. (5.34) and (5.35), the LST can be derived by

$$T_s = \frac{1}{\varepsilon_i}\left(\frac{1-\eta\,(1-\varepsilon_i)\left[\tau_i\left(1-\frac{1}{n_i}\right) - \tau_j\left(1-\frac{1}{n_j}\right)\right] + \eta\Delta\varepsilon\tau_j\left(1-\frac{1}{n_j}\right)}{1+\eta\tau_j\Delta\varepsilon/\varepsilon_i}\right)T_i$$
$$+\frac{\eta}{\varepsilon_j}\left(\frac{1-(1-\varepsilon_j)\,\tau_j\left(1-\frac{1}{n_j}\right)}{1+(1+\eta)\,\tau_i\Delta\varepsilon/\varepsilon_j}\right)(T_i - T_j) + \beta \tag{5.44}$$

with

$$\beta = \frac{(1-\tau_i)\,\tau_j\,(1-\varepsilon_j)\,R_{at_j\downarrow}/\left(\frac{\partial B}{\partial T}\right)_{T_j} - (1-\tau_j)\,\tau_i\,(1-\varepsilon_i)\,R_{at_i\downarrow}/\left(\frac{\partial B}{\partial T}\right)_{T_i}}{\varepsilon_i\tau_i\,(1-\tau_j) - \varepsilon_j\tau_j\,(1-\tau_i)} \tag{5.45}$$

Equation (5.44) is a general formulation of temperature-based SW algorithm for LST retrieval over natural land surface.

Similar to the radiance-based SW, if the land surface is a blackbody ($\varepsilon_i = \varepsilon_j = 1$), Eq. (5.44) can then be simplified as

$$T_s = T_i + \eta\,(T_i - T_j) \tag{5.46}$$

This type of SW has been used to retrieve SST from TOA brightness temperatures instead of radiances employed in Eq. (5.42).

If the land surface behaves as a gray body ($\varepsilon_i = \varepsilon_j = \text{constant}$), Eq. (5.44) becomes

$$\begin{aligned} T_s = {} & \frac{1}{\varepsilon}\left(1 - \eta\,(1-\varepsilon)\left[\tau_i\left(1 - \frac{1}{n_i}\right) - \tau_j\left(1 - \frac{1}{n_j}\right)\right]\right) T_i \\ & + \frac{\eta}{\varepsilon}\left(1 - (1-\varepsilon)\,\tau_j\left(1 - \frac{1}{n_j}\right)\right)(T_i - T_j) \\ & + \frac{(1-\varepsilon)}{\varepsilon}\left[\eta\tau_j R_{at_j\downarrow}\Big/\left(\frac{\partial B}{\partial T}\right)_{T_j} - (1+\eta\tau_j)\,R_{at_i\downarrow}\Big/\left(\frac{\partial B}{\partial T}\right)_{T_i}\right] \end{aligned} \tag{5.47}$$

Equation (5.44) shows explicitly that the LST can be determined directly by two TOA channel brightness temperatures.

In case that the absorptions of atmosphere in two channels *i* and *j* are small, the total channel atmospheric transmittance can be approximated by

$$\tau_k \approx 1 - \frac{A_k W}{\cos\theta} \tag{5.48}$$

with A_k being the absorption coefficient in channel k ($k = i$ and j).

One then gets

$$\eta = \frac{A_i}{A_j - A_i} \tag{5.49}$$

This equation implies that η is only dependent on the ratio of the absorption coefficients in two channels. In this case, that is, $W < 2.5$ g/cm^2, coefficient η

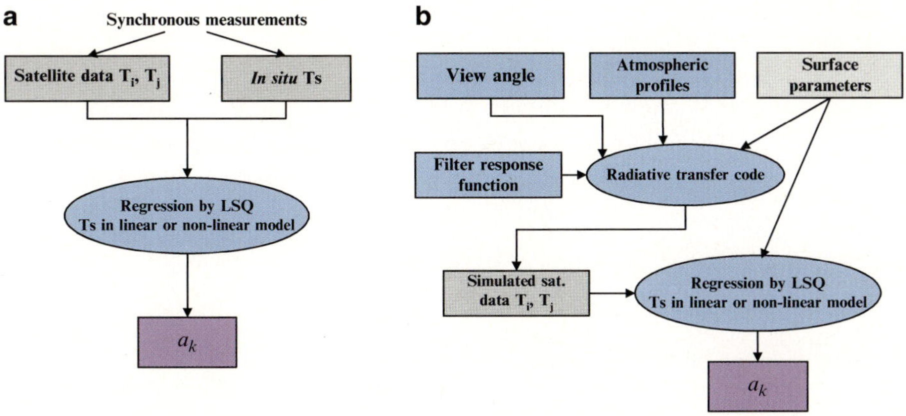

Fig. 5.6 Flowchart for determining coefficients a_k in the SW algorithm (LSQ = least square)

is a constant which is independent of atmospheric parameters and view angle. Otherwise, η is generally in function of W and view zenith angle (VZA).

For simplicity, Eqs. (5.38) and (5.44) are commonly written in typical linear SW forms as

$$B\left(T_{\mathrm{s}}\right)=a_0+a_1 B_i\left(T_i\right)+a_2\left(B_i\left(T_i\right)-B_i\left(T_j\right)\right) \tag{5.50a}$$

$$\mathrm{LST}=a_0+a_1 T_i+a_2\left(T_i-T_j\right) \tag{5.50b}$$

in which a_k ($k = 0$, 1, and 2) are the corresponding coefficients depending primarily on the spectral response function of the two channels $g_i(\lambda)$ and $g_j(\lambda)$, the two channel emissivities ε_i and ε_j, the W, and the VZA. Thus,

$$a_k=f_k\left(g_i, g_j,\ \varepsilon_i, \varepsilon_j,\ W,\ \mathrm{VZA}\right) \tag{5.51}$$

It should be pointed out that the accuracy of this LST retrieval method is dependent on the correct choice of the coefficients a_k, which are predetermined empirically either by comparing the satellite data against in situ LST measurements (see Fig. 5.6a) or by regressing the simulated satellite data with a set of atmospheres and surface parameters (see Fig. 5.6b). Because it is extremely difficult to obtain a representative in situ LST at satellite pixel scale (a few km^2) that is also synchronized with the satellite's measurements (T_i and T_j) over a wide range of surface types and atmospheric conditions, simulation represents an efficient way to predetermine the coefficients a_k by comparing simulated satellite data against the preset LST used in the simulation (see Fig. 5.6b).

5.4.2.2 Linear Split-Window Algorithm

Over the past several decades, many linear SW algorithms have been developed in the literature, all of which have similar forms but different parameterizations of the coefficients a_k. According to the parameterization pattern of the coefficients, the linear SW algorithms are roughly categorized into three groups: (1) parameterization of the coefficients a_k with consideration of LSEs, (2) parameterization of the coefficients a_k with consideration of LSEs and W, and (3) parameterization of the coefficients a_k with consideration of LSEs, W, and VZA.

Parameterization of the Coefficients with Consideration of LSEs

The coefficients a_k in the split window (SW) parameterized variously as functions of the LSEs have been developed previously, and all are somewhat similar in formulation. For instance, Becker and Li (1990b) developed a local SW algorithm and parameterized a_k as functions of the mean ($\varepsilon = (\varepsilon_i + \varepsilon_j)/2$) and difference ($\Delta\varepsilon = \varepsilon_i - \varepsilon_j$) of the two channel LSEs expressed by

$$\begin{aligned} a_0 &= \text{constant}, \\ a_k &= A_{0,k} + A_{1,k}\frac{1-\varepsilon}{\varepsilon} + A_{2,k}\frac{\Delta\varepsilon}{\varepsilon^2} \quad (k = 1, 2) \end{aligned} \tag{5.52}$$

in which the coefficients A are constant and independent of the atmosphere if the atmosphere is relatively dry ($W < 2.5$ g/cm^2). Some authors have proposed modifying the constant coefficient a_0 in Eq. (5.52) to correct for emissivity effects while keeping the other coefficients a_1 and a_2 independent of the LSEs as they are for SST retrieval. Other forms of a_0 have been proposed, including $a_0 = B_0 + B_1(1-\varepsilon) + B_2\Delta\varepsilon$ (Ulivieri and Cannizzaro 1985; Sobrino et al. 1994; Ulivieri et al. 1994) or $a_0 = B_0 + B_1(1-\varepsilon)/\varepsilon + B_2\Delta\varepsilon/\varepsilon^2$ (Vidal 1991), in which the coefficients B are sensor dependent and atmosphere independent. A general form of a_0 can be written as

$$a_0 = f_0\left(g_i, g_j, \varepsilon, \Delta\varepsilon\right) \tag{5.53a}$$

where f_0 is a linear or nonlinear function. Some authors (Sun and Pinker 2003, 2005) have proposed a general parameterization of a_k as

$$a_k = f_k\left(g_i, g_j, \ \varepsilon_i, \varepsilon_j\right) \tag{5.53b}$$

Parameterization of the Coefficients with Consideration of LSEs and W

It is worth noting that the atmosphere-independent coefficients a_k mentioned above are only valid in relatively dry atmospheres ($W < 2.5$ g/cm^2). In order to make the SW algorithm (Eq. 5.50) applicable to more general atmospheric conditions, the total column water vapor content W in the atmosphere has to be explicitly included in the coefficients a_k as shown in Eq. (5.51). Many authors have proposed incorporating both the LSEs and the W into the coefficients a_k, and all of them empirically described the coefficients a_k as linear combinations of the LSEs and the W (Sobrino et al. 1991, 1994, 2004a; Becker and Li 1995). François and Ottlé (1996) proposed another parameterization of a_k, in which the a_k is expressed as quadratic functions of the W at a given LSE, that is,

$$a_k = C_{0,k} + C_{1,k}W + C_{2,k}W^2 \tag{5.54}$$

in which the coefficients C are constant for a given LSE.

Parameterization of the Coefficients with Consideration of LSEs, W, and VZA

It should be noted that the atmospheric transmittance will be decreased at larger VZA because of the increased absorption path length. In order to achieve the required LST accuracy of 1 K over a wide range of atmospheric and surface conditions, error analysis demonstrates that the VZA must also be considered in the LST retrieval algorithm in a hot and humid atmosphere. Many studies have shown that LST retrieval can be significantly improved at VZA larger than 50° by introducing the cosine of the VZA into the parameterization of a_k (Becker and Li 1995; Minnis and Khaiyer 2000; Pinheiro et al. 2004). For instance, Becker and Li (1995) further modified their local split-window algorithm by explicitly incorporating the cosine of VZA and the W in the coefficients a_k as shown in Eq. (5.51). The function f_k in Eq. (5.51) is empirically determined by using a linear combination of LSEs, W, and cos(VZA). Wan and Dozier (1996) developed a generalized split window (GSW) algorithm, which uses a formula similar to that proposed by Becker and Li (1990b), to retrieve LSTs from the Moderate Resolution Imaging Spectroradiometer (MODIS) TIR channels i (channel 31) and j (channel 32). The GSW algorithm can be written as

$$\begin{aligned}\mathrm{LST} = b_0 &+ \left(b_1 + b_2\frac{1-\varepsilon}{\varepsilon} + b_3\frac{\Delta\varepsilon}{\varepsilon^2}\right)\frac{T_i+T_j}{2} \\ &+ \left(b_4 + b_5\frac{1-\varepsilon}{\varepsilon} + b_6\frac{\Delta\varepsilon}{\varepsilon^2}\right)\frac{T_i-T_j}{2}\end{aligned} \tag{5.55}$$

where b_k ($k = 0$–6) are the unknown coefficients that need to be determined for a given VZA and for given subranges of ε, atmospheric W, and surface air temperature, that is, the air temperature at the surface level (T_a) (Wan and Dozier 1996) or LST (Tang et al. 2008). Using the same form as Eq. (5.50), it can be shown that

$$a_0 = b_0, \quad a_1 = b_1 + b_2\frac{1-\varepsilon}{\varepsilon} + b_3\frac{\Delta\varepsilon}{\varepsilon^2}, \quad \text{and}$$

$$a_2 = \frac{b_4 - b_1}{2} + \frac{b_5 - b_2}{2}\frac{1-\varepsilon}{\varepsilon} + \frac{b_6 - b_3}{2}\frac{\Delta\varepsilon}{\varepsilon^2}$$

Similar to the use of piecewise linear functions to approximate nonlinear functions, in the operational algorithm, for each VZA, the atmospheric W, averaged emissivity ε, and T_a or LST are divided into several tractable subranges to improve the LST retrieval accuracy over a wide range of surface and atmospheric conditions. The W is divided into subranges up to 6.5 g/cm^2 with an overlap of 0.5 g/cm^2. The ε is also separated into two groups, one varying from 0.90 to 0.96 and the other varying from 0.94 to 1.0. The T_a subranges are divided by 273, 281, 289, 295, 300, 305, and 310 K. The LST varies within T_a ±16 K, and the LST range of 32 K may be divided into four overlapped subranges (Wan and Dozier 1996). Instead of dividing T_a into several subranges, Tang et al. (2008) proposed to divide LST into several subranges that overlap by 5 K, for example, ≤280.0 K, 275–295 K, 290–310 K, 305–325 K, and ≥320 K. For a given VZA and given subranges of ε, atmospheric W, and T_a or LST, the coefficients b_k ($k = 0$–6) are determined by minimizing Eq. (5.55) using radiative transfer simulation data in ranges wide enough to cover the variations of surface and atmospheric conditions and are then saved in a set of multidimensional look-up tables (LUT). The coefficients b_k can be linearly interpolated using the cosine of the VZA. Accordingly, the coefficients b_k in the GSW algorithm vary with the LSEs, the VZA, the atmospheric W, and T_a or LST. In practice, the LST is estimated in two steps. For Wan and Dozier (1996)'s algorithm, the approximate LST is first estimated with coefficients b_k that cover the entire LST range of $T_a \pm 16$ K in a suitable W subrange, and then a more accurate LST is obtained by the coefficients b_k determined by the difference between the approximate LST and T_a. For Tang et al. (2008)'s algorithm, the approximate LST is first estimated with coefficients b_k that cover the entire LST range from 240 to 330 K in a suitable W subrange, and then a more accurate LST is determined in terms of the coefficients b_k for the LST subrange in which the approximate LST is found.

Wan (1999) compared the viewing-angle-dependent LST algorithm to the viewing-angle-independent algorithm and pointed out that the latter has a LST error of up to 1.6 K if there is an uncertainty of 0.01 in the value of $\Delta\varepsilon/\varepsilon^2$ in the GSW algorithm. The major improvements of the GSW algorithm, as described by Wan et al. (2002, 2004), incorporated in the LST retrieval include (1) VZA dependence, (2) W dependence, and (3) T_a or LST dependence. Validation results showed that an LST accuracy better than 1 K in the range 263–322 K can be obtained using the GSW algorithm given in Eq. (5.55) for certain land covers under clear-sky conditions. One of the advantages of the GSW algorithm that should be highlighted here is that LST retrieval is less sensitive to the uncertainties in the LSEs and the atmospheric conditions. Consequently, several SW formulas similar to the GSW algorithm have recently been developed to estimate LSTs from different satellite instruments, such as the Spinning Enhanced Visible and Infrared Imager (SEVIRI) and FengYun Meteorological satellite (FY-2C) instruments (Jiang and Li 2008; Tang et al. 2008; Atitar and Sobrino 2009). However, the errors in the GSW retrieved LSTs may be larger in bare soil sites in cases where LSTs are larger than T_a by more than 16 K (Wan and Li 2008). A set of new coefficients developed for bare soil lands based on RT simulation data in a much wide LST range has been used in the new version of the GSW algorithm prepared for the reprocessing and forward processing of the collection-6 or version-6 MODIS LST product in the near future.

5.4.2.3 Nonlinear Split-Window Algorithm

Because of the errors introduced by linearizing the RTE and also by some approximations used in the development of SW algorithms, for example, approximating the transmittance as a linear function of the *W*, the linear SW algorithm described by Eq. (5.50) results in large errors in LST retrieval under wet and hot atmospheric conditions. To improve the accuracy of LST retrieval, nonlinear SW algorithms have been developed:

$$\mathrm{LST} = c_0 + c_1 T_i + c_2\left(T_i - T_j\right) + c_3\left(T_i - T_j\right)^2 \tag{5.56}$$

where c_k ($k = 0$–3) are the coefficients predetermined by regressing Eq. (5.56) on simulated satellite data with a set of atmospheres and surface parameters, similar to the a_k in Eq. (5.50).

Many similar forms of nonlinear SW algorithms have been developed in the literature in recent decades (Sobrino et al. 1994; François and Ottlé 1996; Coll and Caselles 1997; Sobrino and Raissouni 2000; Atitar and Sobrino 2009). Similar to the linear SW algorithms, the nonlinear SW algorithms can also be roughly categorized into three groups: (1) coefficients c_k with consideration of LSEs, (2) coefficients c_k with consideration of LSEs and *W*, and (3) coefficients c_k with consideration of LSEs, *W*, and VZA. The following will introduce in detail these three groups of nonlinear SW algorithms.

With Consideration of LSEs

In order to account for the effect of the LSEs on the retrieval of LST, Sobrino et al. (1994) and Coll and Caselles (1997) proposed a nonlinear SW algorithm with a form similar to Eq. (5.56) with c_0 formulated as a linear function of the mean ε and the difference $\Delta\varepsilon$ of the two channels:

$$c_0 = D_0 + D_1(1 - \varepsilon) + D_2\Delta\varepsilon \tag{5.57}$$

in which the coefficients D_k ($k = 0$–2) are the constants that are independent of the atmosphere. Sun and Pinker (2003) also proposed a nonlinear SW algorithm in which the LSEs are implicitly considered by making each parameter c_k in Eq. (5.56) dependent on the land surface type.

With Consideration of LSEs and *W*

To further improve the accuracy of and reduce the influence of wet atmospheric conditions on LST retrieval, Sobrino and Raissouni (2000) and Sobrino et al. (2004a) developed a nonlinear SW algorithm to retrieve the LST from the global land 1-km Advanced Very High Resolution Radiometer (AVHRR) data with an estimated error of 1.3 K compared with a set of three hundred LSTs measured in situ in two regions of Australia (Prata 1994b). They parameterized D_k ($k = 1, 2$) in c_0 (Eq. 5.57) as linear functions of *W*, that is,

$$D_k = E_{0,k} + E_{1,k} W \quad (k = 1, 2)\,, \tag{5.58}$$

in which the coefficients *E* are the sensor-dependent constants.

With Consideration of LSEs, *W*, and VZA

To improve the accuracy of LST retrieval from TIR data measured at larger VZA, the VZA has to be considered when developing the LST retrieval algorithm. Sobrino and Romaguera (2004) and Atitar and Sobrino (2009) proposed a physics-based nonlinear SW algorithm for SEVIRI data in two TIR channels. They set $c_1 = 1$ and parameterized the coefficients c_k ($k = 2, 3$) in Eq. (5.56) and *E* in Eq. (5.58) as linear functions of the square of the secant of the VZA. It has been shown that this type of SW is capable of obtaining LST values with a root-mean-square error (RMSE) of 1.3 K using SEVIRI data at VZAs lower than 50°.

5.4.2.4 Linear or Nonlinear Multichannel Algorithms

When there are three or more TIR channels available, the LST can be retrieved from a linear or nonlinear combination of the TOA brightness temperatures in those channels using methods similar to the SW algorithms described above (Sun and Pinker 2003, 2005, 2007). For instance, Sun and Pinker (2003) developed a three-channel linear algorithm to retrieve nighttime LSTs from the Geostationary Operational Environmental Satellite (GOES) data, assuming that the LSEs in these three channels can be estimated from the land surface types. They proposed to use the characteristics of the MIR channel i_l at 3.9 μm to improve the atmospheric correction at night, and the coefficients in the three-channel linear equation explicitly include the channel LSEs, but neglect the W and the VZA:

$$\mathrm{LST} = d_0 + \left(d_1 + d_2\frac{1-\varepsilon_i}{\varepsilon_i}\right)T_i + \left(d_3 + d_4\frac{1-\varepsilon_j}{\varepsilon_j}\right)T_j + \left(d_5 + d_6\frac{1-\varepsilon_{i1}}{\varepsilon_{i1}}\right)T_{i1} \quad (5.59)$$

where the coefficients d_k ($k = 0$–6) are constants, independent of atmosphere and VZA. Comparisons with some published SW algorithms (Becker and Li 1990b; Wan and Dozier 1996) demonstrated that the proposed three-channel algorithm obtained the best LST values, with an RMSE less than 1 K (Sun and Pinker 2003). Furthermore, Sun and Pinker (2007) also proposed a four-channel nonlinear algorithm to retrieve the LST from SEVIRI data, with the coefficients depending on land surface types to account for LSE effects. For LST retrieval at nighttime, Eq. (5.60) is used:

$$\begin{aligned}\mathrm{LST} = {} & e_0 + e_1 T_i + e_2\left(T_i - T_j\right) + e_3\left(T_{i1} - T_{i2}\right) + e_4\left(T_i - T_j\right)^2 \\ & + e_5\left(\sec(\mathrm{VZA}) - 1\right)\end{aligned} \quad (5.60)$$

where the subscript i_2 represents the TIR channel at 8.7 μm and the coefficients e_k ($k = 0$–5) are dependent on the land surface type. To account for the solar radiation reflected by the earth's surface in the MIR channel i_l during the daytime, a solar correction term $d_6 T_{i1} \cos\theta_s$ was added to Eq. (5.60). When evaluated against ground observations, the results showed that the LSTs retrieved using the four-channel algorithm are more accurate than those obtained using the GSW algorithm.

However, it should be noted that MIR data measured at the TOA during the daytime consists of a combination of reflected solar radiance and emitted radiance from both the surface and the atmosphere, and the error caused by the solar correction term can affect the accuracy of LST retrieval especially in arid and semiarid regions with high reflectance in the MIR. In addition, introducing one more channel comes with the expense of increased measurement errors. The errors related to instrumental noise and other uncertainties, such as in surface emissivity of the 8.7 μm channel in arid regions, might influence the final LST retrieval accuracy.

5.4.3 Multiangle Method

Similar to the rationale of the SW method, the multiangle method is based on differential atmospheric absorption due to the different path-lengths when the same object is observed in a given channel from different viewing angles (Chédin et al. 1982; Prata 1993, 1994a, b; Sobrino et al. 1996, 2004c; Li et al. 2001).

This method was primarily developed based on the first sensor to operate in biangular mode, the Along-Track Scanning Radiometer (ATSR) onboard the first European Remote Sensing Satellite (ERS-1). ATSR can achieve a dual-angle observation of the same region of the earth's surface within about 2 min. One viewing angle is at nadir with a zenithal angle from 0° to 21.6°, and the other is in the forward

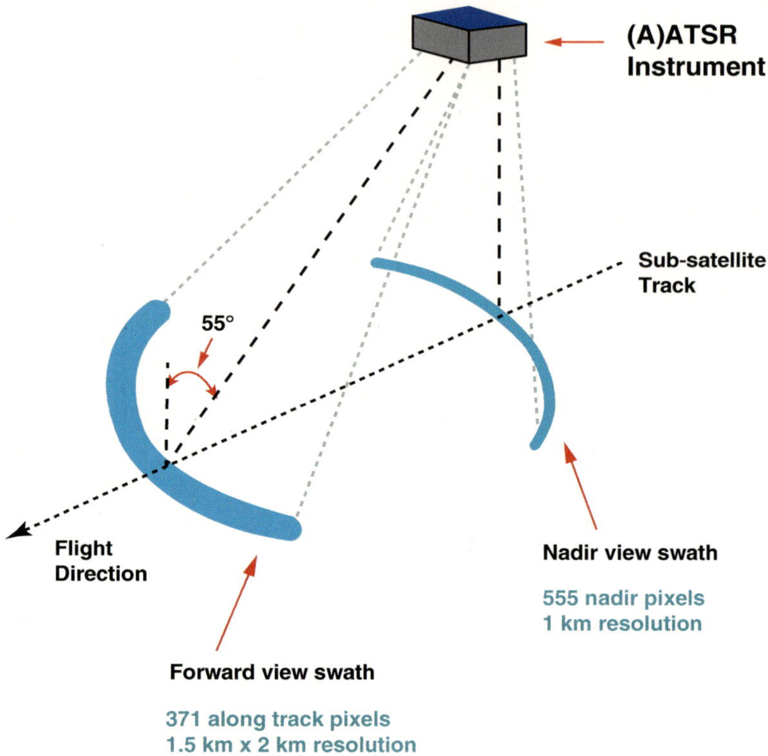

Fig. 5.7 Illustration of the along-track scanning geometry of the Along-Track Scanning Radiometer (ATSR) (http://www.leos.le.ac.uk/aatsr/whatis/instrument.html)

view with a zenithal angle from 52° to 55°. Figure 5.7 illustrates the along-track scanning geometry of this imaging radiometer. Assuming that the LST and SST are independent on the VZA and that the atmosphere is horizontally uniform and stable over the observation time, Prata (1993, 1994a) derived a dual-angle method to retrieve the SST and the LST from ATSR data. Sobrino et al. (1996) later proposed an improved dual-angle algorithm that incorporates the emissivity ε_n at nadir and the emissivity ε_f in the forward view:

$$\text{LST} = T_n + p_1\left(T_n - T_f\right) + p_2 + p_3\left(1 - \varepsilon_n\right) + p_4\left(\varepsilon_n - \varepsilon_f\right) \tag{5.61}$$

where p_k ($k = 1$–4) are the coefficients related to the atmospheric transmittances and mean air temperatures in the nadir and forward views; T_n and T_f are the brightness temperatures measured in the nadir and forward views, respectively. This algorithm includes only emissivity dependence and has no explicit W dependence. A nonlinear dual-angle algorithm has been developed by Sobrino et al. (2004c) to reduce the influence of atmospheric W on the LST retrieval:

$$\begin{aligned}\text{LST} = T_n + q_1\left(T_n - T_f\right) + q_2\left(T_n - T_f\right)^2 + (q_3 + q_4 W)\,(1 - \varepsilon_n) \\ + (q_5 + q_6 W)\,\Delta\varepsilon + q_0\end{aligned} \tag{5.62}$$

where q_k ($k = 0$–6) are the sensor-dependent constants that are predetermined by the simulation method. Using simulated TIR data, Sobrino and Jiménez-Muñoz (2005) compared the LSTs retrieved using the dual-angle algorithm given in Eq. (5.61) and the nonlinear SW algorithm incorporating LSEs, W, and VZA described in Sect. 5.4.2.3. The results showed that the dual-angle algorithm is superior to the SW algorithm provided that the spectral and angular variations of the LSEs are well known.

However, it should be pointed out that although the multiangle (dual-angle) algorithm provides better results than the SW algorithm, the dual-angle algorithm has several practical difficulties when applied to satellite data (Sobrino and Jiménez-Muñoz 2005). A critical phenomenon in the multiangle method is the angular dependence of the emissivity, as the angular behavior of natural surfaces such as soils and rocks is not well known at the scales of satellite's spatial resolution (Sobrino and Cuenca 1999). The angular dependence of the LST is also an issue. In addition to the requirement that the atmosphere is free of clouds and horizontally uniform, the multiangular measurements must have significant differences in slant path-lengths. Otherwise, the measurements will be highly correlated, and the algorithm will be unstable and extremely sensitive to measurement noise (Prata 1993, 1994a). Furthermore, "the same" object observed at different viewing angles can cover different sensor areas (in terms of pixels). Even though the same pixel size may be obtained, an object observed at different observation angles may also appear totally different because of the three-dimensional structure of the object. Finally, misregistration of pixels under different viewing angles can result in drastic errors in the LST retrieval results. Consequently, multiangle methods can only be applied to homogeneous areas (e.g., the surface of the sea or densely vegetated forest) in ideal atmospheric conditions but not to heterogeneous areas.

5.5 LST Retrieval with Unknown LSEs

All of the methods described above estimate the LST by the assumption that the LSEs are known a priori. In practice, the heterogeneity of the surface and the angular and spectral variation of the LSEs (Li et al. 2013b) make it challenging to exactly determine the LSEs at the satellite pixel scale in advance. Generally, an uncertainty of 1 % in the LSEs will result in about 0.5 K errors in the LST under normal conditions. To ensure the target of 1 K accuracy in LST retrieval, methods to estimate the LSEs from space must also be developed. To date, there have been at least three distinct methods to estimate the LST from space when the LSEs are not known, namely, the stepwise retrieval methods, the simultaneous LST and LSE retrieval methods with known atmospheric information, and the simultaneous retrieval methods for LSEs, LST, and atmospheric profiles. The stepwise retrieval methods determine the LSE and the LST separately. The LSE is estimated first, and the LST is then retrieved. The simultaneous LST and LSE retrieval methods with known atmospheric information treat both the LST and the LSE as unknowns and resolve both of them simultaneously from the atmospherically corrected radiances or with approximated atmospheric profiles. The simultaneous retrieval methods for LSEs, LST, and atmospheric profiles take all of these surface and atmospheric parameters as unknowns and retrieve them simultaneously.

5.5.1 Stepwise Retrieval Method

The stepwise retrieval method estimates first the LSEs empirically from visible/near-infrared (VNIR) measurements or physically from pairs of atmospherically corrected MIR and TIR radiances at ground level. The LST is then estimated using any of the single, multichannel (SW), or multiangle retrieval methods described in Sect. 5.4. The representative methods include the classification-based emissivity method (CBEM) (Snyder et al. 1998; Sun and Pinker 2003; Peres and DaCamara 2005), the Normalized Difference Vegetation Index (NDVI)-based emissivity method (NBEM) (Van de Griend and Owe 1993; Valor and Caselles 1996; Sobrino and Raissouni 2000), and the temperature-independent spectral indices (TISI)-based method (Becker and Li 1990b; Li and Becker 1993; Li et al. 2000). Compared with the TISI-based method, the CBEM and NBEM are simpler, and LSEs can be obtained at the higher spatial resolution, and the TIR channel and accurate atmospheric correction are not required. However, there are some uncertainties remained in the estimated LSEs using the CBEM and NBEM because some a priori knowledge, such as the knowledge of the dynamic states and the LSEs for soil and vegetation, is often unavailable. The TISI-based method is complex, however, of high accuracy due to its physical basis. More details about this type of stepwise retrieval method can be found in Sect. 4.3.1.

5.5.2 Simultaneous LST and LSE Retrieval Methods with Known Atmospheric Information

Because the accuracy of the retrieved LST is primarily dependent on the accuracy of the LSEs, simultaneous determination of the LSEs and the LST has been proposed to improve the retrieval accuracy. Many simultaneous LSEs and LST retrieval methods with given known atmospheric information have been developed since the 1990s. Usually, those methods make use of measurements at different times under the assumption that the LSEs are time invariant or rely on the intrinsic spectral behavior of the LSE rather than temporal information to retrieve simultaneously the LSEs and the LST. The representatives of these methods include the two-temperature method (TTM) (Watson 1992), the physics-based day/night operational method (D/N) (Wan and Li 1997), the reference channel method (RCM) (Kahle et al. 1980), the normalization emissivity method (NEM) (Gillespie 1985), the temperature emissivity separation method (TES) (Gillespie et al. 1996, 1998), the gray body emissivity method (GEM) (Barducci and Pippi 1996), the iterative spectrally smooth temperature emissivity separation method (ISSTES) (Borel 1997, 2008), and the linear emissivity constraint temperature emissivity separation method (LECTES) (Wang et al. 2011). From the analysis performed in Sect. 4.3.2, TTM and D/N assume LSEs do not change significantly in two observation times, which implies multitemporal data are required and the retrieved accuracies of LSEs and LST would depend on the geometry registration. RCM, NEM, TES, GEM, ISSTES, and LECTES make use of the intrinsic spectral behaviors of the LSEs, such as the empirical relationship between the spectral contrast and the minimum emissivity or the emissivity in a small segment is linearly related to the wavenumber or wavelength. Based on some reasonable assumptions or constraints, those methods can retrieve the LSEs and LST simultaneously from the atmospherically corrected radiances at the ground level either by reducing the number of unknowns or by increasing the number of equations. More details about this type of retrieval method can be found in Sect. 4.3.2.

5.5.3 *Simultaneous Retrieval of LST, LSEs, and Atmospheric Profiles*

Because of the coupling of atmosphere and surface, accuracies of LSEs and LST retrieved by the methods described in Sect. 5.3.2 will be degraded if accurate atmospheric effects cannot be well corrected for. Taking into account the fact that the atmospheric profiles are usually unavailable synchronously with TIR measurements, Ma et al. (2002) proposed to retrieve simultaneously the LSEs, LST, and atmospheric parameters (e.g., atmospheric profiles) for improving the overall accuracies. Subsequently, several methods have been proposed to retrieve simultaneously the surface and atmospheric parameters. The representatives of these methods are the artificial neural network (ANN) method (Wang et al. 2010, 2013) and the two-step physical retrieval method (TSRM) (Ma et al. 2000, 2002). Because of the complexities and limitations, those methods are only applicable for some specific sensors. More details can be found in Sect. 4.3.3.

5.6 Comparison and Analysis of Different Methods

It should be noted that LST retrieval methods reviewed above are developed for use under different conditions with different assumptions, and there is consequently no universal method capable of always accurately retrieving LSTs from all satellite TIR data. It is meaningless to perform a comparison of these algorithms without considering those assumptions. Therefore, it is generally difficult to decide which algorithm is superior to others. The optimal method to retrieve the LST from space in practice can be selected by considering the characteristics of the sensor, the availability of emissivity data and atmospheric information, the complexity of the method, and other considerations.

Because the SW algorithms are simple, effective, and generally suitable for most sensors, many comparative studies evaluating the performance of these methods have been carried out. Vazquez et al. (1997) found that most SW algorithms are statistically indistinguishable. Kerr et al. (2000) performed an algorithm comparison and concluded that the selection of the best LST algorithm may depend on a priori knowledge of the *W* and the LSE. Sòria and Sobrino (2007) reported that the RMSE of retrieved LST in SW algorithms generally decreases as more input parameters are explicitly included. However, more input parameters will introduce more uncertainties and decrease the accuracy of the retrieval LST. Yu et al. (2008) have evaluated nine published SW algorithms to determine which are most capable of generating a consistent LST climate data record across satellite sensors and platforms. The results showed that the SW algorithms that depend on both the mean and the difference of channel emissivities are the most accurate and stable over a wide range of conditions if the emissivities are accurately known. However, Yu et al. (2009) reported that the use of both the mean and the difference of channel emissivities may be too sensitive to the emissivity uncertainty and should not be used in operational practice. As a compromise, the SW algorithms that only use the mean emissivities are recommended by Yu et al. (2009).

Because all of the assumptions and restrictions involved in the LST retrieval methods, such as the required number of TIR channels and the knowledge regarding emissivities or atmospheric quantities, cannot be met simultaneously, comparisons are seldom made except for the SW algorithms. To provide a concise overview, the assumptions, advantages, and limitations of each of these methods with known LSEs are summarized in Table 5.2 to help users select the optimal method in practice (Li et al. 2013a).

5.7 Validation of Satellite-Derived LST

Validation is a process of independently assessing the uncertainty of the data derived from the system outputs. Without validation, no methods, algorithms, or parameters derived from remotely sensed data can be used with confidence. As the retrieved LSTs from satellite TIR data involve corrections to the satellite-observed radiances

Table 5.2 Comparison of different methods for retrieving LSTs with a known emissivity from satellite data

Methods	Assumptions	Superiority	Limitations and disadvantages	References
Single- channel algorithms	No special assumption	Simplicity; applicable to sensors with only one TIR channel	Require a priori knowledge of the pixel emissivity in the TIR channel; require accurate atmospheric profiles and a good RTM to estimate atmospheric quantities; the uncertainty of atmospheric profiles may have strong effects on the accuracy of LST retrieval; forward calculations of atmospheric quantities using an RTM are time-consuming; use of empirical relationships provides poor results at high atmospheric water vapor contents	Jiménez-Muñoz and Sobrino (2003), Cristóbal et al. (2009), Jiménez-Muñoz et al. (2009), and Qin et al. (2001)
Multichannel (SW) algorithms	Radiances involved in the RTE can be approximated by the first order of the Taylor's expansion with respect to a wavelength or a mean temperature; difference of the brightness temperatures at TOA results mainly from the differential absorption of the atmosphere in different channels	Simplicity and high efficiency; atmospheric profiles are not needed; suitable for various sensors with more than one TIR channels within the atmospheric window	Require a priori knowledge of the pixel emissivity in each TIR channel; require a priori determination of the SW coefficients; LST retrieval accuracy is dependent on the correct choice of the SW coefficients	McMillin (1975), Becker and Li (1990b), Sobrino et al. (1994), and Wan and Dozier (1996)
Multiangle algorithms	LSTs are independent of the VZA; the atmosphere is horizontally uniform and stable over the observation time	Simplicity and high efficiency; atmospheric profiles are not required; LST retrieval accuracy is insensitive to the uncertainties in the optical properties of the atmospheric absorbents	Require a priori knowledge of the angular variation of LSEs at satellite pixel scale; require a significant difference in the slant path-lengths; require accurate geometric registration; applicable only to the homogeneous surfaces	Prata (1993), Sobrino et al. (1996), and Sòria and Sobrino (2007)

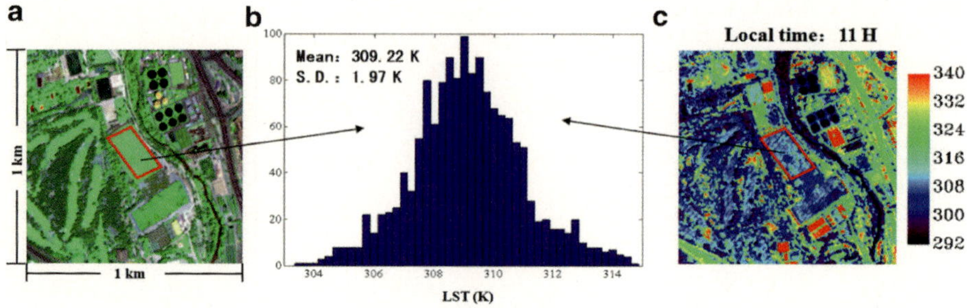

Fig. 5.8 Illustration of LST variation within a MODIS pixel scale (**a**) 1 km × 1 km composited image by visible channels, (**b**) LST histogram corresponding to the homogeneous grass in the *red rectangle* region in (**a**), and (**c**) LST image corresponding to 1 km × 1 km pixel of (**a**)

to account for atmospheric effects and nonunity LSEs, it is necessary to assess the accuracy of the retrieval to provide potential LST users with reliable information regarding the quality of the LST product and to provide feedback to the developers of LST retrieval algorithms for future improvement. However, validation of satellite-derived products is often a challenge in itself. This is true for the sensors having a coarse resolution and even more with the thermal bands over land. Although many algorithms have been proposed and developed over recent decades to retrieve the LST from satellite TIR data, validation of satellite-derived LSTs remains far behind the development of new LST retrieval algorithms due to the difficulties summarized below:

1. It is difficult to conduct in situ LST measurements at local scale, because LST, LSE, and the atmospheric and environmental radiances are coupled together as shown in Eq. (5.19). To obtain LST, one has to measure simultaneously the atmospheric and environmental radiances and the LSE, which is time-consuming and hard to monitor.
2. It is difficult to obtain representative LST data at the scale of a pixel, which usually covers an area of about several square kilometers (1 km^2 for the MODIS and 1.2 km^2 for the AVHRR), due to the spatial variability of landscape and the large spatial–temporal variations in the LST itself. To give a rough idea of this variability, Fig. 5.8c shows the spatial variation of LST within a MODIS pixel, which is composed of different land covers and land uses as displayed in Fig. 5.8a. One notices that even though the green area within a red rectangle in Fig. 5.8a looks like very homogeneous, the spatial variation of LST is very large ranging from 304 to 315 K with a standard deviation of 1.97 K (Fig. 5.8b).
3. The third difficulty concerns the temporal sampling. Surface temperature is characterized by a very high temporal frequency. In a matter of seconds, the surface skin temperature may change by several degrees due to wind, shadow, and etc. Generally, ground radiometers often integrate the temperature over several minutes, but the sensor onboard the satellite acquires the data in a fraction of a second.

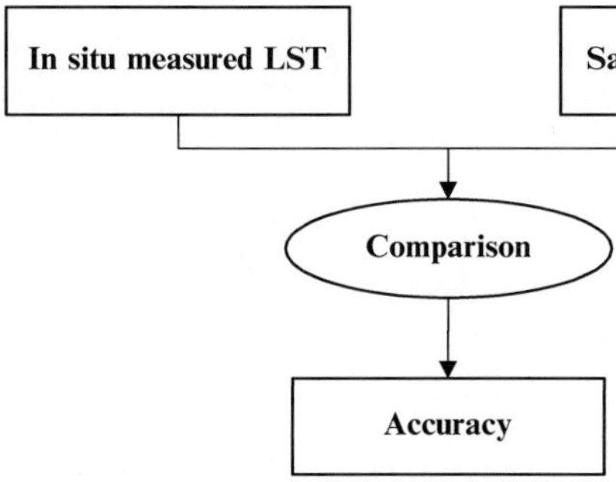

Fig. 5.9 Flowchart of the conventional temperature-based satellite LST validation method

Although there exist many problems and difficulties in the validation of the LST derived from satellite TIR data, in recent years, many efforts and studies have been performed to validate the derived LSTs from ETM + (Enhanced Thematic Mapper), ASTER, AVHRR, AATSR (Advanced Along-Track Scanning Radiometer), MODIS, and SEVIRI data (Prata 1994b; Wan et al. 2002, 2004; Sawabe et al. 2003; Coll et al. 2005, 2010, 2012b; Hook et al. 2005; Sòria and Sobrino 2007; Sobrino et al. 2007; Wan 2008; Wan and Li 2008; Hulley and Hook 2009a; Sabol et al. 2009; Wang and Liang 2009; Niclòs et al. 2011). The methods utilized to validate LST values retrieved from space may be roughly grouped into four categories: the temperature-based method (*T*-based), the radiance-based method (*R*-based), the cross-validation method, and the indirect validation method. The following will introduce each validation method and discuss their advantages and disadvantages.

5.7.1 Temperature-Based Method (T-Based)

The temperature-based (*T*-based) method is a conventional ground-based method for satellite-derived LST validation, which compares directly the satellite-derived LSTs with ground measurements performed at a field site concurrently with the satellite overpass (Prata 1994b; Slater et al. 1996; Wan et al. 2002; Coll et al. 2005; Pinker et al. 2009) (see Fig. 5.9). The measurements observed by the ground instruments at one or several points over the test site are regarded as representatives of the average LST over the ground instantaneous field of view of the satellite sensor. It is required that the spatial variability of temperature and emissivity be very small within one satellite pixel, so that ground point measurements could be compared with satellite area-averaged measurements. However, performing LST measurements in the field is a complex and difficult task due to the difference of the scales probed by satellite pixels (a few km^2) and field sensors (a few m^2 or cm^2). Moreover, natural land cover and the corresponding LST and LSE values are quite variable at the scale of km^2. Snyder et al. (1997) pointed out that

Fig. 5.10 Some validation study sites of the MODIS LST products (Courtesy of Dr. Z. Wan)

homogeneous and flat surfaces that can be easily instrumented and characterized, including inland water, sand, snow, and ice, can serve as validation sites (Sobrino et al. 2004c; Coll et al. 2005; Wan 2008; Guillevic et al. 2012). The size of the area that needs to be viewed by the validation instrument depends on the intra-pixel variability of the surface and on how well measurements of several "end-members" can be combined to obtain a representative value for the satellite pixel. This process remains challenging due to the difficulties inherent to finding adequate surfaces in the image and performing a representative thermal sampling on the ground.

Because most of the earth's surface is heterogeneous at the satellite pixel scale, high-quality ground LST validation data are scarce and are limited to a few homogeneous surface types such as lakes, silt playas, grasslands, and agricultural fields collected during dedicated campaigns (Wan et al. 2002, 2004; Coll et al. 2005, 2009, 2010). Once a thermally homogeneous area is identified, the average LST measured at several points by ground instruments within the validation site is considered to be the true LST and is compared with the satellite-derived LST at the pixel scale. Using this method, many authors have performed validation studies of the LST values produced using different sensors (Wan et al. 2002; Coll et al. 2005, 2009, 2010; Peres et al. 2008; Wan 2008; Sabol et al. 2009). Figure 5.10 shows some appropriate validation sites at which the MODIS LST validation campaigns were conducted. Figure 5.11 gives some comparisons between MODIS V5-derived LSTs and those measured in situ in field campaigns in 2000 to 2003. The results show that MODIS LST accuracy is better than 1 K in most clear-sky cases for the LST ranging from 263 to 331 K and the water vapor content varying from 0.4 to 4.0 cm.

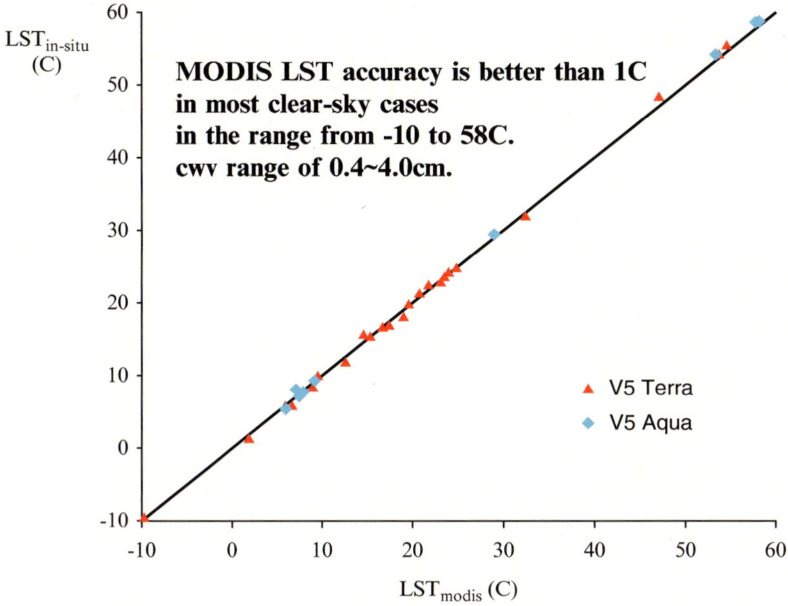

Fig. 5.11 Comparisons between V5 MODIS-derived LSTs and in situ LST measurements in field campaigns in 2000–2003 (Reproduced from Wan (2008), with permission from Taylor & Francis)

The main advantage of the *T*-based method is that it provides a direct evaluation of the radiometric quality of the satellite sensor and the ability of the LST retrieval algorithm to correct for atmospheric and emissivity effects. However, the success of *T*-based validations depends crucially on the accuracy of the ground LST measurements and how well they represent the LST at the satellite pixel scale. Because the spatial and temporal variations of the LST at daytime might be 10 K or more over a few meters or over short time intervals depending on the nature of the surface, the solar irradiation level, and the local meteorological conditions, *T*-based validation activities are often restricted to nighttime and homogeneous surfaces such as lakes, dense grasslands, and vegetated regions. Moreover, even if ground-level LST measurements can be performed, there is still a problematic difficulty in scaling up from the ground point measurements to the pixel scale under the field of view of the satellite sensor, especially over heterogeneous surfaces (Wan et al. 2002). As a result, only a few surface types are suitable for *T*-based validation within an uncertainty of ± 1 K for the ground-measured LST at the pixel scale. The collection of in situ measurements is also a demanding task and is often limited to short-term, dedicated field campaigns. Therefore, the *T*-based method is not appropriate for global validation of satellite-derived LST measurements.

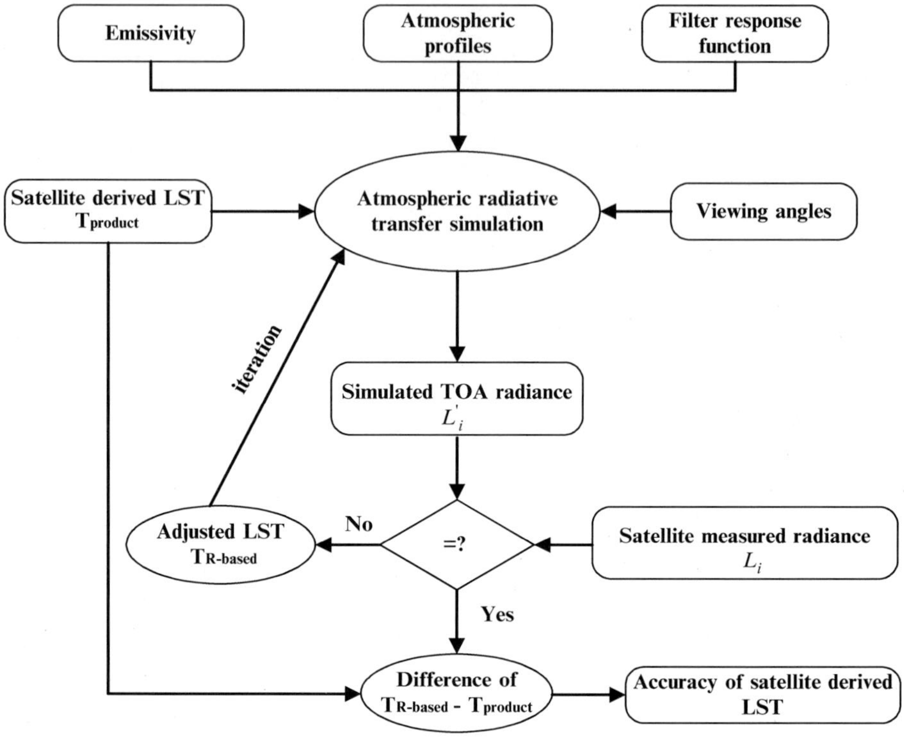

Fig. 5.12 Illustration of the radiance-based method for validation of satellite-derived LST

5.7.2 Radiance-Based Method (R-Based)

The radiance-based (*R*-based) method is an advanced alternative method for validating space-based LST measurements (Wan and Li 2008; Coll et al. 2012b). This method does not rely on ground-measured LST values, but does require both LSE spectra, which can be measured in the field or estimated from land cover types or from other auxiliary data, and measured atmospheric profiles over the validation site at the time of the satellite overpass (Wan 2008; Wan and Li 2008). This method uses the satellite-derived LST and the aforementioned in situ atmospheric profiles and LSEs as initial input parameters to an atmospheric RTM that simulates the TOA radiance at the moment of the satellite overpass. Using the difference between the simulated TOA radiance and the measured radiance, the initial LST will be adjusted and the simulated radiance will be iteratively recalculated to match the satellite-measured radiance. The difference between the adjusted LST and the initial satellite-derived LST is the accuracy of the retrieved LST. Figure 5.12 depicts the flowchart of the *R*-based method. More details about the *R*-based method are

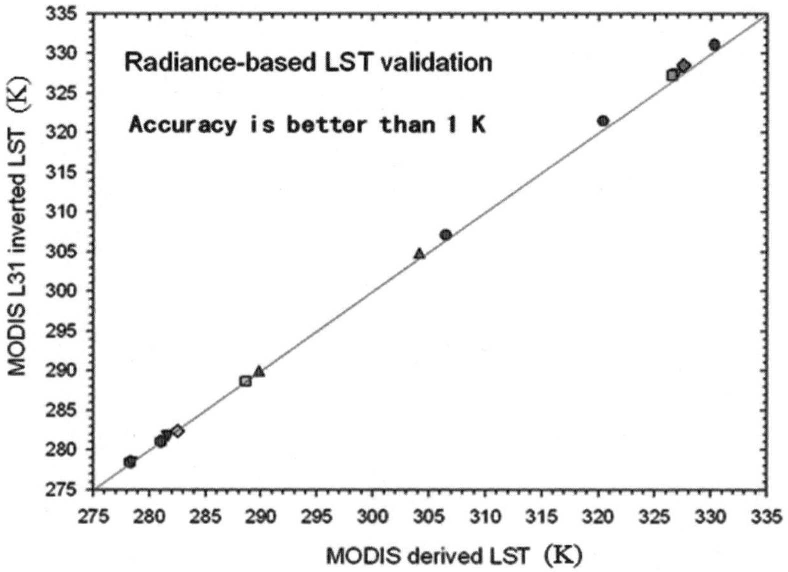

Fig. 5.13 Comparison between MODIS-derived LST and that inverted from MODIS band 31 using the radiance-based method

provided by Wan and Li (2008). Figure 5.13 shows the comparison between the MODIS-derived LST and that inverted from MODIS band 31 data by using the radiance-based validation method, which indicates that the accuracy of the MODIS-derived LST is better than 1 K.

The *R*-based method does not require ground LST measurements, and it can therefore be applied to the surfaces on which ground LST measurements are unfeasible (such as forests, surfaces with partial vegetation cover, semiarid areas, deserts, remote regions) and extended to homogeneous and nonisothermal surfaces. The promising performance of the *R*-based method offers the possibility of validating satellite-derived LST values during the daytime and nighttime over homogeneous and nonisothermal surfaces where the *T*-based method is not feasible. However, the strongest limitations of the *R*-based method are the use of measured or estimated LSEs representative at pixel scale, how to check the actual atmosphere really free of clouds, and how well the profiles used in simulations represent the actual atmosphere at the time of observations (Coll et al. 2012b). The success of the *R*-based method depends on the accuracies of the atmospheric RTM, the atmospheric profiles, and the LSEs at pixel scale. Consequently, this alternative method should be evaluated at more validations, especially at soil sites under various atmospheric conditions.

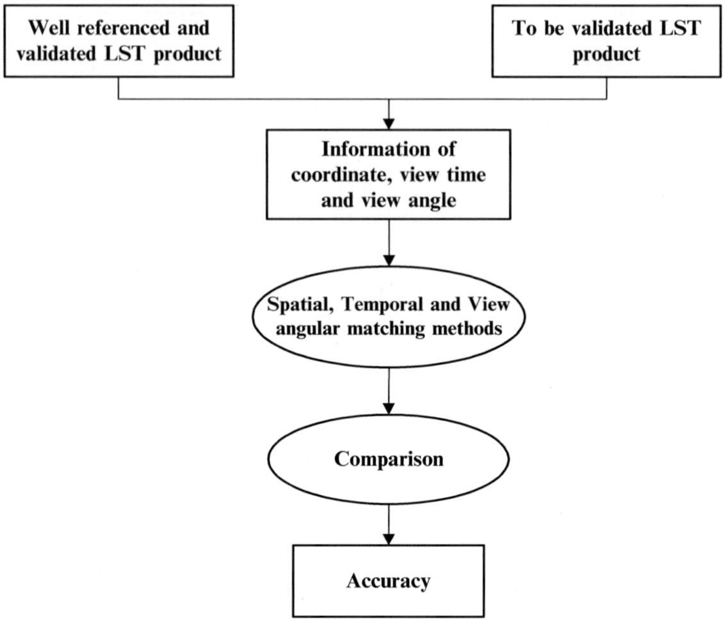

Fig. 5.14 Flowchart of the cross-validation method for cross-validating the satellite-derived LST

5.7.3 Cross-Validation Method

This method involves cross-validating the LST values retrieved by the method under test with well-documented and validated LST values retrieved from other satellite data (Trigo et al. 2008a). This technique represents an alternative method for LST validation if there are no atmospheric profiles or ground LST measurements available or if the *T*- and *R*-based validations cannot be conducted.

The cross-validation method uses a well-referenced or validated LST product derived from other satellites to compare with the satellite-derived LST to be validated. Figure 5.14 displays the flowchart of this validation method for cross-validating the satellite-derived LST. Due to the large spatial and temporal variations in the LST, geographic coordinate matching, temporal matching, and VZA matching have to be performed before the two satellite-derived LST products can be compared (Trigo et al. 2008a; Qian et al. 2013). The main advantage of this method is that the LST can be validated without any ground measurements, and it can be used anywhere in the world if well-validated LST products are available. As mentioned above, the accuracy of this method is sensitive to the spatial and temporal mismatches of the two LST measurements. The observation time interval between the two measurements should be as short as possible. Considering that the LSE also depends on the viewing zenith angle and that the pixels of the two sensors cover different areas and contain different land surface information under different viewing angles, only pixels with the same or nearly same viewing zenith angles should be used for cross-validation.

5.7.4 Indirect Validation

In contrast to the direct validation of LST, this method validates the LST derived from satellite data by using assimilated or predicted LST through land surface process models such as the Community Land Model (CLM) (Oleson et al. 2004) and the Common Land Model (CoLM) (Dai et al. 2003, 2004) or climate models such as the National Center for Atmospheric Research (NCAR) Community Climate Model (CCM3), and Atmospheric General Circulation Model (AGCM). As land data assimilation systems or climate models combine the complementary information from modeled and observed land surface fields by appropriately weighting the sources of the errors and produce dynamically consistent, spatially complete, and temporally continuous estimates of global land surface conditions, and consequently give better description of the evolving state of the LST, it may be an operational way to indirectly validate the satellite-derived LST.

It is well known that the LST varies with the change of the spatial and temporal variations. The scale mismatch issue in both space and time, between the satellite measurements and the in situ observations, is inevitable. The accuracy of the validation is more or less affected by the mismatch. Errors in the forcing fields, however, along with the imperfect parameterization of land–atmosphere interactions can lead to considerable drifts in modeled land surface states and consequently affect the accuracy of the assimilated LST. Moreover, the spatial resolution of the assimilated or predicted LST is very coarse (larger than 10 km), which may affect the validation accuracy of satellite derived LST. Anyway, the indirect validation is an alternative to the direct validation if no ground measurement or atmospheric data is collected when satellite overpasses.

References

Adams, J. B., Smith, M. O., & Gillespie, A. R. (1989). Simple models for complex natural surfaces: A strategy for the hyperspectral era of remote sensing. *IEEE International Geosciences Remote Sensing Symposium,* 16–21. New York.

Arnfield, A. J. (2003). Two decades of urban climate research: A review of turbulence, exchanges of energy and water, and the urban heat island. *International Journal of Climatology, 23*, 1–26.

Atitar, M., & Sobrino, J. A. (2009). A split-window algorithm for estimating LST from Meteosat 9 data: Test and comparison with in situ data and MODIS LSTs. *IEEE Geoscience and Remote Sensing Letters, 6*, 122–126.

Barducci, A., & Pippi, I. (1996). Temperature and emissivity retrieval from remotely sensed images using the "grey body emissivity" method. *IEEE Transactions on Geoscience and Remote Sensing, 34*, 681–695.

Barton, I. J., Zavody, A. M., O'Brien, D. M., Cutten, D. R., Saunders, R. W., & Llewellyn-Jones, D. T. (1989). Theoretical algorithms for satellite-derived sea surface temperatures. *Journal of Geophysical Research, 94*, 3365–3375.

Bastiaanssen, W. G. M., Menenti, M., Feddes, R. A., & Holtslag, A. A. M. (1998). A remote sensing surface energy balance algorithm for land (SEBAL). 1. Formulation. *Journal of Hydrology, 212*, 198–212.

Becker, F. (1981). Angular reflectivity and emissivity of natural media in the thermal infrared bands. In G. Guyot & M. Verbrugghe (Eds.), *First International Colloquium on Spectral Signatures of Objects in Remote Sensing* (Vol. 5, pp. 57–72). Avignon: INRA.

Becker, F. (1987). The impact of spectral emissivity on the measurement of land surface temperature from a satellite. *International Journal of Remote Sensing, 8*, 1509–1522.

Becker, F., & Li, Z.-L. (1990a). Towards a local split window method over land surfaces. *International Journal of Remote Sensing, 11*, 369–393.

Becker, F., & Li, Z.-L. (1990b). Temperature-independent spectral indices in thermal infrared bands. *Remote Sensing of Environment, 32*, 17–33.

Becker, F., & Li, Z.-L. (1995). Surface temperature and emissivity at various scales: Definition, measurement and related problems. *Remote Sensing Reviews, 12*, 225–253.

Berk, A., Anderson, G. P., Acharya, P. K., Hoke, M. L., Chetwynd, J. H., Bernstein, L. S., Shettle, E. P., Matthew, M. W., & Adler-Golden, S. M. (2003). MODTRAN4 Version 3 Revision 1 USER'S MANUAL. In *Hanscom Air Force Base*. St, Bedford: Air Force Res. Lab.

Borel, C. (1997). Iterative retrieval of surface emissivity and temperature for a hyperspectral sensor. In *First JPL workshop on Remote Sensing of Land Surface Emissivity* (pp. 1–5), Pasadena, CA, held at JPL.

Borel, C. (2008). Error analysis for a temperature and emissivity retrieval algorithm for hyperspectral imaging data. *International Journal of Remote Sensing, 29*, 5029–5045.

Brunsell, N. A., & Gillies, R. R. (2003). Length scale analysis of surface energy fluxes derived from remote sensing. *Journal of Hydrometeorology, 4*, 1212–1219.

Bussieres, N., Louie, P. Y. T., Hogg, W. (1990, February 13-14). Progress report on the implementation of an algorithm to estimate regional evapotranspiration using satellite data. In *Proceeding of the workshop on Applications of Remote Sensing in Hydrology*, Saskaton Saskatchewan.

Caselles, V., Sobrino, J. A., & Becker, F. (1988). Determination of the effective emissivity and temperature under vertical observation of a citrus orchard: Application to frost nowcasting. *International Journal of Remote Sensing, 9*, 715–727.

Chaumat, L., Standfuss, C., Tournier, B., Armante, R., & Scott, N. A. (2009). 4A/OP reference documentation. In NOV-3049-NT-1178-v4. 0, NOVELTIS, LMD/CNRS, CNES.

Chédin, A., Scott, N. A., & Berroir, A. (1982). A single-channel, double-viewing angle method for sea surface temperature determination from coincident Meteosat and TIROS-N radiometric measurements. *Journal of Applied Meteorology, 21*, 613–618.

Chédin, A., Scott, N. A., Wahiche, C., & Moulinier, P. (1985). The improved initialization inversion method: A high resolution physical method for temperature retrievals from satellites of the TIROS-N series. *Journal of Climate and Applied Meteorology, 24*, 128–143.

Coll, C., & Caselles, V. (1997). A split-window algorithm for land surface temperature from advanced very high resolution radiometer data: Validation and algorithm comparison. *Journal of Geophysical Research, 102*, 16697–16713.

Coll, C., Caselles, V., Sobrino, J. A., & Valor, E. (1994). On the atmospheric dependence of the split-window equation for land surface temperature. *International Journal of Remote Sensing, 15*, 105–122.

Coll, C., Caselles, V., Galve, J. M., Valor, E., Niclòs, R., Sanchez, J. M., & Rivas, R. (2005). Ground measurements for the validation of land surface temperatures derived from AATSR and MODIS data. *Remote Sensing of Environment, 97*, 288–300.

Coll, C., Hook, S. J., & Galve, J. M. (2009). Land surface temperature from the Advanced Along-Track Scanning Radiometer: Validation over inland waters and vegetated surfaces. *IEEE Transactions on Geoscience and Remote Sensing, 47*, 350–360.

Coll, C., Galve, J. M., Sanchez, J. M., & Caselles, V. (2010). Validation of Landsat-7/ETM + thermal-band calibration and atmospheric correction with ground-based measurements. *IEEE Transactions on Geoscience and Remote Sensing, 48*, 547–555.

Coll, C., Caselles, V., Valor, E., & Niclòs, R. (2012a). Comparison between different sources of atmospheric profiles for land surface temperature retrieval from single channel thermal infrared data. *Remote Sensing of Environment, 117*, 199–210.

Coll, C., Valor, E., Galve, J. M., Mira, M., Bisquert, M., García-Santos, V., Caselles, E., & Caselles, V. (2012b). Long-term accuracy assessment of land surface temperatures derived from the Advanced Along-Track Scanning Radiometer. *Remote Sensing of Environment, 116*, 211–225.

Cooper, D. I., & Asrar, G. (1989). Evaluating atmospheric correction models for retrieving surface temperatures from the AVHRR over a tallgrass prairie. *Remote Sensing of Environment, 27*, 93–102.

Cristóbal, J., Jiménez-Muñoz, J. C., Sobrino, J. A., Ninyerola, M., & Pons, X. (2009). Improvements in land surface temperature retrieval from the Landsat series thermal band using water vapor and air temperature. *Journal of Geophysical Research, 114*, D08103.

Dai, Y., Zeng, X., Dickinson, R. E., Baker, I., Bonan, G. B., Bosilovich, M. G., Scott Denning, A., Dirmeyer, P., Houser, P., Niu, G., Oleson, K., Schlosser, C., & Yang, Z. L. (2003). The common land model. *Bulletin of the American Meteorological Society, 84*(8), 1013–1023.

Dai, Y., Dickinson, R. E., & Wang, Y.-P. (2004). A two-big-leaf model for canopy temperature, photosynthesis and stomatal conductance. *Journal of Climate, 17*, 2281–2299.

Dash, P., Gottsche, F. M., Olesen, F. S., & Fischer, H. (2002). Land surface temperature and emissivity estimation from passive sensor data: Theory and practice-current trends. *International Journal of Remote Sensing, 23*, 2563–2594.

Deschamps, P. Y., & Phulpin, T. (1980). Atmospheric correction of infrared measurements of sea surface temperature using channels at 3.7, 11 and 12 um. *Boundary Layer Meteorology, 18*, 131–143.

França, G. B., & Carvalho, W. S. (2004). Sea surface temperature GOES-8 estimation approach for the Brazilian coast. *International Journal of Remote Sensing, 25*, 3439–3450.

François, C., & Ottlé, C. (1996). Atmospheric corrections in the thermal infrared: Global and water vapor dependent split-window algorithms-applications to ATSR and AVHRR data. *IEEE Transactions on Geoscience and Remote Sensing, 34*, 457–470.

Gillespie, A. R. (1985). Lithologic mapping of silicate rocks using TIMS. In *TIMS data users' workshop*. Pasadena: Jet Propul. Lab., JPL Publication 86-38, 29–44.

Gillespie, A. R. (2005). Extending surface temperature and emissivity retrieval to the mid-infrared (3–5mm) using the Multispectral Thermal Imager. *Remote Sensing of Environment, 98*, 141–151.

Gillespie, A. R., Rokugawa, S., Hook, S. J., Matsunaga, T., & Kahle, A. B. (1996). Temperature/emissivity separation algorithm theoretical basis document, Version 2.4 (pp. 1–64). Maryland: NASA/GSFC.

Gillespie, A. R., Rokugawa, S., Matsunaga, T., Cothern, J. S., Hook, S., & Kahle, A. B. (1998). A temperature and emissivity separation algorithm for Advanced Spaceborne Thermal Emission and Reflection Radiometer (ASTER) images. *IEEE Transactions on Geoscience and Remote Sensing, 36*, 1113–1126.

Gillespie, A. R., Abbott, E. A., Gilson, L., Hulley, G., Jiménez-Muñoz, J. C., & Sobrino, J. A. (2011). Residual errors in ASTER temperature and emissivity standard products AST08 and AST05. *Remote Sensing of Environment, 115*, 3681–3694.

Guillevic, P. C., Privette, J. L., Coudert, B., Palecki, M. A., Demarty, J., Ottlé, C., & Augustine, J. A. (2012). Land Surface Temperature product validation using NOAA's surface climate observation networks-Scaling methodology for the Visible Infrared Imager Radiometer Suite (VIIRS). *Remote Sensing of Environment, 124*, 282–298.

Hansen, J., Ruedy, R., Sato, M., & Lo, K. (2010). Global surface temperature change. *Reviews of Geophysics, 48*, RG4004.

Hook, S. J., Clodius, W. B., Balick, L., Alley, R. E., Abtahi, A., Richards, R. C., & Schladow, S. G. (2005). In-flight validation of mid- and thermal infrared data from the Multispectral Thermal Imager (MTI) using an automated high-altitude validation site at Lake Tahoe CA/NV, USA. *IEEE Transactions on Geoscience and Remote Sensing, 43*, 1991–1999.

Hulley, G. C., & Hook, S. J. (2009a). Intercomparison of versions 4, 4.1 and 5 of the MODIS Land Surface Temperature and Emissivity products and validation with laboratory measurements of sand samples from the Namib desert, Namibia. *Remote Sensing of Environment, 113*, 1313–1318.

Jiang, G. M., & Li, Z.-L. (2008). Split-window algorithm for land surface temperature estimation from MSG1-SEVIRI data. *International Journal of Remote Sensing, 29*, 6067–6074.

Jiang, G. M., Li, Z.-L., & Nerry, F. (2006). Land surface emissivity retrieval from combined mid-infrared and thermal infrared data of MSG-SEVIRI. *Remote Sensing of Environment, 105*, 326–340.

Jiménez-Muñoz, J. C., & Sobrino, J. A. (2003). A generalized single-channel method for retrieving land surface temperature from remote sensing data. *Journal of Geophysical Research, 108*, 4688–4695.

Jiménez-Muñoz, J. C., & Sobrino, J. A. (2010). A single-channel algorithm for land-surface temperature retrieval from ASTER data. *IEEE Geoscience and Remote Sensing Letters, 7*, 176–179.

Jiménez-Muñoz, J. C., Cristobal, J., Sobrino, J. A., Soria, G., Ninyerola, M., & Pons, X. (2009). Revision of the single-channel algorithm for land surface temperature retrieval from Landsat thermal-infrared data. *IEEE Transactions on Geoscience and Remote Sensing, 47*, 339–349.

Jiménez-Muñoz, J. C., Sobrino, J. A., Mattar, C., & Franch, B. (2010). Atmospheric correction of optical imagery from MODIS and Reanalysis atmospheric products. *Remote Sensing of Environment, 114*, 2195–2210.

Kahle, A. B., Madura, D. P., & Soha, J. M. (1980). Middle infrared multispectral aircraft scanner data: Analysis for geological applications. *Applied Optics, 19*, 2279–2290.

Kalma, J. D., McVicar, T. R., & McCabe, M. F. (2008). Estimating land surface evaporation: A review of methods using remotely sensed surface temperature data. *Surveys in Geophysics, 29*, 421–469.

Kealy, P. S., & Hook, S. J. (1993). Separating temperature and emissivity in thermal infrared multispectral scanner data: Implications for recovering land surface temperatures. *IEEE Transactions on Geoscience and Remote Sensing, 31*, 1155–1164.

Kerr, Y. H., Lagouarde, J. P., & Imbernon, J. (1992). Accurate land surface temperature retrieval from AVHRR data with use of an improved split window algorithm. *Remote Sensing of Environment, 41*, 197–209.

Kerr, Y. H., Lagouarde, J. P., Nerry, F., & Ottlé, C. (2000). Land surface temperature retrieval techniques and applications. In D. A. Quattrochi & J. C. Luvall (Eds.), *Thermal remote sensing in land surface processes* (pp. 33–109). Boca Raton: CRC Press.

Kogan, F. N. (2001). Operational space technology for global vegetation assessment. *Bulletin of the American Meteorological Society, 82*, 1949–1964.

Li, Z.-L., & Becker, F. (1993). Feasibility of land surface temperature and emissivity determination from AVHRR data. *Remote Sensing of Environment, 43*, 67–85.

Li, Z.-L., Petitcolin, F., & Zhang, R. H. (2000). A physically based algorithm for land surface emissivity retrieval from combined mid-infrared and thermal infrared data. *Science in China, Series E Technological Sciences, 43*, 23–33.

Li, Z.-L., Stoll, M. P., Zhang, R. H., Jia, L., & Su, Z. B. (2001). On the separate retrieval of soil and vegetation temperatures from ATSR data. *Science in China Series D: Earth Sciences, 44*, 97–111.

Li, F., Jackson, T. J., Kustas, W. P., Schmugge, T. J., French, A. N., Cosh, M. H., & Bindlish, R. (2004a). Deriving land surface temperature from Landsat 5 and 7 during SMEX02/SMACEX. *Remote Sensing of Environment, 92*, 521–534.

Li, Z.-L., Tang, B. H., Wu, H., Ren, H., Yan, G., Wan, Z., Triggo, I. F., & Sobrino, J. A. (2013a). Satellite-derived land surface temperature: Current status and perspectives. *Remote Sensing of Environment, 131*, 14–37.

Li, Z.-L., Wu, H., Wang, N., Qiu, S., Sobrino, J. A., Wan, Z., Tang, B., & Yan, G. J. (2013b). Land surface emissivity retrieval from satellite data. *International Journal of Remote Sensing, 34*(9–10), 3084–3127.

Liu, Y., Hiyama, T., & Yamaguchi, Y. (2006). Scaling of land surface temperature using satellite data: A case examination on ASTER and MODIS products over a heterogeneous terrain area. *Remote Sensing of Environment, 105*, 115–128.

Llewellyn-Jones, D. T., Minnett, P. J., Saunders, R. W., & Zavody, A. M. (1984). Satellite multichannel infrared measurements of sea surface temperature of the NE Atlantic Ocean using AVHRR/2. *Quarterly Journal of the Royal Meteorological Society, 110*, 613–631.

Ma, X. L., Schmit, T. J., & Smith, W. L. (1999). A nonlinear physical retrieval algorithm-its application to the GOES-8/9 sounder. *Journal of Applied Meteorology, 38*, 501–513.

Ma, X. L., Wan, Z., Moeller, C. C., Menzel, W. P., Gumley, L. E., & Zhang, Y. L. (2000). Retrieval of geophysical parameters from Moderate Resolution Imaging Spectroradiometer thermal infrared data: Evaluation of a two-step physical algorithm. *Applied Optics, 39*, 3537–3550.

Ma, X. L., Wan, Z., Moeller, C. C., Menzel, W. P., & Gumley, L. E. (2002). Simultaneous retrieval of atmospheric profiles, land-surface temperature, and surface emissivity from Moderate-Resolution Imaging Spectroradiometer thermal infrared data: extension of a two-step physical algorithm. *Applied Optics, 41*, 909–924.

Mannstein, H. (1987). Surface energy budget, surface temperature, and thermal inertia. In R. A. Vaughan, & D. Reidel (Eds.), *Remote sensing applications in meteorology and climatology*. Dordrecht (NATO ASI Series C: Math. Phys. Sci., Vol. 201).

McClain, E. P., Pichel, W. G., & Walton, C. C. (1985). Comparative performance of AVHRR-based multichannel sea surface temperatures. *Journal of Geophysical Research, 90*, 11587–11601.

McMillin, L. M. (1975). Estimation of sea surface temperature from two infrared window measurements with different absorptions. *Journal of Geophysical Research, 80*, 5113–5117.

Minnis, P., & Khaiyer, M. M. (2000). Anisotropy of land surface skin temperature derived from satellite data. *Journal of Applied Meteorology, 39*, 1117–1129.

Mushkin, A., Balick, L. K., & Gillespie, A. R. (2005). Extending surface temperature and emissivity retrieval to the mid-infrared (3-5um) using the Multispectral Thermal Imager (MTI). *Remote Sensing of Environment, 98*, 141–151.

Neteler, M. (2010). Estimating daily land surface temperatures in mountainous environments by reconstructed MODIS LST data. *Remote Sensing, 2*, 333–351.

Niclòs, R., Caselles, V., Coll, C., & Valor, E. (2007). Determination of sea surface temperature at large observation angles using an angular and emissivity-dependent split-window equation. *Remote Sensing of Environment, 111*, 107–121.

Niclòs, R., Galve, J. M., Valiente, J. A., Estrela, M. J., & Coll, C. (2011). Accuracy assessment of land surface temperature retrievals from MSG2-SEVIRI data. *Remote Sensing of Environment, 115*, 2126–2140.

Norman, J. M., & Becker, F. (1995). Terminology in thermal infrared remote sensing of natural surfaces. *Agricultural and Forest Meteorology, 77*, 153–166.

Oleson, K., Dai, Y., Bonan, G. B., Bosilovichm, M., Dickinson, R., Dirmeyer, P., Hoffman, F., Houser, P., Levis, S., Niu, G. -Y., Thornton, P., Vertenstein, M., Yang, Z., & Zeng, X. (2004). *Technical description of the Community Land Model (CLM)* (NCAR Technical Note NCAR/TN-461+STR). doi:10.5065/D6N877R0.

Ottlé, C., & Stoll, M. (1993). Effect of atmospheric absorption and surface emissivity on the determination of land surface temperature from infrared satellite data. *International Journal of Remote Sensing, 14*, 2025–2037.

Ottlé, C., & Vidal-Madjar, D. (1992). Estimation of land surface temperature with NOAA9 data. *Remote Sensing of Environment, 40*, 27–41.

Peres, L. F., & DaCamara, C. C. (2005). Emissivity maps to retrieve land-surface temperature from MSG/SEVIRI. *IEEE Transactions on Geoscience and Remote Sensing, 43*, 1834–1844.

Peres, L. F., Sobrino, J. A., Libonati, R., Jiménez-Muñoz, J. C., Dacamara, C. C., & Romaguera, M. (2008). Validation of a temperature emissivity separation hybrid method from airborne hyperspectral scanner data and ground measurements in the SEN2FLEX field campaign. *International Journal of Remote Sensing, 29*, 7251–7268.

Perry, E. M., & Moran, M. S. (1994). An evaluation of atmospheric corrections of radiometric surface temperatures for a semiarid rangeland watershed. *Water Resources Research, 30*, 1261–1269.

Pinheiro, A. C. T., Privette, J. L., Mahoney, R., & Tucker, C. J. (2004). Directional effects in a daily AVHRR land surface temperature dataset over Africa. *IEEE Transactions on Geoscience and Remote Sensing, 42*, 1941–1954.

Pinker, R. T., Sun, S., Hung, M.-P., & Li, C. (2009). Evaluation of satellite estimates of land surface temperature from GOES over the United States. *Journal of Applied Meteorology and Climatology, 48*, 167–180.

Pozo Vazquez, D., Olmo Reyes, F. J., & Alados Arboledas, L. (1997). A comparative study of algorithms for estimating land surface temperature from AVHRR data. *Remote Sensing of Environment, 62*, 215–222.

Prabhakara, C., Dalu, G., & Kunde, V. G. (1974). Estimation of sea surface temperature from remote sensing in the 11-13 mm window region. *Journal of Geophysical Research, 79*, 5039–5044.

Prata, A. J. (1993). Surface temperatures derived from the advanced very high resolution radiometer and the along track scanning radiometer. 1. Theory. *Journal of Geophysical Research, 98*, 16689–16702.

Prata, A. J. (1994a). Land surface temperature determination from satellites. *Advances in Space Research, 14*, 15–26.

Prata, A. J. (1994b). Land surface temperatures derived from the advanced very high resolution radiometer and the along-track scanning radiometer 2. Experimental results and validation of AVHRR algorithms. *Journal of Geophysical Research, 99*, 13025, 13013,13058.

Prata, A. J., Caselles, V., Coll, C., Sobrino, J. A., & Ottlé, C. (1995). Thermal remote sensing of land surface temperature from satellites: Current status and future prospects. *Remote Sensing Reviews, 12*, 175–224.

Price, J. C. (1982). On the use of satellite data to infer surface fluxes at meteorological scales. *Journal of Applied Meteorology, 21*, 1111–1122.

Price, J. C. (1983). Estimating surface temperatures from satellite thermal infrared data – A simple formulation for the atmospheric effect. *Remote Sensing of Environment, 13*, 353–361.

Price, J. C. (1984). Land surface temperature measurements from the split window channels of the NOAA 7 AVHRR. *Journal of Geophysical Research, 89*, 7231–7237.

Price, J. C. (1990). The potential of remotely sensed thermal infrared data to infer surface soil moisture and evaporation. *Water Resources, 16*, 787–795.

Qian, Y. G., Li, Z.-L., & Nerry, F. (2013). Evaluation of land surface temperature and emissivities retrieved from MSG-SEVIRI data with MODIS land surface temperature and emissivity products. *International Journal of Remote Sensing, 34*(9–10), 3140–3152.

Qin, Z., Karnieli, A., & Berliner, P. (2001). A mono-window algorithm for retrieving land surface temperature from Landsat TM data and its application to the Israel-Egypt border region. *International Journal of Remote Sensing, 22*, 3719–3746.

Running, S. W., Justice, C., Salomonson, V., Hall, D., Barker, J., Kaufman, Y., Strahler, A., Huete, A., Muller, J.-P., Vanderbilt, V., Wan, Z., & Teillet, P. (1994). Terrestrial remote sensing science and algorithms planned for EOS/MODIS. *International Journal of Remote Sensing, 17*, 3587–3620.

Sabol, D. E., Gillespie, A. R., Abbott, E., & Yamada, G. (2009). Field validation of the ASTER Temperature-Emissivity Separation algorithm. *Remote Sensing of Environment, 113*, 2328–2344.

Sawabe, Y., Matsunaga, T., Rokugawa, S., & Hoyano, A. (2003). Temperature and emissivity separation for multi-band radiometer and validation ASTER TES algorithm. *Journal of Remote Sensing Society of Japan, 23*, 364–375.

Schmugge, T. J., & André, J. C. (1991). *Land surface evaporation: Measurements and parameterization*. New York: Springer-Verlag.

Schroedter, M., Olesen, F., & Fischer, H. (2003). Determination of land surface temperature distributions from single channel IR measurements: an effective spatial interpolation method for the use of TOVS, ECMWF and radiosonde profiles in the atmospheric correction scheme. *International Journal of Remote Sensing, 24*, 1189–1196.

Sellers, P. J., Hall, F. G., Asrar, G., Strebel, D. E., & Murphy, R. E. (1988). The first ISLSCP field experiment (FIFE). *Bulletin of the American Meteorological Society, 69*, 22–27.

Serafini, V. V. (1987). Estimation of the evapotranspiration using surface and satellite data. *International Journal of Remote Sensing, 8*, 1547–1562.

Slater, P. N. (1980). *Remote sensing, optics and optical system* (pp. 246–247). Reading, MA: Addison-Wesley.

Slater, P. N., Biggar, S. F., Thome, K. J., Gellman, D. I., & Spyak, P. R. (1996). Vicarious radiometric calibrations of EOS sensors. *Journal of Atmospheric and Oceanic Technology, 13*, 349–359.

Snyder, W. C., Wan, Z., Zhang, Y., & Feng, Y. Z. (1997). Requirements for satellite land surface temperature validation using a silt playa. *Remote Sensing of Environment, 61*, 279–289.

Snyder, W. C., Wan, Z., Zhang, Y., & Feng, Y. Z. (1998). Classification-based emissivity for land surface temperature measurement from space. *International Journal of Remote Sensing, 19*, 2753–2774.

Sobrino, J. A., & Cuenca, J. (1999). Angular variation of thermal infrared emissivity for some natural surfaces from experimental measurements. *Applied Optics, 38*, 3931–3936.

Sobrino, J. A., & Jiménez-Muñoz, J. C. (2005). Land surface temperature retrieval from thermal infrared data: An assessment in the context of the Surface Processes and Ecosystem Changes Through Response Analysis (SPECTRA) mission. *Journal of Geophysical Research, 110*, D16103.

Sobrino, J. A., & Raissouni, N. (2000). Toward remote sensing methods for land cover dynamic monitoring: application to Morocco. *International Journal of Remote Sensing, 21*, 353–366.

Sobrino, J. A., & Romaguera, M. (2004). Land surface temperature retrieval from MSG1-SEVIRI data. *Remote Sensing of Environment, 92*, 247–254.

Sobrino, J. A., Coll, C., & Caselles, V. (1991). Atmospheric correction for land surface temperature using NOAA-11 AVHRR channels 4 and 5. *Remote Sensing of Environment, 38*, 19–34.

Sobrino, J. A., Caselles, V., & Coll, C. (1993). Theoretical split-window algorithms for determining the actual surface temperature. *IL Nuovo Cimento C, 16*, 219–236.

Sobrino, J. A., Li, Z. L., Stoll, M. P., & Becker, F. (1994). Improvements in the split-window technique for land surface temperature determination. *IEEE Transaction on Geosciences and Remote Sensing, 32*, 243–253.

Sobrino, J. A., Li, Z.-L., Stoll, M. P., & Becker, F. (1996). Multi-channel and multi-angle algorithms for estimating sea and land surface temperature with ATSR data. *International Journal of Remote Sensing, 17*, 2089–2114.

Sobrino, J. A., Jiménez-Muñoz, J. C., El-Kharraz, J., Gómez, M., Romaguera, M., & Sòria, G. (2004a). Single-channel and two-channel methods for land surface temperature retrieval from DAIS data and its application to the Barrax site. *International Journal of Remote Sensing, 25*, 215–230.

Sobrino, J. A., Jiménez-Muñoz, J. C., & Paolini, L. (2004b). Land surface temperature retrieval from LANDSAT TM 5. *Remote Sensing of Environment, 90*, 434–440.

Sobrino, J. A., Sòria, G., & Prata, A. J. (2004c). Surface temperature retrieval from Along Track Scanning Radiometer 2 data: Algorithms and validation. *Journal of Geophysical Research, 109*, D11101.

Sobrino, J. A., Jiménez-Muñoz, J. C., Balick, L., Gillespie, A. R., Sabol, D. A., & Gustafson, W. T. (2007). Accuracy of ASTER level-2 thermal-infrared standard products of an agricultural area in Spain. *Remote Sensing of Environment, 106*, 146–153.

Sòria, G., & Sobrino, J. A. (2007). ENVISAT/AATSR derived land surface temperature over a heterogeneous region. *Remote Sensing of Environment, 111*, 409–422.

Su, Z. (2002). The Surface Energy Balance System (SEBS) for estimation of turbulent heat fluxes. *Hydrology and Earth System Sciences, 6*, 85–100.

Sun, D., & Pinker, R. T. (2003). Estimation of land surface temperature from a Geostationary Operational Environmental Satellite (GOES-8). *Journal of Geophysical Research, 108*, 4326.

Sun, D., & Pinker, R. T. (2005). Implementation of GOES-based land surface temperature diurnal cycle to AVHRR. *International Journal of Remote Sensing, 26*, 3975–3984.

Sun, D., & Pinker, R. T. (2007). Retrieval of surface temperature from the MSG-SEVIRI observations: Part I. Methodology. *International Journal of Remote Sensing, 28*, 5255–5272.

Susskind, J., Rosenfield, J., Reuter, D., & Chahine, M. T. (1984). Remote sensing of weather and climate parameters from HIRS2/MSU on TIROS-N. *Journal of Geophysical Research, 89*, 4677–4697.

Sutherland, R. A., & Bartholic, J. F. (1977). Significance of vegetation in interpreting thermal radiation from a terrestrial surface. *Journal of Applied Meteorology, 16*, 759–764.

Tang, B. H., & Li, Z.-L. (2008). Retrieval of land surface bidirectional reflectivity in the mid-infrared from MODIS channels 22 and 23. *International Journal of Remote Sensing, 29*, 4907–4925.

Tang, B. H., Bi, Y., Li, Z.-L., & Xia, J. (2008). Generalized Split-Window algorithm for estimate of land surface temperature from Chinese geostationary FengYun meteorological satellite (FY-2C) data. *Sensors, 8*, 933–951.

Tonooka, H. (2001). An atmospheric correction algorithm for thermal infrared multispectral data over land – A water-vapor scaling method. *IEEE Transactions on Geoscience and Remote Sensing, 39*, 682–692.

Townshend, J. R. G., Justice, C. O., Skole, D., Malingreau, J. P., Cihlar, J., Teillet, P., Sadowski, F., & Ruttenberg, S. (1994). The 1 km resolution global data set: Needs of the international geosphere biosphere programme. *International Journal of Remote Sensing, 15*, 3417–3441.

Trigo, I. F., Monteiro, I. T., Olesen, F., & Kabsch, E. (2008a). An assessment of remotely sensed land surface temperature. *Journal of Geophysical Research, 113*, D17108.

Ulivieri, C., & Cannizzaro, G. (1985). Land surface temperature retrievals from satellite measurements. *Acta Astronautica, 12*, 977–985.

Ulivieri, C., Castronuovo, M. M., Francioni, R., & Cardillo, A. (1994). A split window algorithm for estimating land surface temperature from satellites. *Advances in Space Research, 14*, 59–65.

Valor, E., & Caselles, V. (1996). Mapping land surface emissivity from NDVI: Application to European, African, and South American areas. *Remote Sensing of Environment, 57*, 167–184.

Van de Griend, A. A., & Owe, M. (1993). On the relationship between thermal emissivity and the normalized difference vegetation index for natural surfaces. *International Journal of Remote Sensing, 14*, 1119–1131.

Vauclin, M., Vieira, R., Bernard, R., & Hatfield, J. L. (1982). Spatial variability of surface temperature along two transects of a bare. *Water Resources Research, 18*, 1677–1686.

Vidal, A. (1991). Atmospheric and emissivity correction of land surface temperature measured from satellite using ground measurements or satellite data. *International Journal of Remote Sensing, 12*, 2449–2460.

Voogt, J. A., & Oke, T. R. (2003). Thermal remote sensing of urban climates. *Remote Sensing of Environment, 86*, 370–384.

Wan, Z. (1999). MODIS land-surface temperature algorithm theoretical basis document. In Greenbelt MD, USA: NASA/GSFC.

Wan, Z. (2008). New refinements and validation of the MODIS land-surface temperature/emissivity products. *Remote Sensing of Environment, 112*, 59–74.

Wan, Z., & Dozier, J. (1996). A generalized split-window algorithm for retrieving land-surface temperature from space. *IEEE Transactions on Geoscience and Remote Sensing, 34*, 892–905.

Wan, Z., & Li, Z.-L. (1997). A physics-based algorithm for retrieving land-surface emissivity and temperature from EOS/MODIS data. *IEEE Transactions on Geoscience and Remote Sensing, 35*, 980–996.

Wan, Z., & Li, Z.-L. (2008). Radiance-based validation of the V5 MODIS land-surface temperature product. *International Journal of Remote Sensing, 29*, 5373–5395.

Wan, Z., Zhang, Y., Zhang, Q., & Li, Z.-L. (2002). Validation of the land-surface temperature products retrieved from Terra Moderate Resolution Imaging Spectroradiometer data. *Remote Sensing of Environment, 83*, 163–180.

Wan, Z., Zhang, Y., Zhang, Q., & Li, Z.-L. (2004). Quality assessment and validation of the MODIS global land surface temperature. *International Journal of Remote Sensing, 25*, 261–274.

Wang, K., & Liang, S. (2009). Evaluation of ASTER and MODIS land surface temperature and emissivity products using long-term surface longwave radiation observations at SURFRAD sites. *Remote Sensing of Environment, 113*, 1556–1565.

Wang, N., Tang, B. H., Li, C., & Li, Z.-L. (2010). A generalized neural network for simultaneous retrieval of atmospheric profiles and surface temperature from hyperspectral thermal infrared data. In *IEEE International Geoscience and Remote Sensing Symposium* (pp. 1055–1058). Honolulu: IEEE.

Wang, N., Wu, H., Nerry, F., Li, C. R., & Li, Z.-L. (2011). Temperature and emissivity retrievals from hyperspectral thermal infrared data using linear spectral emissivity constraint. *IEEE Transactions on Geoscience and Remote Sensing, 49*, 1291–1303.

Wang, N., Li, Z.-L., Tang, B. H., Zeng, F. N., & Li, C. R. (2013). Retrieval of atmospheric and land surface parameters from satellite-based thermal infrared hyperspectral data using a neural network technique. *International Journal of Remote Sensing, 34*(9–10), 3485–3502.

Watson, K. (1992). Two-temperature method for measuring emissivity. *Remote Sensing of Environment, 42*, 117–121.

Weng, Q. (2009). Thermal infrared remote sensing for urban climate and environmental studies: Methods, applications, and trends. *ISPRS Journal of Photogrammetry and Remote Sensing, 64*, 335–344.

Weng, Q., Lu, D., & Schubring, J. (2004). Estimation of land surface temperature-vegetation abundance relationship for urban heat island studies. *Remote Sensing of Environment, 89*, 467–483.

Wu, H., & Li, Z.-L. (2009). Scale issues in remote sensing: A review on analysis, processing and modeling. *Sensors, 9*, 1768–1793.

Yu, Y., Privette, J. L., & Pinheiro, A. C. (2008). Evaluation of split-window land surface temperature algorithms for generating climate data records. *IEEE Transactions on Geoscience and Remote Sensing, 46*, 179–192.

Yu, Y., Tarpley, D., Privette, J. L., Goldberg, M. D., Rama Varma Raja, M. K., Vinnikov, K. Y., & Xu, H. (2009). Developing algorithm for operational GOES-R land surface temperature product. *IEEE Transactions on Geoscience and Remote Sensing, 47*, 936–951.

Zhang, R., Tian, J., Su, H., Sun, X., Chen, S., & Xia, J. (2008). Two improvements of an operational two-layer model for terrestrial surface heat flux retrieval. *Sensors, 8*, 6165–6187.

Chapter 6
Estimation and Validation of Evapotranspiration from Thermal Infrared Remote Sensing Data

6.1 Background

As one of the most significant components in the hydrological processes (accounting for approximately 60 % of the mean precipitation globally), land surface evapotranspiration (ET, or latent heat flux, LE, in W/m^2) controls the water and energy transfer in the interface between land surface and atmosphere (Idso et al. 1975; Brutsaert 1986; Su 2002; Oki and Kanae 2006). Nowadays, measuring directly the actual ET remains unresolved over large heterogeneous areas in practical applications (Brutsaert 1986). Conventional ET measurements (i.e., sap flow, weighing lysimeter, pan measurement, Bowen ratio system, eddy correlation system, and scintillometer) are generally of limited use because of their spatial unrepresentativeness (with a fetch of 10^0–10^3 m) and have difficulties in spatial extrapolation due to the significant surface heterogeneity. To date, remote sensing technology has been recognized as the most effective means to capture surface information at various spatial scales. Advantages in the remote sensing technology over the conventional "point" measurements include (1) its large and continuous spatial coverage within a few minutes, (2) its less cost when same spatial information is required, and (3) its benefit in ungauged areas (Engman and Gurney 1991; Rango 1994). Different combinations of the remote sensing spectral information from visible and near-infrared bands to mid and thermal infrared bands can produce surface characteristic parameters that are indispensable to ET models. These parameters consist of Normalized Difference Vegetation Index (NDVI), Leaf Area Index (LAI), fractional vegetation coverage (Fr), surface albedo, surface emissivity, and radiometric surface temperature (Mauser and Schädlich 1998).

The possibility of using remote sensing technology to evaluate ET can be traced back to as early as the 1970s when the Heat Capacity Mapping Mission (HCMM) and polar orbiting weather satellites TIROS-N were launched (Price 1980). Significant progress has been made during the subsequent 40 years on the use of satellite thermal infrared data to retrieve regional and local scale ET (Seguin and Itier 1983). Both empirical and physical models, which differ greatly in the

H. Tang and Z.-L. Li, *Quantitative Remote Sensing in Thermal Infrared: Theory and Applications*, Springer Remote Sensing/Photogrammetry,
DOI 10.1007/978-3-642-42027-6_6, © Springer-Verlag Berlin Heidelberg 2014

model structure, input, and output, have been developed with varying degrees of complexity. In these models, there exists a contradiction between complex models that are physically based and simple models that are short of mechanistic realism (Cleugh et al. 2007). The trend seems to increase the model complexity rather than applicability (Fisher et al. 2008), which may require detailed ground-based measurements as input. To better understand the mechanisms of the water and energy interactions between the atmosphere and land surface, it is urgently required and indispensable to summarize and intercompare the different remote sensing-based ET approaches.

This chapter inherits and updates the overview given by Li et al. (2009) of a variety of remote sensing ET methods and models that have been developed at field, regional, and continental scales. The main theory and assumption are detailed and the advantages, drawbacks, and potentiality are highlighted for each method or model. In the latter part, representative methods of upscaling instantaneous ET to daily values are presented, several exercises in validating remote sensing estimates of ET outlined, and the problems and issues finally addressed.

6.2 Methodologies for Evapotranspiration Retrieval

Current remote sensing ET models can be roughly categorized into the (semi)empirical method and the analytical method. The (semi)empirical method is developed by establishing empirical functions of remote sensing observations and minimum ground-based measurements, while the analytical method considers the physical processes at the scale of interest and generally requires a more variety of direct and indirect measurements.

6.2.1 Empirical Regression Method

The main theory of the simplified empirical regression method is to statistically fit the ET at different temporal scales to remote sensing- or ground-based variables, for example, surface temperature (T_s), surface net radiation (R_n), near-surface air temperature (T_a), and vegetation index. For example, Jackson et al. (1977) related the daily ET from the irrigated wheat at Phoenix, Arizona, to the difference between instantaneous T_s and T_a measured near midday at local time about 13 h30 to 14 h00. Wang et al. (2007) constructed a linear relationship to estimate ET using the vegetation index, daytime-averaged R_n and T_a data collected over the Southern Great Plains (SGP) area of the USA from January 2002 to May 2005. The most general form of the simplified regression method can be expressed mathematically as

$$\mathrm{ET} = f_{\mathrm{regress}}\left(R_n, T_s, T_a, \cdots\right) \tag{6.1}$$

where the independent variable can be either R_n, T_s, T_a, wind speed, relative humidity, NDVI, or a combination of several of these parameters. The regression method can be applied at daily or monthly scale.

The simplified regression method proposed by Jackson et al. (1977) and its refinements have attracted great attentions in the subsequent operational applications of ET mapping. The implications of the regression coefficients were theoretically analyzed by Carlson et al. (1995) in Jackson's simplified equation. All contributions to the Jackson's simplified equation have shown that the error is on the magnitude of about 1 mm/day in the calculated daily ET, and the simplified equation can provide reliable information to the water availability in the practical applications (Seguin et al. 1994).

The main advantage of the empirical regression method is its simplicity and relatively high accuracy. Thus, it is very convenient for the simplified empirical equation to be applied. However, the site-specific coefficients in the regression equation have more or less limited the applications of this kind of method at regional scales with variable surface conditions.

6.2.2 Surface Energy Balance Method

Surface energy balance controls the exchange and partition of the surface turbulent fluxes into sensible heat flux and latent heat flux. In the surface energy balance method, ET is calculated as the residual of the surface energy budget equation. Depending on whether land surface is treated uniformly as a big leaf or separately as a combination of soil and canopy components, surface energy balance method can be subdivided into single-source and dual-source models. Both are the most widely applied approaches to mapping ET at different temporal and spatial scales. Figure 6.1 presents a schematic diagram of the energy exchange between land surface and atmosphere based on the single-source and dual-source models.

6.2.2.1 Single-Source Model

Neglecting horizontal advective heat flow and the heat storage of photosynthetic vegetation and surface residuals, the one-dimensional form of surface energy budget equation at instantaneous time scale can be expressed mathematically as

$$\mathrm{LE} = R_n - G - H \tag{6.2}$$

where R_n is the surface net radiation [W/m^2], G is the soil heat flux [W/m^2], and H is the sensible heat flux [W/m^2]. The three components at the right-hand side of the energy budget equation can be estimated by combining remote sensing parameters and ancillary ground-based surface and meteorological variables (including air temperature, wind speed, and humidity).

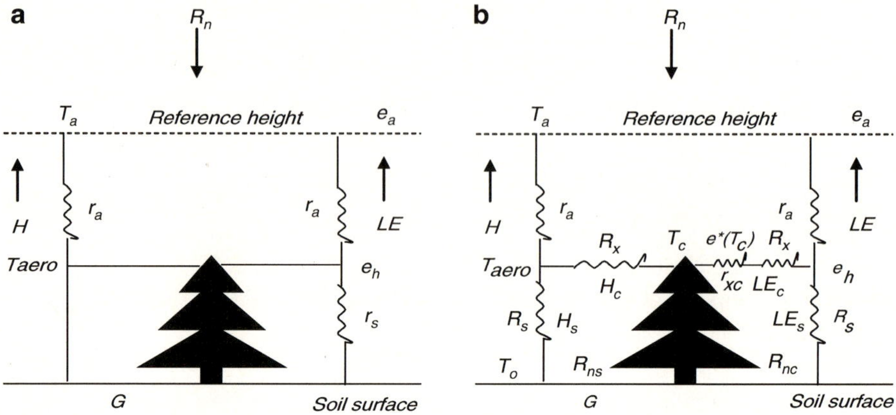

Fig. 6.1 Schematic diagram of the energy exchange between land surface and atmosphere with the resistance analogy to (**a**) single-source and (**b**) dual-source models. R_n is the surface net radiation; G is the soil heat flux; H is the sensible heat flux; LE is the latent heat flux; T_a is the air temperature at reference height; e_a is the actual vapor pressure at reference height; T_{aero} is the aerodynamic temperature at heat source height; e_h is the actual vapor pressure at water source height; r_a is the aerodynamic resistance; r_s is the bulk surface resistance; H_c and H_s are the sensible heat flux for canopy and soil components, respectively; LE_c and LE_s are the latent heat flux for canopy and soil components, respectively; T_c and T_o are the canopy and soil temperatures, respectively; R_s is the soil boundary layer resistance; r_{xc} is the stomatal resistance; $e^*(T_c)$ is the vapor pressure at the leaf surface

Net Radiation Equation (R_n)

Surface net radiation (R_n), which is the sum of the difference between incoming (R_g) and reflected outgoing shortwave solar radiation (0.15–5 μm) and the difference between downwelling and surface emitted and reflected longwave radiation (3–100 μm), represents the total heat energy that can be used for the partitioning into G, H, and LE (Jackson 1985; Kustas and Norman 1996):

$$R_n = (1 - \alpha_s) R_g + \varepsilon_s \varepsilon_a \sigma T_a^{\ 4} - \varepsilon_s \sigma T_s^{\ 4} \tag{6.3}$$

where α_s is the surface shortwave albedo, usually calculated from a combination of narrow band spectral reflectances (Liang 2005) [–]; R_g is the global solar radiation, determined by a number of factors (e.g., solar constant, solar inclination angle, geographic location, time of year, atmospheric transmissivity, and ground elevation) (Allen et al. 2007) [W/m^2]; ε_s is the surface emissivity [–]; ε_a is the atmospheric emissivity, estimated as a function of actual vapor pressure and air temperature (Brutsaert 1975) [–]; and σ is the Stefan–Boltzmann constant, [5.67×10^{-8} W m^{-2} K^{-4}].

The uncertainty of various remote sensing methods of surface net radiation estimation was found by Kustas and Norman (1996) to be generally within 5–10 % when comparisons were made with ground-based observations at meteorologically temporal scales. Note that over the mountainous area, surface slope and aspect information derived from a digital elevation model should be taken into account (Allen et al. 2007).

Soil Heat Flux (G)

Soil heat flux (G) represents the heat energy that is used to warm or cool substrate soil volume. It is proportional to thermal conductivity and vertical temperature gradient of the topsoil and is traditionally measured using sensors buried beneath the surface soil. Many papers have expressed G as a fraction of R_n (from 0.05 for full vegetation cover or wet bare soil to 0.5 for dry bare soil) (Price 1982; Jackson 1985; Reginato et al. 1985; Daughtry et al. 1990; Choudhury 1994; Kustas and Norman 1996; Li and Lyons 1999) and related this fraction as a function of LAI (Choudhury 1989), NDVI (Moran et al. 1989; Bastiaanssen et al. 1998b; Allen et al. 2007), T_s (Bastiaanssen 2000; Allen et al. 2007), and/or solar zenith angle (Gao et al. 1998) based on field observations:

$$\frac{G}{R_n} = f\,(\mathrm{NDVI}, \mathrm{LAI}, T_s, \ldots) \tag{6.4}$$

Soil heat flux varies in both diurnal and yearly cycles over diverse surface conditions (Kustas and Daughtry 1990). Remote sensing satellite platforms, however, are not able to directly measure G under current circumstances. Fortunately, the fraction in Eq. (6.4) is generally relatively small at the satellite overpass. Moreover, daily G can be assumed equal to 0 and thus can be negligible in the daily energy balance (Price 1982). An uncertainty of 20–30 % is found when G estimated from the simplified techniques is compared with observations at micrometeorological scales (Kustas and Norman 1996).

Sensible Heat Flux (H)

Sensible heat flux (H) represents the heat energy that drives to warm/cool the air above the surface. It can be calculated in analogy to Ohm's law by combining the difference between aerodynamic (T_{aero}) and air temperatures with the aerodynamic resistance (r_a) in the single-source energy balance model:

$$H = \frac{\rho c_p\,(T_{aero} - T_a)}{r_a} \tag{6.5}$$

where ρ is the air density [kg/m^3], and c_p is the specific heat at constant pressure [J/(kg · K)].

A number of factors, including surface roughness (vegetation height, vegetation structure), wind speed, and atmospheric stability, are found to affect the aerodynamic resistance, and extremely elementary (a function of wind speed only) to quite rigorous methods (accounting for atmospheric stability, wind speed, surface "aerodynamic" roughness, etc.) have been developed to estimate r_a (Seguin et al. 1982; Hatfield 1983; Choudhury et al. 1986; Moran et al. 1994). The one that has been most commonly applied is (Brutsaert 1982)

$$r_a = \frac{\ln\left[(z_a - d)/z_{om} - \psi_1\right] \ln\left[(z_a - d)/z_{oh} - \psi_2\right]}{k^2 u} \quad (6.6)$$

where z_a is the reference height for the wind speed and humid measurements [m]; d is the zero plane displacement height [m]; z_{om} and z_{oh} are the momentum and heat transfer lengths, respectively [m]; ψ_1 and ψ_2 are the stability correction functions for momentum and heat transfer, respectively [−]; k is the von Karman's constant [−]; and u is the wind speed [m/s]. Under neutral atmospheric stability conditions, $\psi_1 = \psi_2 = 0$.

Under stable and unstable stability conditions, Monin–Obukhov length (L_{MO}) was introduced to solve H iteratively (Monin and Obukhov 1954):

$$L_{MO} = \frac{u*^3 \rho c_p T_a}{kgH} \quad (6.7)$$

where $u*$ is the friction velocity [m/s], and g is the acceleration due to gravity of the earth [m/s^2]. Under unstable stability condition, $L_{MO} < 0$; under stable stability condition, $L_{MO} > 0$.

Under unstable conditions that usually prevail at daytime and with no significant free convection, ψ_1 and ψ_2 are expressed as (Paulson 1970)

$$\psi_1 = 2\ln\left(\frac{1+x}{2}\right) + \ln\left(\frac{1+x^2}{2}\right) - 2\arctan(x) + \frac{\pi}{2} \quad (6.8)$$

$$\psi_2 = 2\ln\left(\frac{1+x^2}{2}\right) \quad (6.9)$$

with

$$x = \left(1 - 16\frac{z_a - d}{L_{MO}}\right)^{0.25} \quad (6.10)$$

Under stable conditions that usually prevail at nighttime, ψ_1 and ψ_2 are expressed on the basis of work done by Webb (1970) and Businger et al. (1971):

$$\psi_1 = \psi_2 = -5\frac{z_a - d}{L_{MO}} \quad (6.11)$$

The study made by Hatfield et al. (1983) shows that no correction for atmospheric stability to the aerodynamic resistance could lead to an overestimation of ET if the absolute canopy–air temperature difference is greater than ~2 °C.

Surface roughness has a significant impact on the estimation of aerodynamic resistance and changes apparently with leaf size, the flexibility of petioles, and plant stems (Jackson 1985). The effective roughness for momentum (z_{om}) in some papers has been specified to equal that for heat (z_{oh}). It can be either expressed as a fraction of vegetation height (typically 5–15 % depending on vegetation characteristics) (Soer 1980; Gurney and Camillo 1984; Monteith and Unsworth 2007) or estimated by extrapolating the standard log-linear wind profiles to zero wind speed (Gurney and Camillo 1984). Because of the different mechanisms between heat and momentum transfer (molecular diffusion drives the heat transfer while both viscous shear and pressure forces control the momentum transfer) (Brutsaert 1982), there is a difference between z_{om} and z_{oh}, causing an additional resistance (often expressed as aerodynamic definition of kB^{-1}) (Garratt and Hicks 1973) or an excess (extra) resistance (Norman and Becker 1995) to heat transfer. The kB^{-1} ($\mathrm{kB}^{-1} = \ln(z_{om}/z_{oh})$) was related by Kustas et al. (1989) to wind speed and surface–air temperature difference:

$$\mathrm{kB}^{-1} = S_{\mathrm{kB}} u \left(T_{\mathrm{s}} - T_{\mathrm{a}}\right) \tag{6.12}$$

where S_{kB} is an empirical coefficient, varying from 0.05 to 0.25. The kB^{-1} is found to be sensitive to the errors in the micrometeorological variables and in the roughness length for momentum (Verhoef et al. 1997). It is challenging to determine surface roughness for the large-scale retrieval of sensible heat flux and latent heat flux despite the efforts made in the past (Massman 1999; Su 2002).

Aerodynamic temperature (T_{aero}) is recognized as the temperature of the apparent sources or sinks of sensible heat flux (Kalma and Jupp 1990), and by definition, it may be estimated by extrapolating temperature profile down to $z = d + z_{oh}$. Remotely sensed surface temperature T_{s} instead of T_{aero} has been used in a number of papers to calculate H because of the difficulty in measuring T_{aero} (Blad and Rosenberg 1976; Seguin et al. 1982; Moran et al. 1989; Kalma and Jupp 1990). However, problems arise if one assumes that measured T_{s} is identical to T_{aero} (Kalma and Jupp 1990). Under stable (unstable) atmospheric conditions, T_{aero} is shown to be lower (higher) than T_{s}. Only under neutral conditions, they approximate to each other (Choudhury et al. 1986; Kalma and Jupp 1990). The difference between T_{aero} and T_{s} could vary from 2 °C over uniform vegetation cover to 10 °C for partially vegetated area (Kustas and Norman 1996).

The single-source bulk transfer equation (resistance-based model) has been widely applied since the 1970s to various vegetation covers at local and regional scales (Blad and Rosenberg 1976; Hatfield 1983; Moran and Jackson 1991). Over homogeneous areas, H and LE can be retrieved from the single-source models with a relatively high accuracy. The average difference of H is found to be about 15–20 %, which is around the magnitude of uncertainty in the measurements from eddy correlation system and Bowen ratio system (Seguin 1984; Kustas and Norman 1996). However, over partially vegetated areas, a dual-source model, which can separate the heat and water exchange of soil from that of vegetation, is strongly needed. This kind of model often deals with a decomposition of surface heat flux and radiometric surface temperature to soil and vegetation components. Moreover, sensible heat flux in the single-source surface energy balance models is more sensitive to errors in near-surface meteorological and ground-based inputs (e.g., surface–air temperature, wind speed, vegetation height). There also seems no unique relationship between surface roughness characteristics and the roughness length for heat transfer (Cleugh et al. 2007). The weakness for the ET estimates from surface energy balance equation is primarily that both the physics-based single- and dual-source models require detailed ground-based meteorological and vegetation measurements as model input (e.g., air temperature, wind speed, vegetation height).

6.2.2.2 Dual-Source Model (Also Called Two-Source Model)

To produce reliable estimates of sensible heat flux and latent heat flux, single-source models that treat the earth's surface as a single, uniform layer are generally required to be calibrated, which thus hinders their direct application to a diverse range of surface conditions. For example, over a partial vegetation cover, one may need to adjust r_a, which unfortunately cannot be applicable over other circumstances (Kustas and Daughtry 1990). Because single-source bulk transfer models are dependent upon absolute surface temperature or surface–air temperature difference, they suffer from the errors in the sensor calibration, the atmospheric corrections, the specification of surface emissivity (Mecikalski et al. 1999), and the interpolation of shelter-level air temperature over large heterogeneous areas. Dual-source models on the contrary do not need a priori calibration, nor do they need more ground-based information than single-source models. It is believed that dual-source models are an advance over single-source surface models and have a wider range of applicability (Anderson et al. 1997). In the dual-source models, the divergence of surface energy balance inside the canopy needs to be estimated and the clumping considered (affecting the wind speed profile, the radiation penetration, and the radiative surface temperature partitioning between soil and vegetation) (Kustas and Norman 1997).

The energy balance components of the total soil and vegetation system in the representative dual-source N95 model (Norman et al. 1995) are calculated from the following formulae: Soil and canopy energy budgets:

$$R_{\mathrm{n,s}} = H_{\mathrm{s}} + \mathrm{LE}_{\mathrm{s}} + G \tag{6.13a}$$

$$R_{\mathrm{n,c}} = H_{\mathrm{c}} + \mathrm{LE}_{\mathrm{c}} \tag{6.13b}$$

Sensible heat flux:

$$H = H_{\mathrm{s}} + H_{\mathrm{c}} = \rho c_p \frac{T_0 - T_{\mathrm{AC}}}{R_{\mathrm{S}}} + \rho c_p \frac{T_{\mathrm{c}} - T_{\mathrm{AC}}}{R_{\mathrm{X}}} \tag{6.13c}$$

Latent heat flux:

$$LE = LE_s + LE_c \tag{6.13d}$$

$$LE_c = \alpha f_g \frac{\Delta}{\Delta + \gamma} R_{n,c} \tag{6.13e}$$

Soil heat conduction flux:

$$G = \Gamma_s R_{n,s} \tag{6.13f}$$

where subscripts "s" and "c" represent the soil and canopy components, respectively; T_c and T_o are the vegetation canopy and soil component temperatures, respectively [K]; R_X and R_S are the total boundary layer resistance of the complete canopy of leaves and resistance to heat transfer immediately above the soil surface, respectively [s/m]; f_g is the fraction of the total LAI that is green; Γ_s is the fraction of the ratio of soil heat flux to surface net radiation at bare soil surface; α is the Priestley–Taylor coefficient for the potentially transpiring canopy with an initial approximation of 1.3; Δ is the slope of the saturated vapor pressure versus air temperature curve [kPa/°C]; and γ is the psychrometric constant [kPa/°C].

Over the past years, temperature and energy components of both soil and vegetation over various landscapes have been investigated by a series of papers through a set of applications, validations, and modifications to the dual-source energy balance models (Shuttleworth and Wallace 1985; Norman et al. 1995, 2000; Kustas and Norman 1997, 1999; Anderson et al. 2005; Li et al. 2005; Sánchez et al. 2008; Tang 2011; Cammalleri et al. 2012). For example, based on remote sensing data from single/multiple view angles, a dual-source (N95) model was developed by Norman et al. (1995) to partition temperature and energy into soil and vegetation components, which can accommodate the difference between radiometric surface and aerodynamic temperatures. On the basis of N95, the dual-source time integrated model (TSTIM, subsequently named as ALEXI, Atmosphere-Land Exchange Inverse (Mecikalski et al. 1999)), which related the morning rise of surface temperature acquired at 1.5 and 5.5 h past sunrise to the growth of a planetary boundary layer, was later developed by Anderson et al. (1997) for an estimate of temperature and energy from soil and vegetation components. The ALEXI removes the need for the measurements of near-surface air temperature and is relatively insensitive to uncertainties in surface emissivity and atmospheric corrections on the remotely sensed surface temperatures. It is believed to be a practical means of operational estimates of surface fluxes over continental scales with the spatial resolution of 5–10 km. Using time rate of change in T_s and T_a, Norman et al. (2000) developed the dual-temperature-difference (DTD) method that is simpler than other modifications of N95. This method does not require modeling boundary layer development. By combining low- and high-resolution remotely sensed data, a two-step disaggregated ALEXI (DISALEXI) was proposed without a need for local observations (Norman et al. 2000; Kustas et al. 2003). Li et al. (2005) compared the series and parallel resistance (with and without consideration of the soil and vegetation interaction, respectively) network formulations in a dual-source model that were used to parameterize soil and canopy energy exchanges and found that both resistance formulations produced similar estimates compared with tower-based flux observations, but the parallel resistance formulation seemed more able to achieve the temperature and energy balance of soil and canopy components.

Dual-source energy balance models have been shown to be more robust than single-source models under a wide range of landscape and hydrometeorological conditions (Kustas and Norman 1997). Several advantages make dual-source models (take N95 as an example) superior to single-source models, including (1) precise atmospheric corrections, emissivity estimations, and high accuracy in sensor calibration are not required, (2) ground-based measurement of T_a is avoided when coupled with a PBL (Kustas and Norman 1996; Anderson et al. 1997), (3) effects of view geometry are incorporated, and (4) empirical corrections for the "excess resistance" are circumvented. However, applications of the aforementioned DTD and ALEXI models generally require data from a geostationary satellite. The suboptimal viewing orientation at high latitudes and the coarse spatial resolution from a geostationary satellite may constrain its ability to provide a series of cloud-free images (Van den Hurk 2001). With its relatively smallpixel size and high observation frequency, the new Meteosat Second Generation/Spinning Enhanced Visible and Infrared Imager (MSG/SEVIRI) sensor provides a good promise for applications in Europe and Africa.

6.2.2.3 End-Member-Based Model

Depending on whether the contextual information (emergence of representative dry and wet pixels) of land surface characteristics in the area of interest is used, single-source surface energy balance models are further subdivided into end-member- and non-end-member-based models. When surface parameters derived from remote sensing data and ground-based variables are readily available, the end-member-based models use the contrast between dry and wet limits to derive pixel-by-pixel ET and EF over local/regional scale surfaces (Gowda et al. 2008). The dry (wet) limit, no matter how it was specifically defined, often has the following characteristics: (1) generally maximum (minimum) surface temperature and (2) usually low or no (high or maximum) ET. In this section, several representative end-member-based single-source models are reviewed.

SEBI (Surface Energy Balance Index) and SEBS (Surface Energy Balance System)

The Surface Energy Balance Index (SEBI), first developed by Menenti and Choudhury (1993), along with its derivatives like the Surface Energy Balance

Algorithm for Land (SEBAL), the Simplified Surface Energy Balance Index (S-SEBI), the Surface Energy Balance System (SEBS), and the Mapping EvapoTranspiration at high Resolution with Internalized Calibration (METRIC), is typically an end-member-based single-source energy balance model.

The SEBI has a dry limit that is characterized by a zero surface ET (latent heat flux) for a given set of boundary layer characteristics, which indicates that sensible heat flux at the dry limit is equal to the surface available energy. Surface temperature ($T_{s,max}$) at this limit is estimated by inverting the bulk transfer equation (Van den Hurk 2001):

$$T_{s,max} = T_{pbl} + r_{a,dry}\frac{H}{\rho c_p} \tag{6.14}$$

where T_{pbl} is the average planetary boundary layer temperature [K] and $r_{a,dry}$ is the aerodynamic resistance to sensible heat transfer at the dry limit [s/m].

At the wet limit, the surface evaporates potentially and the minimum surface temperature ($T_{s,min}$) can be estimated similarly (with potential ET calculated from Penman–Monteith equation with a zero internal resistance):

$$T_{s,min} = T_{pbl} + \frac{r_{a,wet}\frac{R_n - G}{\rho c_p} - \mathrm{VPD}/\gamma}{1 + \Delta/\gamma} \tag{6.15}$$

where $r_{a,wet}$ is the aerodynamic resistance to sensible heat transfer at the wet limit [s/m], and VPD is the vapor pressure deficit [kPa].

By interpolating the observed surface temperature between the maximum and minimum surface temperatures, the relative evaporation fraction ($\mathrm{EF_r}$) can then be calculated:

$$\mathrm{EF_r} = \frac{\mathrm{LE}}{\mathrm{LE_p}} = 1 - \frac{r_a{}^{-1}\left(T_s - T_{pbl}\right) - r_{a,wet}{}^{-1}\left(T_{s,min} - T_{pbl}\right)}{r_{a,dry}{}^{-1}\left(T_{s,max} - T_{pbl}\right) - r_{a,wet}{}^{-1}\left(T_{s,min} - T_{pbl}\right)} \tag{6.16}$$

where the second part of the right-hand side of Eq. (6.16) corresponds to the so-called SEBI, which varies between 0 (wet limit) and 1 (dry limit).

By defining theoretical pixel-wise ranges for LE and T_s, the SEBI approach, which is essentially a modification from crop water stress index (CWSI) proposed by Idso et al. (1981) and Jackson et al. (1981, 1988), was proposed to reveal the spatial variability of actual evaporation (Menenti and Choudhury 1993). Menenti and Choudhury (1993) accounted for the effect of atmospheric stratification on external resistance to calculate the pixel-wise maximum and minimum surface temperature and redefined CWSI as a pixel-wise SEBI to estimate regional ET. Based on the conception of the SEBI, the SEBAL, the SEBS, and the S-SEBI models are subsequently developed. The main distinction between these models lies in the difference in defining the dry (characterized by maximum sensible heat flux and minimum latent heat flux) and wet (characterized by maximum latent heat flux and minimum sensible heat flux) limits and interpolating the sensible heat flux between the defined upper and lower limits. Note that atmospheric conditions (mainly the surface available energy) should be uniform in space and sufficient hydrological contrasts are required for several models (i.e., the SEBAL) to ensure that dry and wet limits exist in the area of interest.

With a dynamic model for the thermal roughness, the Bulk Atmospheric Similarity Theory for the planetary boundary layer (PBL) scaling and the Monin–Obukhov atmospheric surface layer (ASL) similarity for surface layer scaling, Su (2002, 2005) and Su et al. (2003) developed the Surface Energy Balance System (SEBS) in a more complex framework based on an extension of the SEBI to estimate land surface energy balance. This framework consists of (1) a set of tools for calculating land surface physical parameters, (2) a new parameterization of roughness length for heat transfer, and (3) an estimation of the evaporative fraction using energy balance at limiting cases. The dry limit in the SEBS has a zero ET and maximum H (i.e., $H_{\mathrm{dry}} = R_{\mathrm{n}} - G$), whereas ET ($\mathrm{LE_{wet}}$) evaporates potentially (limited only by the energy available under the given surface and atmospheric conditions and can be calculated from the Penman–Monteith combination equation assuming a zero bulk internal resistance) at the wet limit and sensible heat flux reaches its minimum value. Sensible heat flux at the dry and wet limits are expressed as

$$H_{\mathrm{dry}} = R_{\mathrm{n}} - G \tag{6.17}$$

$$H_{\mathrm{wet}} = \frac{\left((R_{\mathrm{n}} - G) - \frac{\rho C_p}{r_{\mathrm{a}}} \frac{\mathrm{VPD}}{\gamma}\right)}{\left(1 + \frac{\Delta}{\gamma}\right)} \tag{6.18}$$

The $\mathrm{EF_r}$ is introduced to express EF:

$$\mathrm{EF_r} = 1 - \frac{H - H_{\mathrm{wet}}}{H_{\mathrm{dry}} - H_{\mathrm{wet}}}, \tag{6.19}$$

$$\mathrm{EF} = \frac{\mathrm{EF_r} \cdot \mathrm{LE_{wet}}}{R_{\mathrm{n}} - G}. \tag{6.20}$$

SEBS makes a distinction between the atmospheric boundary layer (ABL) or PBL and the ASL similarity. Its inputs consist of remote sensing land parameters (land surface temperature, LAI, fractional vegetation cover, albedo) and ground-based meteorological measurements (wind speed, humidity, air temperature). It could be used to estimate daily, monthly, and annual evaporation for both local and regional scales under all atmospheric stability regimes (Su 2002; Jia et al. 2003; Su et al. 2003; Wood et al. 2003).

A significant improvement to the SEBS model was recently made by Gokmen et al. (2012) who incorporated the level of soil water stress into the parameterization of kB^{-1} to mitigate the overestimation of daily ET in sparsely vegetated semiarid

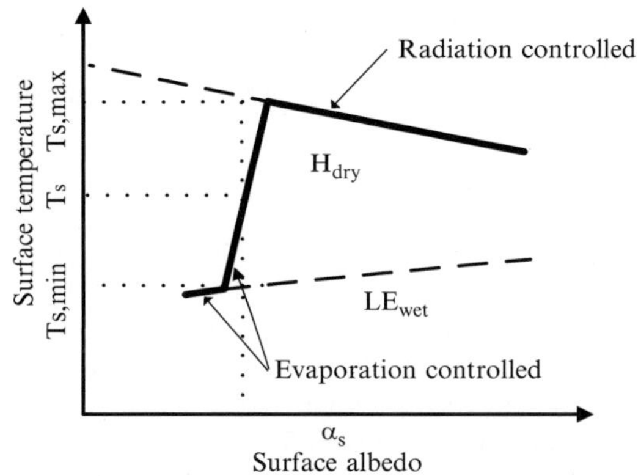

Fig. 6.2 A schematic illustration of the relationship between surface temperature and albedo in the S-SEBI (Reproduced from Li et al. (2009))

regions. This improvement emphasizes the significance of accounting for the dynamics of surface roughness length induced by the soil water stress in the use of single- or dual-source bulk transfer equation to estimate sensible heat flux in the future work.

Advantages of the SEBS are that (1) the use of the energy balance at the limiting cases constrains the uncertainty from the surface temperature or meteorological variables, (2) new formulation of the roughness height rather than a fixed value is adopted, and (3) no a priori knowledge of the actual turbulent heat fluxes is required. However, inconveniences may arise because of the too many parameters required as input.

S-SEBI

Based on the contrast between a reflectance (or albedo)-dependent maximum surface temperature at dry limit and a reflectance (or albedo) dependent minimum surface temperature at wet limit, Roerink et al. (2000) developed the Simplified Surface Energy Balance Index (S-SEBI) to partition available energy into sensible and latent heat fluxes. The S-SEBI has been tested and validated using data from a small field campaign conducted during August 1997.

Given a range of surface characteristics changing from dry/dark soil to wet/bright pixels, a theoretical interpretation to the S-SEBI can be given: (1) at low reflectance (albedo), such as over open water or irrigated lands, because of the sufficient water available, surface temperature remains almost unchanged; (2) at higher reflectance (albedo), because of the less water availability for ET, surface temperature increases with the increase of reflectance to a turning point, which is termed as "evaporation controlled"; and (3) after the turning point, surface temperature decreases with the increase of surface reflectance (albedo), which is called the "radiation controlled" (see Fig. 6.2).

Bounded by the dry and wet limits, EF in the S-SEBI is estimated by linearly interpolating the reflectance (albedo) dependent surface temperatures between the upper and the lower limits, which can be expressed as

$$\mathrm{EF} = \frac{T_{\mathrm{s,max}} - T_{\mathrm{s}}}{T_{\mathrm{s,max}} - T_{\mathrm{s,min}}} \tag{6.21}$$

where $T_{\mathrm{s,max}}$ represents the temperature at the dry limit [K], corresponding to minimum latent heat flux ($\mathrm{LE_{dry}} = 0$) and maximum sensible heat flux ($H_{\mathrm{dry}} = R_{\mathrm{n}} - G$) (the upper decreasing envelope when T_{s} is plotted against surface reflectance); and $T_{\mathrm{s,min}}$ represents the temperature at the wet limit, [K], indicative of the maximum latent heat flux ($\mathrm{LE_{wet}} = R_{\mathrm{n}} - G$) and minimum sensible heat flux ($H_{\mathrm{wet}} = 0$) (the lower increasing envelope when T_{s} is plotted against surface reflectance). $T_{\mathrm{s,max}}$ and $T_{\mathrm{s,min}}$ are linearly fit to the surface reflectance (albedo):

$$T_{\mathrm{s,max}} = a_{\mathrm{max}} + b_{\mathrm{max}}\alpha_{\mathrm{s}} \tag{6.22}$$

$$T_{\mathrm{s,min}} = a_{\mathrm{min}} + b_{\mathrm{min}}\alpha_{\mathrm{s}} \tag{6.23}$$

where a_{max}, b_{max}, a_{min}, and b_{min} are the site-specific and time-dependent regression coefficients.

Substituting Eqs. (6.22) and (6.23) into Eq. (6.21), EF can be obtained as

$$\mathrm{EF} = \frac{a_{\mathrm{max}} + b_{\mathrm{max}}\alpha_{\mathrm{s}} - T_{\mathrm{s}}}{a_{\mathrm{max}} - a_{\mathrm{min}} + (b_{\mathrm{max}} - b_{\mathrm{min}})\,\alpha_{\mathrm{s}}} \tag{6.24}$$

To apply the S-SEBI to estimate EF and LE, it should be ensured that the atmosphere is uniform in space and extreme hydrological contrast exists. A maximum relative difference of 8 % was found by Roerink et al. (2000) when estimated EF was compared with measured EF obtained from a small field campaign during 1997 in Italy.

The major advantage of the S-SEBI is that (1) no ground-based measurements other than purely remote sensing surface temperature and reflectance (albedo) are needed and (2) unlike the methods that have fixed temperatures at the dry and wet pixels (e.g., SEBAL), S-SEBI has different extreme temperatures at the wet and dry limits associated with different reflectances (albedos). When atmospheric conditions over larger areas are not constant, it is recommended that the study area be split into several subareas to ensure the uniform atmosphere over each of the subareas. The extreme temperatures at the wet and dry limits may also be subarea specific.

SEBAL and METRIC

Assuming that the near-surface temperature difference is linearly related to remote sensing surface temperature, Bastiaanssen (1995) and Bastiaanssen et al. (1998a) developed the Surface Energy Balance Algorithm for Land (SEBAL) with minimum ground-based measurements to estimate regional ET. The SEBAL has been used to estimate field crop coefficients and basin-wide irrigation performance under several climatic conditions in more than 30 countries worldwide (Bastiaanssen et al. 2005, 2010; Timmermans et al. 2007; Singh et al. 2008; Teixeira et al. 2009; Long and Singh 2013). The test of SEBAL shows that the typical accuracy at field scale is 85 %, 95 %, and on average 96 % at daily, seasonal, and annual scales, respectively (Bastiaanssen et al. 1998b, 2005; Bastiaanssen 2000; Allen et al. 2001).

The core of the SEBAL model is to establish the linear relationship between T_s and the surface–air temperature difference (dT) for the area of interest based on the surface characteristics at dry and wet points. The dT is linearly regressed to T_s:

$$\mathrm{d}T = a + bT_s \tag{6.25}$$

where a and b are the regression coefficients derived from two anchor points (dry and wet points).

Similar to other end-member-based models, the SEBAL model has a dry pixel of zero LE and maximum LE at the wet pixel. Both the sensible heat flux and the surface-air temperature difference ($\mathrm{d}T_{\mathrm{wet}}$) at the wet pixel equal zero. The surface-air temperature difference ($\mathrm{d}T_{\mathrm{dry}}$) at the dry pixel is obtained by inverting the single-source bulk transfer equation:

$$\mathrm{d}T_{\mathrm{dry}} = \frac{H_{\mathrm{dry}} \times r_{\mathrm{a}}}{\rho C_p} \tag{6.26}$$

where H_{dry} is the sensible heat flux at the dry pixel and equal to $R_{\mathrm{n}} - G$ [W/m^2].

Coefficients a and b in Eq. (6.25) can be obtained when ($T_{\mathrm{s,dry}}$, $\mathrm{d}T_{\mathrm{dry}}$) at the dry pixel and ($T_{\mathrm{s,wet}}$, $\mathrm{d}T_{\mathrm{wet}}$) at the wet pixels are linearly regressed. Provided that a and b are known, dT at each pixel over the study area can be figured out with T_s, and H can be obtained with an iterative procedure and the stability correction. This procedure requires a blending-height wind speed (100–200 m above the ground) that is not affected by surface horizontal variation and is obtained by extrapolating the near-surface measurements.

Three major refinements have been made to the most current version of the SEBAL model (SEBAL2008) through (1) incorporating the effect of land surface slope and aspect in mountainous areas, (2) making correction to the advection effect, and (3) improving estimates of surface albedo and soil heat flux (Bastiaanssen et al. 2010). For the SEBAL model, recent progress was primarily made on the investigation of the spatial-scale effect caused by the variation of the sizes of both the area of interest (AOI) and/or the satellite pixel in estimating regional sensible heat flux and latent heat flux (Tang et al. 2013a). The effect of the size of the satellite pixel was generally evaluated by comparing ET estimated at a targeted coarse resolution (e.g., MODIS) with the spatially averaged value that is derived by aggregating either output (output upscaling) or input (input upscaling) of SEBAL over high-resolution data (e.g., TM/ETM+ or ASTER) (Compaoré et al. 2008;

Hong et al. 2009; Gebremichael et al. 2010; Long et al. 2011; Long and Singh 2013). In contrast to the previous studies that evaluated the spatial-scale effect of the SEBAL model using a simple comparison, generalized analytical equations were proposed by Tang et al. (2013a), which were verified to be able to reveal the mechanism of the scaling transfer of the SEBAL model.

One prerequisite in successfully applying the SEBAL model is to ensure that the full hydrological contrast (i.e., wet and dry pixels) should be present in the area of interest. The selection of dry pixel and wet pixel can have a significant impact on the heat flux distribution from SEBAL. Therefore, it is very important to correctly identify the dry and wet pixels. The large calm water surface or well-watered agriculture areas is a good candidate of wet pixel in the practical applications. The advantages of the SEBAL are (1) minimum ancillary ground-based data are required and (2) its automatic internal calibration within each analyzed image makes the accurate atmospheric corrections on surface temperature unnecessary. However, SEBAL also has several drawbacks: (1) the specification of representative dry and wet pixels within the scene is generally subjective and is influenced by the area of interest and satellite pixels, (2) it is often applied over flat surfaces and needs to be temperature and wind speed corrected based on a digital elevation model when applied over the mountainous area, (3) the estimate of H is sensitive to the errors in surface temperature or surface–air temperature difference, (4) the view angle effect of radiometer is not taken into account, which can lead to a bias in T_s of several degrees in some cases, and (5) the assumption of linearity between surface temperature and the air temperature gradient used in defining the sensible heat fluxes does not generally hold true for strongly heterogeneous landscape (Norman et al. 2006).

To overcome the problems encountered in applying the SEBAL over complicated surfaces, Allen et al. (2005a, b, 2007) developed the Mapping EvapoTranspiration at high Resolution with Internalized Calibration (METRIC) to derive regional ET. Same as the derivative of the SEBAL model, the METRIC also adopts an automatic internal calibration in calculating the sensible heat flux and latent heat flux.

Main distinctions between METRIC and SEBAL are the following: (1) METRIC assumes the ET at the wet pixel to be 1.05 times tall reference crop (e.g., alfalfa) ET rather than $R_n - G$. The selected wet pixel should have similar biophysical characteristics to the reference crop. (2) The alfalfa reference EF (ET_rF, defined as the ratio of instantaneous actual ET to reference ET), which can better account for the effects of variations in environmental variables and horizontal advection during the day, is applied instead of the actual EF to upscale the instantaneous ET to daily value.

T_s-VI Triangle/Trapezoidal Feature Space

Based on the negative relationship between remotely sensed surface temperature T_s and vegetation index (VI) that was first proposed by Goward et al. (1985), the T_s-VI triangle feature space was subsequently established under the conditions

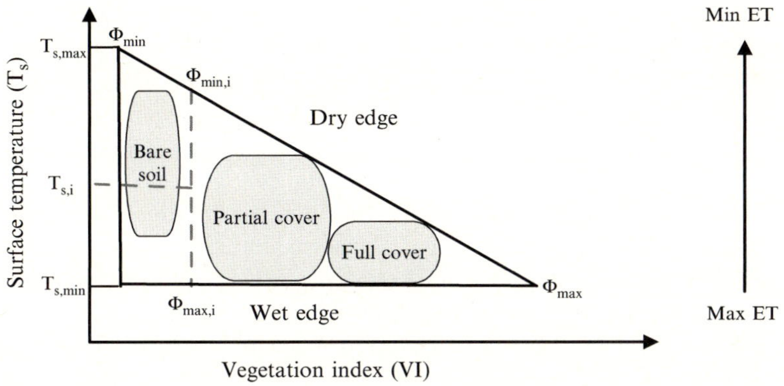

Fig. 6.3 A schematic illustration of the T_s-VI triangular space (Reproduced from Li et al. (2009))

of full ranges of soil moisture content and fractional vegetation cover to study surface resistance, the soil water content, land use and land cover change, drought monitoring, and regional ET (Nemani and Running 1989; Nemani et al. 1993; Jiang and Islam 1999, 2001, 2003), while the trapezoidal space reviewed below originates from a simple CWSI (Idso et al. 1981; Jackson et al. 1981). The T_s-VI triangle/trapezoidal feature space is bounded by an upper decreasing dry edge (representing pixels with unavailability of soil moisture content) and a lower nearly horizontal wet edge (representing pixels with sufficient water supply) with the increase of vegetation cover and the two edges ultimately intersect at a (truncated) point at full vegetation cover (see Fig. 6.3). Two implications are incorporated in constructing the triangle/trapezoid, including (1) canopy and soil components have different sensitivities to the variation of surface temperature, and canopy temperature is less sensitive to surface/deep-layer soil moisture content than soil temperature and (2) variation of surface temperature within a NDVI interval is caused by the variation of available soil water content rather than the difference in atmospheric conditions.

The principal rationale of the T_s-VI triangle and trapezoidal space in the evaluation of ET at regional scale is addressed, respectively, as follows:

1. Triangle Method

By interpreting the remotely sensed T_s-NDVI triangle feature space, Jiang and Islam (1999, 2001, 2003) adapted the Priestley–Taylor equation to dry surface to estimate regional actual EF and ET:

$$\mathrm{LE} = \phi \left[(R_n - G) \frac{\Delta}{\Delta + \gamma} \right] \tag{6.27}$$

where ϕ ranges from 0 to 1.26 [−]. All terms in the right-hand side of Eq. (6.27) can be estimated using fully remotely sensed data (Jiang and Islam 1999).

Some simplicity and assumptions are made in order to solve the parameter ϕ for intermediate pixels in the T_s-NDVI triangle feature space, including (1) full range of soil moisture and vegetation coverage at satellite pixel scale is present in the area of interest and (2) contaminations of clouds and atmospheric effects are removed. A two-step linear interpolation scheme (Jiang and Islam 1999; Carlson 2007; Stisen et al. 2008; Tang et al. 2010) as displayed in Fig. 6.3 is used to get the value of ϕ in Eq. (6.27) in the following manner: (1) First, set the global minimum ϕ to $\phi_{\min} = 0$ on the driest bare soil pixel and global maximum ϕ to $\phi_{\max} = 1.26$ on the pixel with largest NDVI and lowest T_s, respectively. (2) Second, interpolate linearly between $\phi_{\min}$ and $\phi_{\max}$ with NDVI to get local minimum ϕ ($\phi_{\min,i}$) for each NDVI interval (i) and local maximum ϕ ($\phi_{\max,i}$) for each NDVI interval (i), assuming $\phi_{\max,i} = \phi_{\max} = 1.26$. (3) ϕ (ϕ_i) for an intermediate pixel within each NDVI interval (i) is linearly interpolated between $\phi_{\min,i}$ and $\phi_{\max,i}$ with the decrease of T_s.

Based on the data collected from the MONSOON 90 (Kustas et al. 1991) and FIFE 1987 and 1989 field programs (Sellers et al. 1992), the triangular (trapezoidal) feature space constructed from a combination between surface temperature (or temperature difference or a scaled surface temperature) and vegetation indices (e.g., NDVI, SAVI (Soil-Adjusted Vegetation Index), a scaled NDVI, or F_r, fractional vegetation cover) has been applied and verified to derive surface soil moisture and surface fluxes (Hope 1988; Price 1990; Gillies and Carlson 1995; Moran et al. 1996; Venturini et al. 2004; Batra et al. 2006; Wang et al. 2006; Carlson 2007; Stisen et al. 2008). The emergence of the triangle shape under full range of soil moisture availability was found by Carlson et al. (1995) to depend more on the number of pixels rather than the spatial resolutions of the satellite pixels, which indicates that remote sensing data from coarse-resolution satellite sensors (GOES) can also be used to construct the triangle/trapezoid feature space (Diak et al. 1995).

One of the significant recent advances in the use of T_s-VI triangle to estimate regional ET comes from the work of Tang et al. (2010), in which an automatic technique was developed to determine the dry and wet edges. The automatic technique significantly reduces the subjectivity in the determination of dry and wet edges that is encountered in the prior studies. Similar to the SEBAL model, the T_s-VI triangle method is also subject to the spatial-scale effect of the sizes of the area of interest and the satellite pixel (Long and Singh 2013). To address the issue on the scale dependence of the triangle, Long and Singh (2012) figured out the theoretical dry and wet edges based on the detailed ground-based meteorological data, which makes the use of the T_s-VI triangle method immune from the spatial-scale effect.

The advantages of the T_s-VI triangle method are that (1) no ground-based measurements are required as input except remote sensing T_s and VI, (2) accurate correction of atmospheric effects is not indispensable, and (3) EF can be derived without calculating the surface energy balance. The limitations are that (1) the dry and wet edges are in general subjectively determined, (2) to make sure that the dry and wet limits exist, a large number of pixels over a flat area with a full range of soil wetness and fractional vegetation cover are required, and (3) the ET estimation suffers from the spatial-scale effect from the sizes of both satellite pixel and spatial extent of the study area.

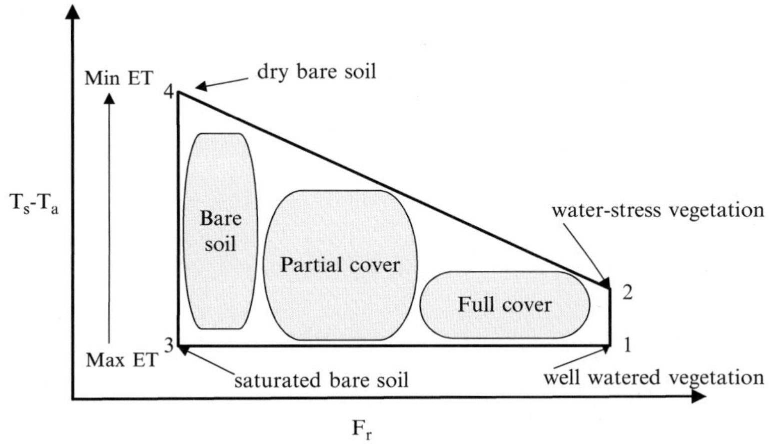

Fig. 6.4 A schematic illustration of the trapezoidal space between $T_s - T_a$ and F_r (Reproduced from Li et al. (2009))

2. Trapezoid Method

Following the conception of the CWSI (Idso et al. 1981; Jackson et al. 1981, 1988) that was developed for the detection of plant water stress over full vegetation cover and bare soil surface conditions, Moran et al. (1994, 1996) extended its application to partially vegetated surfaces by defining the vegetation index/temperature (VI-T_s) trapezoid and introducing the water deficit index (WDI, defined as 1 minus the ratio of actual to potential ET). The defined trapezoid requires as input the ground-based vapor pressure, air temperature, wind speed, physiological parameters of maximum and minimum stomatal resistances, etc. Four physical vertices (see Fig. 6.4), corresponding to (1) well-watered full-cover vegetation, (2) water-stress full-cover vegetation, (3) saturated bare soil, and (4) dry bare soil, are defined in the VI-T_s trapezoid to calculate the WDI value of intermediate pixels. The trapezoid assumes that $T_s - T_a$ varies linearly with vegetation cover along the dry and wet edges, and WDI within each VI interval varies linearly between maximum and minimum temperature differences ($T_s - T_a$). For a partially vegetated surface, WDI can be defined as

$$\mathrm{WDI} = 1 - \mathrm{LE}/\mathrm{LE_P} = \frac{[(T_s - T_a)_{\min} - (T_s - T_a)_i]}{[(T_s - T_a)_{\min} - (T_s - T_a)_{\max}]} \tag{6.28}$$

The trapezoid method is in essence a method to measure the surface water stress. Accurate estimates of ET and relative water deficit can be obtained from Eq. (6.28) with potential ET ($\mathrm{LE_p}$) estimated from the Penman–Monteith equation for both full-cover and partially vegetated surfaces (Moran et al. 1994).

The advantage of the VI-T_s trapezoidal space over the triangular space is that the dry and wet edges are controlled by the four physical limiting vertices that are inverted by combining the energy balance equation and the Penman–Monteith equation. In such case, it is not required that a large number of pixels under full range of soil moisture availability and vegetation cover are present in the area of interest. The weakness of the trapezoid space lies in its requirement of detailed ground-based measurements, which may constrain its broad application. The new WDI index also has some limitations (Luquet et al. 2004), including that (1) heat exchange between soil and vegetation is not accounted for, which may pose a challenge to the validity of the WDI index when soil and vegetation are at different temperatures, (2) the effect of water stress on vegetation cover cannot be reflected at an instantaneous scale, and (3) plant transpiration is not separated from soil evaporation.

6.2.3 Penman–Monteith Equation

The biased estimates of ET in the commonly applied surface energy balance models generally result from the use of remote sensing land surface temperature as a surrogate of aerodynamic temperature and from the use of surface available energy as the upper limit of surface energy balance. To overcome the two problems encountered in the surface energy balance models, Cleugh et al. (2007) used the Penman–Monteith equation (Monteith 1965) to estimate regional ET. Input to their model consists of the MODIS-derived vegetation data and the daily surface meteorological variables. Cleugh's model was subsequently modified by Mu et al. (2007) to estimate ET at the global scale. Input to the algorithm of Mu et al. (2007) (hereafter referred as the old algorithm) includes the MODIS land cover, albedo, Leaf Area Index, and Enhanced Vegetation Index and a daily meteorological reanalysis data set from NASA's Global Modeling and Assimilation Office (GMAO).

In the old algorithm, ET was estimated from the sum of the evaporation from moist soil surface and the transpiration from the vegetation during daytime. The evaporation from the intercepted precipitation was not accounted for and no distinction was made between the evaporation from the saturated soil surface and that from the moisture soil surface. Soil heat flux and nighttime ET were assumed to be negligible. Note that although soil heat flux is relatively a small fraction of surface net radiation compared to sensible heat flux and latent heat flux in forest and grassland areas, errors may be induced when neglecting soil heat flux for tundra.

To deal with these problems, Mu et al. (2011) made a further improvement to the old algorithm to produce the MOD16 product at the global scale. In the improved ET algorithm, ET is estimated as the sum of the evaporation from the intercepted water by the canopy (accounting for a substantial amount of ET in ecosystems with high Leaf Area Index after rainfall/sprinkler irrigation events), the transpiration from the dry canopy surface, and the evaporation from the moisture soil surface and saturated soil surface. Specifically, these improvements primarily include (1) vegetation cover fraction is surrogated by the 8-day 1-km MOD15A2 FPAR product; (2) nighttime ET is added; (3) soil heat flux is calculated; (4) estimates of stomatal conductance, aerodynamic resistance, and boundary layer resistance are improved; (5) dry canopy surface is separated from wet canopy surface; and (6) soil surface is divided into saturated soil surface and moisture surface.

Both the old and improved MOD16 ET algorithms use the Penman–Monteith equation to estimate ET at global scale:

$$\mathrm{LE} = \frac{\Delta(R_\mathrm{n} - G) + \rho c_p(e_\mathrm{s} - e_\mathrm{a})/r_\mathrm{a}}{\Delta + \gamma(1 + r_\mathrm{s}/r_\mathrm{a})} \tag{6.29}$$

where e_a is the actual vapor pressure [kPa], and r_s is the bulk surface resistance (s/m).

The Penman–Monteith equation combines both the energy driver and atmospheric demand of evaporation and is a more robust and theoretically sound approach. Evapotranspiration from the Penman–Monteith equation is not overly sensitive to any of the inputs. When surface conductance is high (e.g., the energy-limited condition), ET is sensitive to wind speed (a determinant of aerodynamic resistance) and the sensitivity depends upon the value of surface conductance (Cleugh et al. 2007). The key and difficulty of the use of Penman–Monteith equation is to properly determine the surface bulk conductance, which needs to account for the effect of soil water availability, vapor pressure deficit, carbon dioxide concentration, atmospheric temperature, and canopy physiology through a coupled biophysical model (Tuzet et al. 2003) or an empirical discount function (Jarvis 1976).

Inputs to the improved MOD16 ET algorithm include (1) global 1-km^2 Collection 4 MODIS land cover type 2 (MOD12Q1), (2) global 1-km^2 MODIS Collection 5 LAI/FPAR (MOD15A2), (3) the 10th band of MODIS White-Sky Albedo from Collection 4 0.05-degree CMG MOD43C1 product, and (4) the global GMAO reanalysis meteorological data at $1.00° \times 1.25°$ resolution. The output from the MOD16 ET algorithm consists of the 8-day, monthly, and annual ET, LE, Potential ET, and Potential LE.

Validation results from the eddy covariance (EC) measurements at 46 AmeriFlux sites in Mu et al. (2011) showed that mean absolute bias of daily ET was reduced from 0.39 mm day^{-1} (25 % of the observed mean) in the old algorithm to 0.33 mm day^{-1} (24 % of the observed mean) in the improved algorithm when both algorithms were driven by tower meteorological data and from 0.40 to 0.31 mm day^{-1} when driven by a global meteorological reanalysis data set (GMAO data). Validations on MOD16 product with EC measurements collected at 17 sites in Asia showed that the algorithm of Mu et al. (2011) overestimated the eight-day averaged ET at nine sites, underestimated the eight-day averaged ET at one site, agreed with the eight-day averaged ET at five sites, and mismatched the eight-day averaged ET at two sites (Kim et al. 2012).

As pointed out by Mu et al. (2011), the bias in the MOD16 ET product primarily resulted from (1) the uncertainty from the input data, (2) the uncertainty from the EC measurements, (3) the inconsistent source areas between the EC measurements and MODIS footprint, and (4) the uncertainty from the MOD16 ET algorithm.

For example, uncertainties in MODIS LAI/FPAR and daily GMAO meteorological data may result in the biases in the ET estimates. Overestimates of LAI may contribute to the overestimates of ET even if other input data are relatively accurate. For the ground-based EC measurements, an uncertainty of about 10–30 % has been reported in many studies even though they have been widely applied to validate the remote sensing-based ET estimates. Moreover, the energy imbalance between the surface available energy and the sum of H and LE has been found in a number of papers. However, to date, there has been still no consensus on how to best reconcile this imbalance. Several factors including the assumption of the stomata closure at night, the same biophysical parameters used for a given biome type globally, and no account for the stand age, disturbance history, or species composition in the current MOD16 ET product algorithm can all make the accurate estimates of ET a challenge.

6.2.4 Modified Priestley–Taylor Equation

Although the Penman–Monteith equation can provide in theory accurate ET estimates, it also has difficulty in preparing the input variables for the parameterization of aerodynamic resistance and stomatal resistance, especially when applied globally. To overcome the problems encountered in the PM equation, Fisher et al. (2008) adapted the Priestley–Taylor equation (Priestley and Taylor 1972) to soil water stress conditions by exerting plant physiological limitations and soil drought constraint. Inputs to Fisher's model include surface net radiation (R_n), NDVI, soil adjusted vegetation index, maximum air temperature (T_{max}), and actual vapor pressure (e_a). Their model has been validated at a monthly scale using eddy covariance data from 16 FLUXNET sites ranging from tropical to boreal environments. These sites represent a wide range of plant functional types.

Total LE in Fisher et al. (2008) is composed of canopy transpiration (LE_c), wet and moisture soil evaporation (LE_s), and canopy interception evaporation (LE_i). Each of the three (in fact it is four) components is estimated from the adjusted Priestley–Taylor equation:

$$\mathrm{LE} = \mathrm{LE_s} + \mathrm{LE_c} + \mathrm{LE_i} \tag{6.30}$$

$$\mathrm{LE_c} = (1 - f_{\mathrm{wet}}) f_{\mathrm{g}} f_{\mathrm{T}} f_{\mathrm{M}} \alpha \frac{\Delta}{\Delta + \gamma} R_{\mathrm{n,c}} \tag{6.31}$$

$$\mathrm{LE_s} = [f_{\mathrm{wet}} + f_{\mathrm{SM}}(1 - f_{\mathrm{wet}})]\, \alpha \frac{\Delta}{\Delta + \gamma} (R_{\mathrm{n,s}} - G) \tag{6.32}$$

$$\mathrm{LE_i} = f_{\mathrm{wet}} \alpha \frac{\Delta}{\Delta + \gamma} R_{\mathrm{n,c}} \tag{6.33}$$

where f_{wet} is the relative surface wetness, which can be expressed as the fourth power atmospheric relative humidity (RH^4); f_g is the green canopy fraction, expressed as the ratio of f_{APAR} (fraction of PAR absorbed by green vegetation cover) to f_{IPAR} (fraction of PAR intercepted by total vegetation cover); f_T is the plant temperature constraint factor; f_M is the plant moisture constraint factor; f_{SM} is the soil moisture constraint factor; $R_{n,c}$ is the canopy net radiation [W/m^2]; and $R_{n,s}$ is the soil net radiation [W/m^2].

This modified Priestley–Taylor equation has been used by Fisher et al. (2008) to produce global LE at a monthly temporal scale and a 1 degree spatial scale during the years 1986–1993 using data sets from ISLSCP-II and AVHRR. Although Fisher et al. (2008) have claimed that their model can be potentially applied based solely on remote sensing inputs at finer spatial and temporal scales, to circumvent the difficulty of the satellite-based estimation of relative humidity and vapor pressure deficit, Yao et al. (2013) subsequently excluded the plant soil moisture constraint on the canopy transpiration and introduced the temperature diurnal range to parameterize the soil moisture constraint. The original inputs are also changed to only include surface net radiation, NDVI, air temperature, and diurnal temperature range. Validation results at 16 sites located in China revealed that Yao et al. (2013) improved the monthly LE estimates from Fisher et al. (2008) when their models were applied to MODIS data at a 0.05 degree spatial resolution.

The advantages of this type of method are that very limited number of variables is required as input and all of the input variables can be obtained from remote sensing data (Fisher et al. 2008). Therefore, this method is easy to operate for routine, long-term mapping of LE at a global scale. Limitations in this type of method can be summarized as (1) the threshold value for some of the parameters does not account for the difference in the biome types, (2) the effect of CO_2 concentration is not incorporated as a constraint in the canopy transpiration, (3) soil heat flux is roughly dealt with and may need to be improved, and (4) energy for canopy photosynthesis and storage is not included at a monthly scale.

Literatures conducted over the past several years have shown that there appears a trend in the scientific community to estimate continental/global ET in the context of climate change using the modified Penman–Monteith equation and Priestley–Taylor equation (Cleugh et al. 2007; Mu et al. 2007, 2011; Fisher et al. 2008; Yuan et al. 2010; Zhang et al. 2010; Vinukollu et al. 2011; Yao et al. 2013), although the efficacy of these two equations still needs to be further investigated. This trend may be provoked by the limitation of the single- and dual-source energy balance models that absorb the remote sensing surface temperature as one of the inputs. One prerequisite for developing a global land surface ET algorithm using optical/thermal remote sensing data is that the model should not be sensitive to the constraints imposed by the once-daily overpass of the polar orbiting satellite and the necessary cloud screening and compositing procedures (Cleugh et al. 2007). Using the retrieved vegetation biophysical parameters other than surface temperature from remote sensing data as part of the inputs has made the modified Penman–Monteith equation and Priestley–Taylor equation good alternatives to the conventional energy balance models for the global ET estimation. Lack of reliable meteorological data and quantification of the biome-specific limiting factors are nowadays the main constraints for the application of these two equations to a global scale.

6.2.5 Data Assimilation Method

Most of the aforementioned remote sensing physical models generally produce regional ET at instantaneous scale, which may not meet the demand by the water resources managers and engineers in the practical applications who have more interest in the temporally continuous and long-time integrated values. However, data assimilation technique when combined with a dynamic model can overcome this weakness and can take full advantage of the synergy of multisensor/multiplatform observations (Boni et al. 2001; Reichle 2008).

Data assimilation technique was first proposed by meteorologist to display the variation of daily weather variables in space and time (McLaughlin 1995). Its essence is simply to assimilate all available information to estimate objective variables as accurately as possible by minimizing errors of the estimate that is derived by merging noisy observations and model uncertainty under the fundamental physical constraints (Talagrand 1997; Bouttier and Courtier 2002). A data assimilation system generally consists of a set of observations, a dynamic model, and a data assimilation technique (Robinson and Lermusiaux 2000). For ET estimate from the data assimilation technique, detailed atmospheric forcing data are generally required to drive the dynamic (numerical) model, which may be computationally demanding than remote sensing ET models (McLaughlin et al. 2006; Kumar et al. 2008). A trade-off between optimality, computational efficiency, flexibility, and robustness should be made to select an appropriate data assimilation technique. However, because these criteria often conflict (Crow and Kustas 2005), one has to make some compromise for a specific goal.

Data assimilation techniques are roughly classified into the sequential assimilation (e.g., ensemble Kalman filter and optimal interpolation) (Anderson 2001; Reichle et al. 2002a, b; Caparrini et al. 2004; Crow and Kustas 2005; Margulis et al. 2005; Huang et al. 2008) and the un-sequential/variational/retrospective assimilation (e.g., 4-dimentional variational assimilation) (Courtier et al. 1998; Margulis and Entekhabi 2003; Seo et al. 2003; Caparrini et al. 2004). In the sequential assimilation, each individual observation only influences the estimated state of the flow at later times whereas the model solution is adjusted globally to all available observations over the assimilation period in the variational assimilation. Data assimilation techniques have been used to estimate regional sensible heat flux and latent heat flux in several papers (Caparrini et al. 2003; Margulis et al. 2005). For example, a land data assimilation system was developed by Boni et al. (2001) to estimate latent heat flux by assimilating remote sensing surface temperature sequences with surface energy balance equation as physical constraints. This system has been tested using data collected from the International Satellite Land Surface Climatology Project (ISLSCP FIFE) (Caparrini et al. 2004).

Advantages of the data assimilation technique over traditional physical methods are (1) both LE and other intermediate variables can be obtained via the assimilation procedure in a numerical model, (2) these estimates are continuous in time and space and may be produced at a much finer resolution, (3) data sources with different resolutions, coverage, and uncertainties can be organically merged (Margulis et al. 2002). The drawback is that the data assimilation technique is relatively more computationally demanding than the remote sensing ET models.

6.3 Upscaling of Instantaneous ET to Daily Value

Though data assimilation technique when combined with a numerical model can produce spatially and temporal continuous ET, the requirement of detailed ground-based measurements has limited its applicability to some extent. To meet the demand in the practical applications, it is necessary to upscale the instantaneous ET derived at the satellite overpass time to daily or longer-time integrated values. A number of representative methods have thus been proposed to extrapolate the instantaneous ET through the assumed relationship between instantaneous ET (ET_i) and daily values (ET_d). The upscaling scheme is essentially to find a factor that can be nearly constant over a diurnal cycle:

$$\frac{LE_i}{LE_d} = \frac{ET_i}{ET_d} = \frac{\omega_i}{\omega_d} \tag{6.34}$$

where ω can be either the surface available energy, the extraterrestrial solar radiation, the global solar radiation, or the reference ET, corresponding to the constant EF method (Shuttleworth et al. 1989; Brutsaert and Sugita 1992), the constant extraterrestrial solar radiation ratio method (Ryu et al. 2012), the constant global solar radiation ratio method, or the constant reference EF method (Trezza 2002; Allen et al. 2007), respectively. Subscripts “i” and “d” indicate the instantaneous and daily values of the variable, respectively.

With the assumption that the EF during the daytime is relatively stable, a variety of studies has applied the so-called constant EF method to upscale the instantaneous ET to daily or daytime values (Brutsaert and Sugita 1992; Lhomme and Elguero 1999; Gentine et al. 2007). For the constant global solar radiation ratio method, Jackson et al. (1977) modeled the daytime global solar radiation using a sinusoidal function and found this energy flux ratio could produce reliable estimates of daytime ET under cloud-free conditions. Under partially cloudy days, the effect of the amount and temporal duration of the cloud cover should be further taken into account. Similarly, the extraterrestrial solar radiation ratio method was applied by Ryu et al. (2012) to extrapolate the MODIS-derived ET to daily and eight-day values. Furthermore, Trezza (2002) proposed the constant reference EF method to consider the effect of horizontal advection and variation of environmental factors (e.g., wind speed, vapor pressure) on the EF in a diurnal cycle. The performance of a given ET upscaling method depends on how closely the instantaneous flux ratio at the clear-sky satellite overpass time can represent its daily averages. The presence of intermittent cloud covers during daytime questions the validity of Eq. (6.34) and consequently has an effect on the accuracy of the upscaling method.

Both the actual global solar radiation and actual surface available energy are less affected (if not independent) by the variation of daytime air temperature, wind speed, and vapor pressure, and the modeled extraterrestrial solar radiation is only controlled by the sun–earth geometry. Therefore, these three upscaling methods cannot effectively consider the effect of the horizontal advection and the variation of meteorological variables (e.g., wind speed, vapor pressure deficit). On the contrary, the reference ET for a crop, which is estimated from the Penman–Monteith equation (Monteith 1965) and thus closely depends upon the meteorological variables, is used as the upscaling factor in the constant reference EF method. Therefore, this upscaling method can incorporate the variations of horizontal advection and the variation of meteorological variables. However, the estimation of reference ET requires as input the detailed ground-based meteorological measurements of air temperature, vapor pressure deficit, global solar radiation, and wind speed. This may introduce a degree of uncertainty into the upscaled daily ET and hinders the broad application of the reference EF method.

6.4 Comparison of Different Evapotranspiration Estimation and Upscaling Methods

The aforementioned sections overview the theory, advantages, and weaknesses of the various remote sensing ET estimation and upscaling methods. This section aims to summarize and make a comparison of the representative remote sensing ET estimation models and ET upscaling methods, as tabulated in Tables 6.1 and 6.2.

6.5 Validation of Evapotranspiration from EC and LAS Measurements

Surface energy components measured by eddy covariance (EC) system and large aperture scintillometer (LAS) in a number of experiments have been used to validate the instantaneous and daily surface energy components and EF estimated by the various aforementioned models and upscaling methods in the past decades. This section highlights several examples that have been previously presented in Tang et al. (2010, 2011a, b, 2012, 2013b).

Table 6.1 Comparisons of a variety of commonly applied remote sensing ET methods

Methods	References	Main inputs	Main assumptions	Advantages	Disadvantages
Empirical regression equation	Jackson et al. (1977) and Wang et al. (2007)	R_n, T_s, T_a,	–	Simplicity	Site specific
VI-T_s triangle	Jiang and Islam (1999)	R_n, G, T_s, VI	Full range of both soil moisture and vegetation coverage exists; EF varies linearly with T_s for a given VI	No ground-based measurements are needed	Difficult to determine the dry and wet edges; VI-T_s triangle form is not easy recognized with coarse spatial-resolution data
VI-T_s trapezoid	Moran et al. (1994)	T_a, VPD, u, T_s, VI, R_n, G	Dry and wet edges vary linearly with VI; EF varies linearly with T_s for a given VI	Full range of VI and soil moisture in the scene of interest is not required	Lots of ground-based measurements are needed
SEBI	Menenti and Choudhury (1993)	T_{pbl}, h_{pbl}, u, T_s, R_n, G	Dry limit has a zero surface ET; wet limit evaporates potentially	Reducing the sensitivity of the meteorological inputs	Ground-based measurements are needed
SEBAL	Bastiaanssen et al. (1998a)	u, z_0, T_s, VI, R_n, G	Linear relationship exists between T_s and dT; $ET_{dry} = 0$; $ET_{wet} = R_n - G$	Require minimum ground measurements; automatic internal calibration	Applied over flat surfaces; uncertainty in the determination of anchor pixels
S-SEBI	Roerink et al. (2000)	T_s, α_s, R_n, G	EF varies linearly with T_s for a given α_s; $T_{s,max}$ corresponds to minimum LE; $T_{s,min}$ corresponds to maximum LE	No ground-based measurements are needed	Extreme temperatures have to be location specific
SEBS	Su (2002)	T_a, $z_a u$, T_s, R_n, G	At dry limit, ET = 0; at wet limit, ET takes place at potential rate	Uncertainty from the inputs is limited; computing explicitly the roughness height for heat transfer	Require too many parameters; solution of the turbulent heat fluxes is relatively complex
METRIC	Allen et al. (2005a, 2007)	u, z_a, T_s, VI, R_n, G,	At hot pixel, ET = 0; at wet pixel, ET = $1.05ET_r$	Same as SEBAL	Uncertainty in the selection of anchor pixels

(continued)

Table 6.1 (continued)

Methods	References	Main inputs	Main assumptions	Advantages	Disadvantages
N95	Norman et al. (1995)	u, $z_a T_a$, T_s, T_c, F_r or LAI, R_n, G	Priestly–Taylor equation is used to give the first guess of canopy LE	Incorporate the effects of view geometry; no corrections for the "excess resistance"	Require many ground measurements; require component temperatures of soil and vegetation
TSTIM/ ALEXI	Mecikalski et al. (1999)	u, z_a dT, F_r or LAI, R_n, G.	T_s changes linearly with the time during the morning hours of the sensible heating	Errors due to atmospheric corrections and surface emissivity specification are significantly reduced	Determination of an optimal pair of thermal observation times for the linear rise in sensible heating is needed
PM equation	Cleugh et al. (2007) and Mu et al. (2011)	u, z_a, T_a, F_r or LAI, VPD, R_n, G	–	Use of remote sensing T_s is avoided; incorporate the effect of atmospheric evaporative demand	Require detailed ground-based measurements
PT equation	Fisher et al. (2008)	R_n, NDVI, SAVI, T_{max}, e_a	–	Require very limited number of variables; all of the input variables can be obtained from remote sensing data	The threshold value for some of the parameters does not account for the difference in the biome types; energy for canopy photosynthesis and storage is not included at monthly scale

Adapted from Li et al. (2009)

Table 6.2 Comparisons of four commonly applied instantaneous ET upscaling methods

Methods	References	Upscaling factor	Advantages	Disadvantages
Constant evaporative fraction method	Shuttleworth et al. (1989) and Brutsaert and Sugita (1992)	$R_n - G$	Require only the surface available energy as input	The effect of horizontal advection and variation of atmospheric variables (wind speed, vapor pressure deficit) cannot be incorporated
Constant extraterrestrial solar radiation method	Ryu et al. (2012)	E_s	No ground-based measurements are required	The effect of cloud, horizontal advection, and variation of atmospheric variables cannot be incorporated
Constant global solar radiation method	Jackson et al. (1977)	R_g	Require only the global solar radiation as input	The effect of horizontal advection and variation of atmospheric variables (wind speed, vapor pressure deficit) cannot be incorporated
Constant reference evaporative fraction method	Trezza (2002)	ET_rF	The effect of cloud, horizontal advection, and variation of atmospheric variables can be incorporated	Require detailed ground-based measurements as input

6.5.1 Validation of the $T_s - F_r$ Triangle-Derived H Using LAS Measurements Collected at Linze Site

The study area (Fig. 6.5) locates in the middle reach of Heihe river basin, northwest China, with a total area of 38,000 km^2. The latitude and longitude ranges from 38.7° N to 39.8° N and 98.5° to 102° E, respectively. Surface elevation over the most areas varies ~1,200 to ~1,600 m above sea level. The mountainous area with an elevation of ~3,000 m in the southwest was excluded from the study area because pixels used to construct the triangle space should have similar surface elevations in order to ensure that the atmospheric condition was horizontally uniform. The study area is characterized by an arid climate in the temperate zone. Mean annual rainfall is approximately 174 mm and more than 73 % of annual precipitation occurs during the rainfall season from June to September. A LAS instrument was set up in the study area at the Linze site during May 20 to August 21, 2008. This site was covered by sparsely vegetated surfaces (short grass, clover, reed, and agricultural crops).

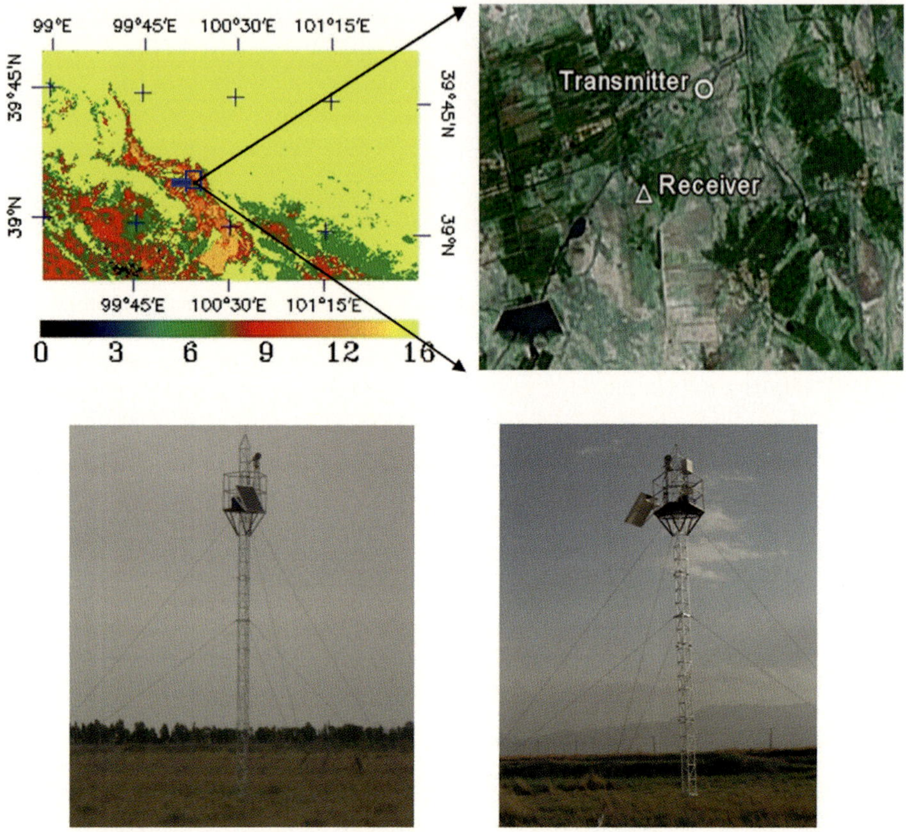

Fig. 6.5 Land cover of the study area and location of the LAS instrument (Reproduced from Tang et al. (2010), with permission from Elsevier) and photos taken for the transmitter (*left*) and receiver (*right*) of the LAS instrument. (In the *upper left* is a yearly IGBP land cover classification map in 2004 from MOD12Q1. *0* water, *1* evergreen needleleaf forest, *2* evergreen broadleaf forest, *3* deciduous needleleaf forest, *4* deciduous broadleaf forest, *5* mixed forests, *6* closed shrubland, *7* open shrublands, *8* woody savannas, *9* savannas, *10* grasslands, *11* permanent wetlands, *12* croplands, *13* urban and built-up, *14* cropland/natural vegetation mosaic, *15* snow and ice, *16* barren or sparsely vegetated. The *upper right* is the zoomed-in map of the LAS instrument site)

Length path between the transmitter and the receiver of the LAS was 1,550 m and the surface elevations of the transmitter site and the receiver sites are 1,384 m and 1,395 m, respectively. The heights of the transmitter and receiver were about 9.25 m and 9.1 m, respectively, above the ground (see Fig. 6.5).

Figure 6.6 compares the instantaneous H estimated from the $T_s - F_r$ triangle method with LAS-measured H over 24 clear-sky MODIS/Terra overpass times from May 23 to August 21, 2008. It was shown that MODIS-derived H varied from ~75 to 226 W/m^2 with a mean of 137 W/m^2. A very good agreement was obtained overall with a root mean square error (RMSE) of ~25 W/m^2. Exceptions

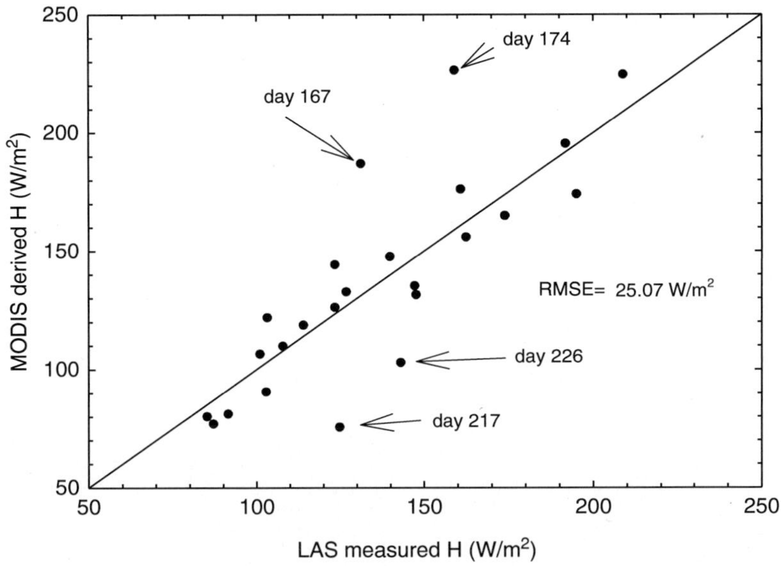

Fig. 6.6 Comparisons of sensible heat flux estimated using the $T_s - F_r$ triangle with that measured by a LAS over 24 clear-sky MODIS/Terra overpass times from May 23 to August 21, 2008 at the Linze site (Reproduced from Tang et al. (2010), with permission from Elsevier)

occurred on day of year (DOY) 167 ($\Delta H = 55.3$ W/m^2), 174 ($\Delta H = 67$ W/m^2), 217 ($\Delta H = -49.7$ W/m^2), and 226 ($\Delta H = -40.6$ W/m^2). The overestimation of H on Julian days 167 and 174 was caused by the underestimation of EF that is a result of the underestimation of the dry edge. The heavy rainfalls taken place on Julian days 165 and 173 over the study area might be responsible for the inaccurate determination of dry edge.

The reason for the relatively large discrepancies found on Julian days 217 and 226 could not be well explored because of the limited available information. It might be contributed by the relatively lower surface net radiation ($R_n = 576$ W/m^2, 518 W/m^2 for DOY 217 and 226, respectively) derived at MODIS overpass time on these days than that estimated on adjacent days ($R_n = 710$ W/m^2, 710 W/m^2 for DOY 216 and 225). Note that the EF results are controlled by the relative position of the dry and wet edges, not by their exact placements.

6.5.2 Validation of the T_s-VI Triangle-Derived EF Using EC Measurements Collected at Two AmeriFlux Sites

The study area locates in the southwest of America (Fig. 6.7), with an area of ~77,000 km^2. The latitude and longitude range from 30.5° to 32.5° N and 108° to

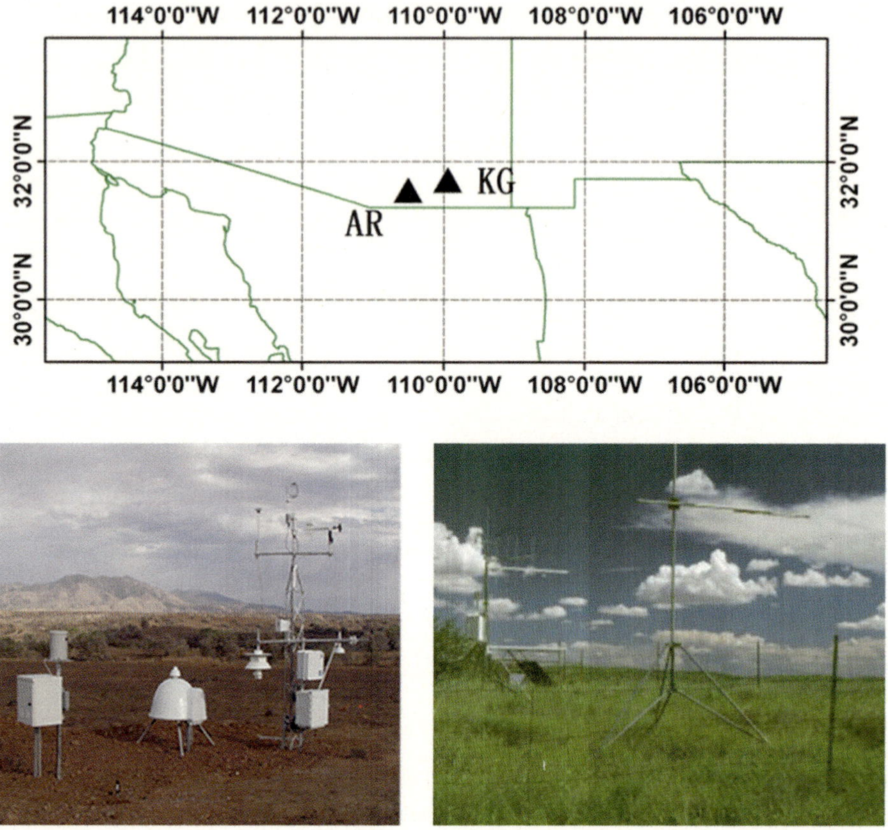

Fig. 6.7 A geographic map encompassing the study area (*upper panel*) (Reproduced from Tang et al. (2011a), with permission from John Wiley and Sons) and photos (*lower panel*) of the two AmeriFlux validation sites, Audubon Research Ranch (*left*) and Kendall Grassland (*right*), taken from the FLUXNET website (http://fluxnet.ornl.gov/)

111.5° W, respectively. This area is characterized by the arid and semiarid climate. It was covered by shrublands, grasslands, and woody savannas. Surface elevation varies from ~1,000 to ~1,800 m above the seal level. Four AmeriFlux sites were in operation over this area and the elevation at these four sites varies from 991 to 1,531 m. Because pixels that have surface elevations far beyond the regional mean should be excluded in the construction of the T_s-VI triangle, only two of the four AmeriFlux sites, namely, Audubon Research Ranch (AR) (31.591° N Lat/110.509° W Lon) and Kendall Grassland (KG) (31.737° N Lat/109.942° W Lon) (see Fig. 6.7), were finally left for further analysis. Attribute of the Audubon Ranch site and the Kendall Grassland site is shown in Table 6.3.

Level 2 surface energy flux measurements of 30-min averaged net radiation, soil heat flux, sensible heat flux, and latent heat flux were downloaded from the Carbon Dioxide Information Analysis Center (CDIAC) (ftp://cdiac.ornl.gov/pub/

Table 6.3 Attribute of the Audubon Ranch site and the Kendall Grassland site (Data involved are measured in situ from year 2004 to 2006 at the Audubon Ranch site and from 2004 to 2007 at the Kendall Grassland site, respectively)

Site name	Latitude longitude	Elevation (m)	Soil type	Land cover	Tower height (m)	Annual precipitation (mm)	Min/max temperature (°C)
Audubon Ranch	31.591° N	1,469	Sandy loam	Desert grassland	4	355	−8.1
	110.509° W						37.3
Kendall Grassland	31.737° N	1,531	Course loam	C4 grassland	6.4	235	−5.0
	109.942° W						35.9

Reproduced from Tang et al. (2011a), with permission from John Wiley and Sons

ameriflux/data/) for the evaluation and validation of the T_s-VI triangle method in the estimates of regional EF and ET. Surface net radiation and soil heat flux were measured by a net radiometer and a soil heat flux plate, respectively, and sensible heat flux and latent heat flux were estimated from data that were measured by an eddy covariance system using a 3-D sonic anemometer and a temperature/humidity probe. Measurements at the AR and the KG sites spanned a period of 3 (2004–2006) and 4 (2004–2007) years, respectively. We applied the Bowen ratio (BR) correction method and the residual energy (RE) correction method (Twine et al. 2000) to close the energy imbalance of the in situ eddy covariance measured turbulent heat fluxes. In the BR correction method, the Bowen ratio measured by the eddy covariance system remained invariant when surface available energy was repartitioned into H and LE, while in the RE correction method, the energy imbalance was assumed to be caused by the underestimation of LE and the imbalance energy totally went into LE.

Totally, 147 and 152 clear-sky (clear sky at the satellite overpass time) daytime MODIS/Terra and Aqua products, respectively, from May to September during the years 2004–2007 over the study area, were used to estimate EF. Figure 6.8 shows the comparison of the $\mathrm{T_s} - \mathrm{F_r}$-derived EF and the in situ EF corrected by the BR and the RE correction methods at both the AR and the KG sites. The performance at the KG site was better than that at the AR site. At both sites, the BR correction method gave the highest R^2 (with an exception for MODIS/Aqua data at the AR site) and the lowest bias (with an exception for MODIS/Aqua data at the AR site) and RMSD. At the MODIS/Terra overpass time, the $T_s - F_r$-derived EF agreed with in situ EF corrected by the BR correction method with a bias of 0.065 and a RMSD of 0.120 at the AR site and a bias of 0.039 and a RMSD of 0.1 at the KG site. At the MODIS/Aqua overpass time, when the BR correction method was applied to correct the in situ measurements, similar performance was obtained with a bias of 0.067 and a RMSD of 0.125 at the AR site and a bias of 0.058 and a RMSD of 0.103 at the KG site. When the in situ EF was corrected by the RE correction method, the performance of the T_s-VI triangle method become a little worse.

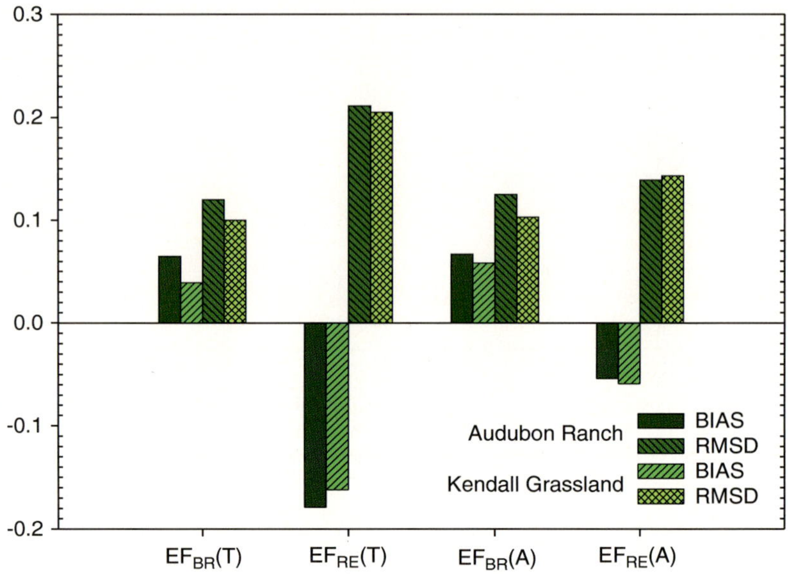

Fig. 6.8 BIAS and RMSD in the comparison of instantaneous evaporative fraction (EF) estimated from the $T_s - F_r$ triangle space using MODIS/Terra (T) and MODIS/Aqua (A) data with the eddy covariance measurement. ($EF_{BR} = LE_{EC}/(H_{EC} + LE_{EC}) = LE_{BR}/(H_{BR} + LE_{BR})$, $EF_{RE} = LE_{RE}/(R_n - G) = LE_{RE}/(H_{EC} + LE_{RE})$; H_{EC} and LE_{EC} are the sensible and latent heat fluxes measured by the eddy covariance system, respectively; H_{BR} and LE_{BR} are the corrected measurements of sensible and latent heat fluxes using the Bowen ratio correction method; LE_{RE} is the corrected measurements of latent heat fluxes using the residual energy correction method) (Reproduced from Tang et al. (2011a), with permission from John Wiley and Sons)

6.5.3 *Validation of the T_s-VI Triangle-Derived EF and H Using LAS Measurements Collected at Changwu Site*

The study area (Fig. 6.9) locates in the North of China, with an area of ~92,000 km^2. The latitude and longitude ranges from 34 to 37° N and from 106 to 109° E, respectively. Validation data throughout the period from late June to late October 2009 were collected from the Changwu agroecological experimental station (35.242° N/107.683° E, hereinafter referred as Changwu station). The Changwu station has a surface elevation of ~1,220 m above sea level and was covered by apple orchard (4.5 m height) with dispersed crops (wheat). It is characterized by a semiarid and subhumid climate with a mean annual precipitation of 584 mm and a mean annual temperature of 9.1 °C. Approximately 68 % of the annual rainfall takes place in the rainfall season from July to October.

From late June to late October 2009, a LAS system (see Fig. 6.9b for its location) was set up along the southeast–northwest direction at the Changwu station to measure the sensible heat flux. The path length between the transmitter and the

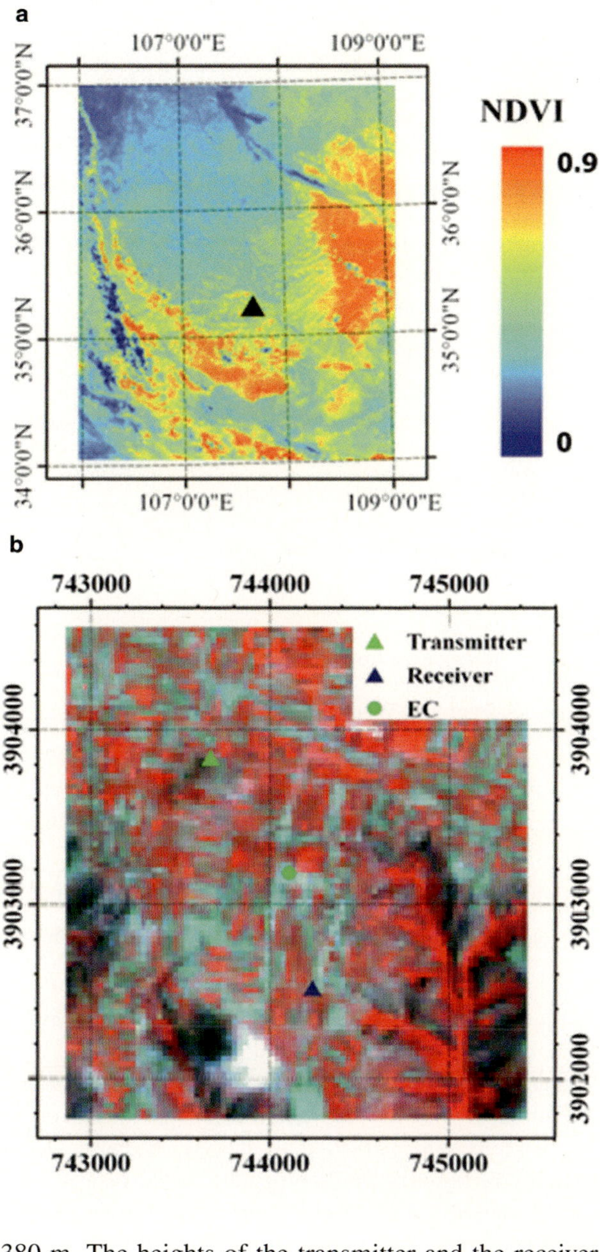

Fig. 6.9 (**a**) A map of MODIS NDVI at 1-km resolution on DOY 183 over the study area and (**b**) a false-color composite image from TM bands 4 (*red*), 3 (*green*), and 2 (*blue*) data on DOY 204 encompassing the Changwu station with the locations of large aperture scintillometer and eddy covariance system indicated (*red*, green crop, apple tree, or forest; *cyan gray*, crop residue filed or bare soil with little vegetation) (Reproduced from Tang et al. (2012), with permission from John Wiley and Sons)

receiver of the LAS was 1,380 m. The heights of the transmitter and the receiver were ~11.36 m and ~11.32 m above the ground, respectively. The transmitter (35.248° N/107.680° E) and the receiver (35.236° N/107.684° E) have surface elevations of ~1,220 m and ~1,219 m, respectively.

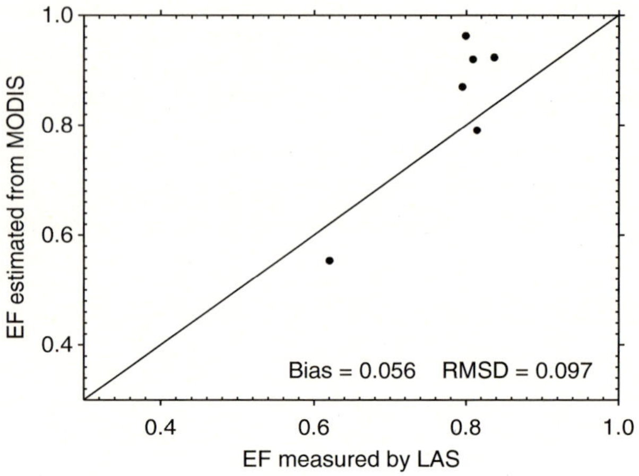

Fig. 6.10 A comparison of the estimated instantaneous evaporative fraction with the LAS measurement ($EF_{LAS} = 1 - H_{LAS}/(R_n - G)$, where $R_n - G$ is derived from the model estimate) at the Changwu station at six MODIS/Terra overpass times (DOY = 183, 199, 204, 224, 225, 226) (Reproduced from Tang et al. (2012), with permission from John Wiley and Sons)

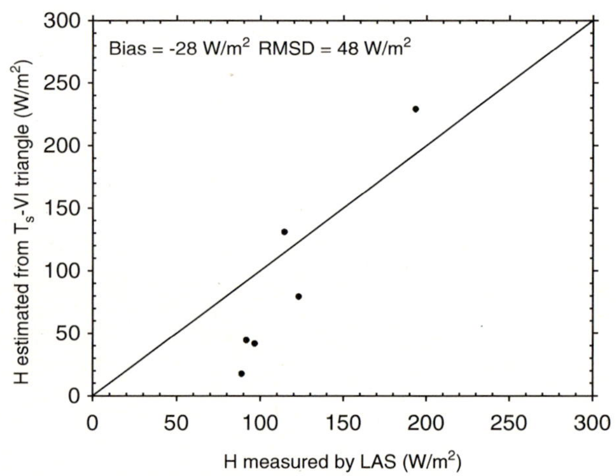

Fig. 6.11 A comparison of the estimated instantaneous sensible heat flux with the LAS measurement at six MODIS/Terra overpass times (DOY = 183, 199, 204, 224, 225, 226) (Reproduced from Tang et al. (2012), with permission from John Wiley and Sons)

Figure 6.10 compares the EF estimated from the $T_s - F_r$ triangle on 6 MODIS/Terra daytime clear-sky days with EF measurement by LAS at the Changwu station. The EF measurement is calculated by using the LAS-measured H (H_{LAS}) and the surface available energy estimated from MODIS products ($EF_{LAS} = 1\text{-}H_{LAS}/(R_n - G)$) because $R_n - G$ estimates were more spatially representative than ground-based measurements. The EF_{LAS} was slightly overestimated from MODIS/Terra products using the $T_s - F_r$ triangle space by 0.056 (RMSD = 0.097). The overestimation could be explained by the underestimation of the dry edge (nonzero surface moisture availability) in a subhumid climate and thus an overestimation of wet edge and EF.

Figure 6.11 compares the H computed from the $T_s - F_r$-derived EF and $R_n - G$ estimation at the Terra overpass time (data on six clear-sky days) with the LAS measurements. LAS-measured H was slightly underestimated by 28 W/m^2 with a

RMSD of 48 W/m^2. Because of the unavailability of the G measurement, the LE estimates could not be evaluated from the residual of surface energy budget. When the estimated $R_n - G$ was used together with the LAS-measured H to derive the LE measurement ($LE_{LAS} = R_n - G - H_{LAS}$), the LE comparison produced a RMSD of 48 W/m^2. The RMSD in the H and LE comparisons was on the similar magnitude to that reported in several studies relevant to the T_s-VI triangle method (Jiang and Islam 1999; Stisen et al. 2008).

6.5.4 Validation of the T_s-VI Triangle-, TSEB-, and SEBS-Derived H Using LAS Measurements Collected at Yucheng Site

The latitude and longitude of the study area range from 35.00 to 38.00° N and from 114.65 to 118.50° E, respectively, with a total area of ~93,000 km^2 (Fig. 6.12a). Surface elevation varies from ~0 to ~400 m above sea level. Data collected at the Yucheng Comprehensive Experimental Station (thereafter referred as the Yucheng station) were used for the evaluation and validation of the model-estimated surface energy fluxes. The Yucheng station was established in 2002 to measure continuously the long-term exchange of carbon dioxide, water vapor, and heat between land and atmosphere and is part of the Chinese terrestrial ecosystem flux research network (ChinaFlux) (http://www.chinaflux.org/). It is geographically located in the southwest of Yucheng County, Shandong Province, and was covered by crops (wheat/corn rotation), bare soil, trees, and water, as shown in the classification map from TM data (see Fig. 6.12b). Winter wheat and summer corn were rotated every year. The soil type is classified as sandy loam. The climate is subhumid and monsoon with a mean annual temperature of 13.1 °C and a mean annual precipitation of 528 mm, respectively.

In late April 2009, a LAS (Kipp & Zonen Inc.) was installed along a northeast–southwest direction at the Yucheng station to measure sensible heat flux at the MODIS pixel scale (see Fig. 6.12b). Path length between the receiver (36.8314° N/116.5717° E) and the transmitter (36.8212° N/116.5661° E) of the LAS was ~1,240 m (Figs. 6.12c, d). The heights of the receiver and the transmitter were about 8.8 m above the ground. The study period was from late April to late September 2009.

Figure 6.13 compares the instantaneous surface net radiation (R_n) estimated from the T_s-VI triangle (TVT), the TSEB, and the SEBS models with ground measurements at the Yucheng station. R_n was overestimated by the SEBS and TVT models with a positive bias (modeled measured) and underestimated by the TSEB model with a smaller bias and RMSD-value. The intermittent cloud cover caused the large deviations in the SEBS and TVT models that occurred on DOYs 132, 239, and 256, which could be verified by the ground-based fluctuating incoming solar radiation measurements surrounding satellite overpass time. The better performance

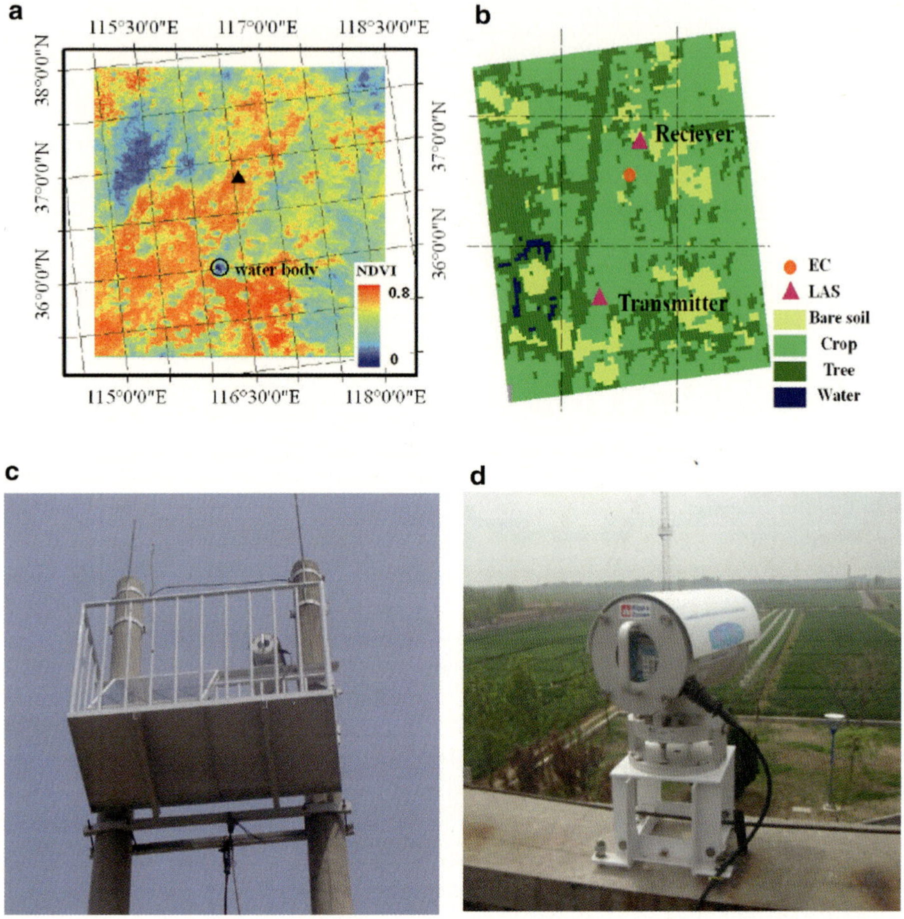

Fig. 6.12 (**a**) The spatial distribution of 1-km resolution MODIS NDVI data on DOY 123 over the study area (the *black-filled triangle* and *circle* represent the Yucheng station and Dongping Lake, respectively), (**b**) a land cover classification map from TM data at the Yucheng station with the locations of the LAS and EC instruments indicated (the grid represents the MODIS pixel on DOY 123) (Reproduced from Tang et al. (2011b), with permission from Elsevier), (**c**) a photo (taken by Chang Liu) of the transmitter of the LAS, and (**d**) a photo (taken by Chang Liu) of the receiver of the LAS

in the TSEB model was primarily contributed by the fact that local ground-based solar radiation measurements were used as input in the estimation of surface net radiation in the TSEB model. A comparison of G was not made because of the lack of multiple measurements of soil heat flux (G) to obtain a spatially representative sample and the no accounting for the heat storage above the plate.

Figure 6.14 compares the H and LE estimated from the TVT, the TSEB, and the SEBS models with LAS-estimated H and LE (derived by using H from LAS and $R_n - G$ from an average of the remotely sensed TSEB and SEBS/TVT model

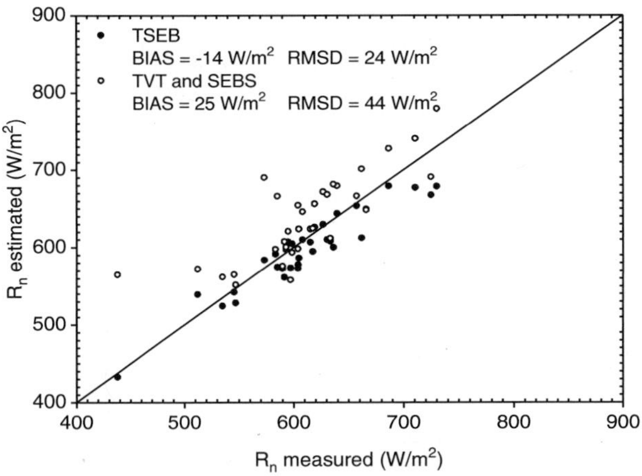

Fig. 6.13 A comparison of instantaneous surface net radiation estimated from the TSEB, the TVT, and the SEBS models using corrected MODIS surface temperature and LAI with ground measurements (Reproduced from Tang et al. (2011b), with permission from Elsevier)

estimates). Both the TSEB and SEBS models underestimated the LAS-measured *H* with relatively small negative bias and a RMSD < 45 W/m^2, whereas the TVT model significantly overestimated the LAS-measured *H* by nearly 80 W/m^2 and a large RMSD-value (Fig. 6.14a). Specifically, when LAS-measured *H* was greater than 100 W/m^2, which corresponds to the corn growth period, both the TSEB and SEBS models systematically underestimated the LAS-measured *H*; when LAS-measured *H* was lower than 100 W/m^2, which corresponds to the wheat growth period, both the TSEB and SEBS models overall overestimated the LAS-measured *H*. The TSEB model produced better LE estimates with a RMSD of ~42 W/m^2 than the SEBS model with a larger RMSD of ~56 W/m^2 when the LAS-measured LE was compared (Fig. 6.14b). The TVT model significantly underestimated the LAS-measured LE with a RMSD of ~112 W/m^2. When LAS-measured LE was less than 500 W/m^2, which occurred primarily during the corn growing season, both the TSEB and the SEBS models systematically overestimated the LAS-derived LE. When LAS measured LE was greater than 500 W/m^2, which occurred primarily during the wheat growing season, the TSEB model typically underestimated LAS-measured LE.

6.5.5 Validation of the Upscaled Daily LE Using EC Measurements Collected at Yucheng Site

The climatic and land information at the Yucheng site has been given in detail in Sect. 6.5.4. Latent heat flux data used for the validation of the upscaled daily LE were measured by an eddy covariance (EC) system which consisted of an open-path

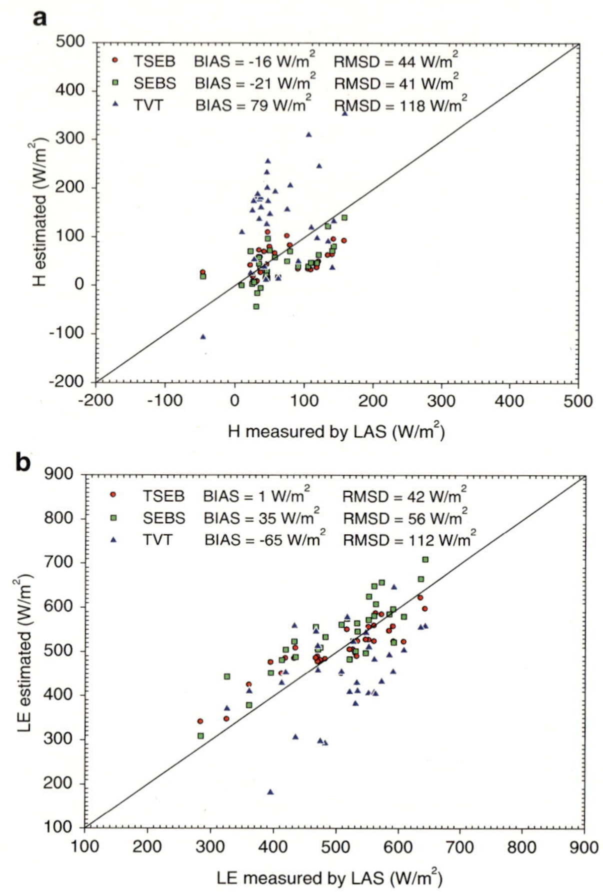

Fig. 6.14 Comparisons of (**a**) instantaneous sensible heat flux and (**b**) instantaneous latent heat flux estimated from the TVT, the SEBS, and the TSEB model with LAS measurements, respectively (LE measured by LAS is computed as the available energy averaged from the TSEB and the SEBS/TVT minus LAS-measured *H*) (Reproduced from Tang et al. (2011b), with permission from Elsevier)

CO_2/H_2O gas analyzer (model LI-7500, Licor Inc., Lincoln, Nebraska, USA) and a 3-D sonic anemometer/thermometer (model CSAT3, Campbell Scientific Inc., Logan, Utah, USA). The measurement height was elevated to ~4.3 m above the ground surface in late July or early August and lowered to ~2.9 m above the ground surface in mid- to late October each year. A series of common corrections following the work of Webb et al. (1980) and Burba and Anderson (2010) were made to obtain reliable EC-measured *H* and LE using the online flux computation and postfield data programs.

Half-hourly averaged near-surface atmospheric variables and ground-based crop height from late April 2009 to late October 2011 were collected as input to drive the N95 model (as presented in Sect. 6.2.2.2) and the daily (24-h) averaged latent heat flux measured by EC to validate the upscaled daily ET. Data quality and completeness were carefully checked through two steps. Spikes and abnormalities (less than -100 W/m^2 or greater than 700 W/m^2) in the measured *H* and LE were first removed. Days when data gaps (e.g., resulting from rainfall, instrument malfunction, or maintenance) were observed in the 30-min averaged atmospheric variables or EC-measured *H* and LE were excluded. Daily measured surface fluxes (surface net radiation, soil heat flux, sensible heat flux, and latent heat flux) were estimated by averaging the 48 half-hourly measurements in a diurnal cycle. The RE correction method, as presented in Sect. 6.5.2, was used to close the energy imbalance of the EC-measured *H* and LE for validation against the satellite-derived LE at both the instantaneous and daily scales (Tang et al. 2013c).

MODIS data used in this study include the land surface temperature and emissivity 5-min L2 Swath 1-km data set product (MOD11_L2 for the Terra satellite and MYD11_L2 for the Aqua satellite), the geolocation data set product (MOD03 and MYD03) at 1-km spatial resolution, and the Leaf Area Index (LAI) and Fractional Photosynthetically Active Radiation (FPAR) product (MOD15 and MYD15). These produces can be freely downloaded from the Land Processes Distributed Active Archive Center (LP DAAC, https://lpdaac.usgs.gov/get_data). To minimize the angular effect of the satellite view geometry, the MODIS data in this study were strictly restricted to those that had a view zenith angle less than 45°. Moreover, data in the harvest stage when senescent and dead leaves dominated were excluded in the upscaling of instantaneous ET because LAI measurements from MOD15 and MYD15 products estimate only the green leaf area of the vegetation. MODIS data at 55 Terra overpass times and 46 Aqua overpass times were finally acquired for use from late April 2009 to late October 2011.

Figure 6.15a compares the MODIS land surface temperature (LST) with that estimated from the simultaneous tower-based upwelling and reflected downwelling longwave radiation measurements. It can be seen that MODIS LST was significantly overestimated and large scatter were found at the Aqua overpass times with a positive BIAS (MODIS LST minus ground LST) of 2.1 K and an RMSE of 4.0 K. Better agreement was obtained at the Terra overpass times with a negative BIAS of −0.1 K and an RMSE of 2.9 K. These biases were probably caused by the inaccurate atmospheric correction, pixel shift or geolocation error, local irrigation, and cirrus cloud. To provide the most accurate inputs for the N95 model, the MODIS LST was corrected by using the linear least square fit, as shown in Fig. 6.15a. Moreover, to eliminate the significant underestimation of LST caused by the linear least square fit, if the difference (ground LST minus corrected MODIS LST) between ground LST and the corrected MODIS LST were larger than 1 K, the corrected MODIS LST would be replaced by the ground LST for further analysis.

Figure 6.15b compares the instantaneous surface net radiation (R_n) estimated from the N95 model using the corrected MODIS LST with ground measurements. The N95 model was shown to underestimate R_n at both the Terra and Aqua overpass times, with a negative BIAS within −22 W/m^2. A comparison with the modeled soil heat flux (*G*) was not made due to the unavailability of multiple measurements of *G* to obtain a representative spatial sample and no account for the heat storage above the plate.

Figure 6.15c compares the N95-modeled instantaneous LE using the corrected MODIS LST with the EC measurement corrected by the RE correction method. The N95-modeled *G* was used together with the R_n measurement to provide reliable

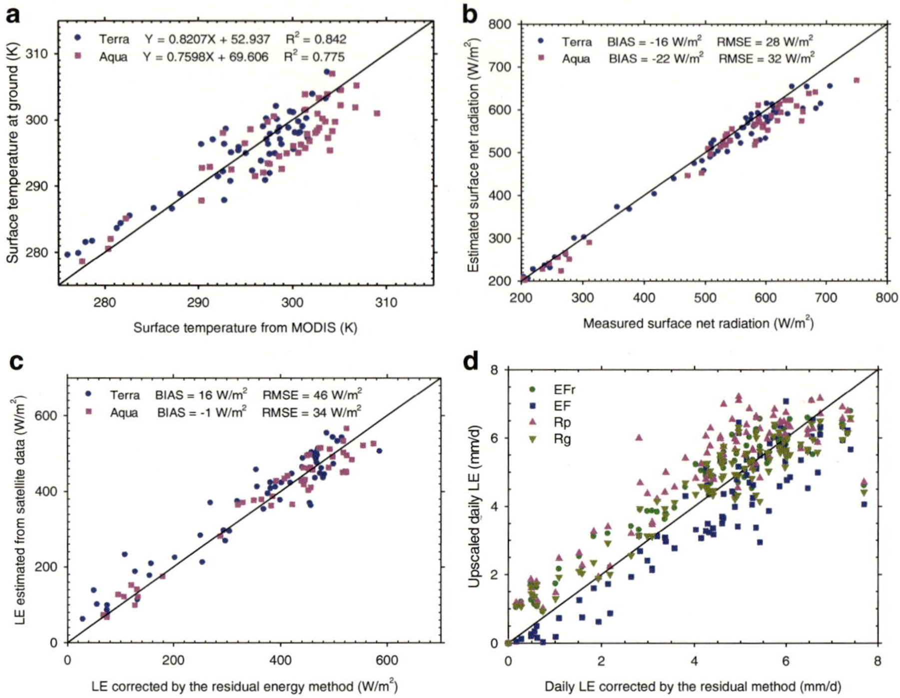

Fig. 6.15 (**a**) Comparison of MODIS land surface temperature (LST) with LST estimated from tower measurements of upwelling and reflected downwelling longwave radiation, (**b**) comparison of the N95-modeled instantaneous surface net radiation using MODIS data with ground-based measurement, (**c–d**) comparison of the N95-modeled instantaneous latent heat flux using MODIS data and the daily latent heat flux extrapolated from the model-derived ET using four upscaling methods with eddy covariance system measurements corrected by the residual energy method (Reproduced from Tang et al. (2013b), with permission from Elsevier)

pixel-averaged $R_n - G$ for the correction of EC-measured LE. At the instantaneous time scale, the model-estimated LE agreed well with the LE measurement at both the Terra and the Aqua overpass times, with a slight overestimation by 16 W/m^2 at the Terra overpass times. At the Terra overpass times, the RMSE and R^2 were 46 W/m^2 (13 % of the observed mean) and 0.916, respectively. At the Aqua overpass times, they were 34 W/m^2 (9 % of the observed mean) and 0.948, respectively. Results from this figure demonstrated the necessity of making corrections to the MODIS LST and the N95 model was able to produce accurate remote sensing instantaneous LE.

Figure 6.15d compares the daily ET converted from the model-derived instantaneous LE using four upscaling methods with the EC measurement corrected using the RE correction method. These four upscaling methods are the constant evaporative fraction (EF) method, the constant extraterrestrial solar radiation ratio (R_p) method, the constant global solar radiation ratio (R_g) method, and the constant

Table 6.4 Statistical measures of the performance of four upscaling methods in converting the N95-modeled instantaneous ET at the 101 satellite (55 MODIS/Terra and 46 MODIS/Aqua) overpass times between late April 2009 and late October 2011, with the eddy covariance system measurements corrected by the residual energy method (R_BIAS = BIAS/$\overline{O}$*100 %, R_RMSE = RMSE/$\overline{O}$*100 %, where $\overline{O}$ represents the average of the EC-measured daily latent heat flux)

	$\overline{O}$ (mm)	EF_r	EF	R_p	R_g
BIAS (mm/day)	4.1	0.4	−0.5	0.7	0.2
RMSE (mm/day)		0.8	0.9	1.1	0.8
R^2		0.902	0.871	0.851	0.905
Slope		0.78	0.84	0.79	0.78
Intercept (mm/day)		1.3	0.1	1.6	1.1
R_BIAS (%)		9	−11	18	5
R_RMSE (%)		19	22	27	19

Reproduced from Tang et al. (2013b), with permission from Elsevier

reference EF (ET_rF) method, as presented in Sect. 6.3. Daily R_n measurements rather than daily $R_n - G$ measurements were used in the constant EF upscaling method to upscale the remote sensing instantaneous ET because in practical applications daily G was generally assumed to be negligible. Table 6.4 presents the statistical measures of the performance of the four upscaling methods in converting the N95-derived instantaneous ET at the 101 satellite (55 MODIS/Terra and 46 MODIS/Aqua) overpass times. Using the R_g factor and R_p factor gave the best and worst performances, respectively. Using the EF factor underestimated the daily ET by 11 % while using the other three factors overestimated the daily ET by 5–18 %. Same magnitude of RMSE but a larger BIAS was obtained when using the EF_r factor was compared with using the R_g factor. Using the R_p factor gave a larger absolute BIAS and RMSE than using the EF factor.

6.6 Problems

In the past decades, a variety of methods ranging from the simplified empirical equation to the physics-based single- and dual-source energy balance models have been developed and great progress made to make use of the remote sensing data to estimate regional ET. However, there also exist several problems that remain unresolved and need further investigation, which are mainly associated with the ET retrieval under the cloudy and nighttime conditions, the ET retrieval over snow- or ice-covered surface, the spatial extrapolation of the meteorological measurements from the limited number of stations/sites, the temporal upscaling of instantaneous ET under cloudy conditions, and the validation of the retrieved ET at the satellite pixel scale.

6.6.1 Instantaneity of the Remote Sensing Data

One of the specialties (or weakness) of optical remote sensing technique is that it only provides instantaneous clear-sky surface information at the satellite overpass time. Land surface characteristic parameters retrieved from the remote sensing data are thus at an instantaneous scale. Based on these remote sensing parameters, most models have estimated regional ET at the same time scale. To meet the requirement of long-time (e.g., daily, weekly, and monthly) integrated ET in the practical applications, the instantaneous ET estimated from optical remote sensing data at the clear-sky satellite overpass time needs to be further temporally upscaled and a number of upscaling methods have therefore been developed. However, this upscaling is often constrained by the presence of cloud at the satellite overpass time. Moreover, the upscaling methods seem to perform better over golden clear-sky days than over the partly cloudy days. The duration and amount of the cloud cover may significantly deteriorate the performance of some of the existing upscaling methods on the partly cloudy days. Though microwave remote sensing can penetrate the cloud and atmosphere to the earth surface, its coarse spatial resolution is perhaps the largest limitation in the retrieval of surface characteristic parameters and ET. The synergy of the high-resolution clear-sky optical data and the coarse-resolution all-weather microwave data should be comprehensively explored in order to get reliable long-time integrated ET in the further work.

6.6.2 Limitation of the Remote Sensing ET Models

Though a variety of remote sensing ET models have been developed in the past decades as reviewed in the aforementioned sections, each model developed has its own advantage and weakness and is generally applied with success under limited surface conditions. Because of the strong heterogeneity of the earth surface, for example, resulting from the differences in the land surface characteristics, in the climate, and in the terrain, no single- or dual-source energy balance model can now be used elsewhere in the world without any calibration or modification.

It needs also to be highlighted that almost none of the single- and dual-source energy balance models are applied during nighttime and most methods have poorly treated the contribution of nocturnal ET to the daily total with the assumption that stomata close during night to prevent water loss in the absence of light. However, past studies over a wide range of species and climate conditions have shown that nocturnal transpiration typically accounts for 10–30 % of the total daily transpiration flux (Novick et al. 2009). Wind speed and vapor pressure deficit are found to be the primary drivers of nocturnal ET in two forested ecosystems when the magnitude and variability of nocturnal ET is explored using multiple years of eddy covariance measurements. Therefore, an important source of error may arise in the long-term ET modeling if ET methods do not effectively account for the contribution of nocturnal transpiration.

Rare studies have been conducted to quantify the evaporation (sublimation) from the snow-covered surface from a remote sensing point of view. ET from the snow-covered surface consists of the evaporation from the land and vegetation and the sublimation from the blowing snow (Vinukollu et al. 2011) and is governed by several factors, for example, aerodynamic resistance, vapor pressure deficit, and radiation (Lundberg and Halldin 2001). Based on reliable, good-quality data on sublimation and precipitation collected over a 90-day period, Suzuki and Nakai (2008) found that 26 % of the total accumulated snowfall was lost by sublimation in a coniferous forest plantation. Because ice and snow pack are two important sources of water in the higher latitude, future studies are recommended to address this issue that has not been well resolved.

6.6.3 Uncertainty of the Retrieved Land Surface Parameters

Radiance measured by the satellite sensor primarily comes from the surface emission and reflection that are disturbed/modulated by the atmosphere absorption, diffusion, and emission. To retrieve land surface characteristic parameters from the satellite radiance, the effect of atmosphere needs to be first corrected for to convert the at-sensor radiance to the at-surface parameters through physical models. The retrieved surface parameters, influenced by the vegetation architecture and solar zenith angle, are also significantly subject to the observational angular effect due to the differing amounts of soil and vegetation in the field of view, especially over heterogeneous surfaces (Norman et al. 1995). Though great progress has been made on quantitatively retrieving land surface parameters from remotely sensed data, accuracy of some of these parameters (e.g., T_s) still needs to be further improved because of the high sensitivity of most ET models. For example, a sensitivity analysis following the work of Zhan et al. (1996) has shown that a 1 K variation of T_s in the N95 dual-source model can incur a maximum bias of 29 and 30 W/m^2 with the average of 17 and 19 W/m^2 in sensible and latent heat fluxes, respectively (Tang 2011). For the single-source SEBS model, maximum (mean) bias of 56 (29) and 61 (34) W/m^2 has been obtained when an uncertainty of 1 K in surface temperature is varied, respectively. An error of 1 K in surface temperature at dry or wet edges in the surface temperature versus vegetation index triangle space, depending on different atmospheric forcing and fractional vegetation cover, will incur an error of less than 0.033 in EF ($T_{\max,i} - T_{\min,i} \approx 30$ K). The difference of surface temperature obtained at nadir and at 60 degrees in zenith angle can reach as large as 5 °C (Anderson et al. 1997), implying that if the angular effect is neglected, a large error would be generated in the ET estimate. Methodologies should be developed in the future work to accurately estimate the component temperatures of vegetation and soil from multispectral and multiangular satellite measurements.

6.6.4 Lack of the Near-Surface Meteorological Measurements

In most remote sensing ET models (e.g., the single- and dual-source energy balance models), near-surface meteorological measurements (e.g., air temperature, wind speed, vapor pressure) at satellite pixel scale are indispensable as input and their accuracies have significant impacts on the model output. For instance, sensitivity analysis by Tang (2011) shows that an uncertainty of 0.5 K in air temperature measurements can approximately lead to maximum bias of ~16 W/m^2 in the N95 dual-source model and 28 W/m^2 in the single-source SEBS model. Because of the limited number of weather stations on the earth, one needs to spatially extrapolate the sparse, discrete, irregular, and essentially point-scale meteorological measurements to satellite pixel scale. However, this extrapolation is subject to the complicated climate and terrain conditions. Approaches to improving the spatial extrapolation may include (1) developing physics-based and dynamic-feedback methods through the use of remote sensing data or atmospheric reanalysis data at high spatial resolution and (2) integrating remote sensing ET models with atmospheric general circulation models or numerical weather forecast models (namely, the data assimilation method). A big challenge in the development of remote sensing ET model is to develop a new parameterization of land surface ET with only land surface variables and parameters directly or indirectly derived from satellite data.

6.6.5 Spatial-Scaling Effects

Spatial-scaling effect in retrieving surface sensible heat flux and latent heat flux is primarily reflected by whether functions of parameters obtained at one scale can be used at another scale (Carlson et al. 1995). Because of the surface heterogeneity and the differing interactions between land and atmosphere at different spatial scales, remote sensing ET models applicable at local scale may be inappropriate for use at regional and global scales. Note that there is a difference in the spatial resolutions of thermal infrared bands and visible/near-infrared bands for a given satellite sensor and also a spatial-resolution difference in the data acquired between different sensors, which may have a significant impact on the current ET estimates. These scale differences, together with the surface heterogeneity, urge a definition and an interpretation of surface parameters independent of the scale. In the past investigation of the spatial-scaling effect, almost all studies have assumed that the remote sensing ET model itself is independent of the spatial scale and the spatial-scaling effect is a result of the scale-dependent inputs of the ET model. To resolve the problems relevant to the spatial-scaling effect, it is recommended to further develop the scaling theory with the fusion of multiscale remote sensing observations (McCabe and Wood 2006; Gowda et al. 2008).

6.6.6 Lack of the Ground-Truth of ET at Satellite Pixel Scale

At this moment, the ET estimates from the remote sensing models are validated primarily with the ground-truth measurements by several conventional techniques/instruments, for example, the Bowen ratio system, the eddy covariance (EC) system, and the weighing lysimeter. A lysimeter is able to directly measure the water lost by the ET over a vegetated surface and its measurement can therefore be used as a standard for the evaluation of the ET estimates. However, the lysimeter measurement is essentially made at the point scale (source area $<10^1$ m^2) and thus cannot be used as the ground-truth over the heterogeneous surface at the satellite pixel scale. The EC technique, assuming the same transport coefficients of water vapor, CO_2, and heat, is particularly applicable over rough surfaces that are characterized by high coefficients of turbulent exchange and has overtaken the Bowen ratio system as the most preferable meteorological technique to gather the ground-truth of ET in the past few decades (Farahani et al. 2007). The EC system and the Bowen ratio system can measure areal ET that originates from an upwind fetch of 10^1–10^3 m (Campbell and Norman 1998), depending on the atmospheric and surface conditions and the sensor height. Though measurements from the EC system and from the Bowen ratio system have been widely applied to evaluate the ET estimates, several uncertainties/limitations still remain to be further investigated. For instance, the accuracy of Bowen ratio-measured LE is found to depend on the range of Bowen ratio values (Angus and Watts 1984) and decreases when soil water becomes less available (Kalma and Jupp 1990). Measurements from the Bowen ratio system and from the large weighing lysimeter can differ by up to 29 % over irrigated alfalfa surface under advective conditions (Blad and Rosenberg 1974; Todd et al. 2000). For the EC system, the energy imbalance can reach up to 20 % over strongly evaporating surfaces even under nonadvective conditions (Gowda et al. 2008). The instrument error and atmospheric stability can contribute to large errors in nighttime measurements from the EC system under low wind-speed stable conditions (Gurney and Camillo 1984; Shuttleworth 2007).

At present, over 500 tower sites are in operation globally in the FLUXNET framework to monitor the long-term exchange of CO_2, water vapor, and energy between terrestrial ecosystems and the atmosphere. Flux measurements from this number of FLUXNET sites are however far from meeting the increasing demand of reliable ground-truth of ET at satellite pixel scale in the scientific community and practical applications because of the diversity of the climates and terrains. Although it may be feasible and reasonable to use these traditional measurements that have limited spatial representativeness to validate the high-resolution ET estimates from remote sensing models over uniform areas, problems will arise over heterogeneous surfaces (Carlson et al. 1995).

The newly developed (extra) large aperture scintillometer (XLAS, LAS) provides the unique possibility of measuring the sensible heat flux averaged over horizontal distances comparable to the grid size of numerical models and satellite pixels (Kohsiek et al. 2002) and thus can be used to validate the remote sensing ET at much larger scales (Hoedjes et al. 2002, 2007; Meijninger et al. 2002; Hemakumara et al. 2003). The limitation lies in the saturation of scintillation and may be overcome by using large, incoherent transmitter and receiver apertures or a longer wavelength (Kohsiek et al. 2006).

References

Allen, R. G., Morse, A., Tasumi, M., Bastiaanssen, W., Kramber, W., & Anderson, H. (2001). Evapotranspiration from Landsat (SEBAL) for water rights management and compliance with multi-state water compacts. IEEE *Geoscience and Remote Sensing Symposium, 2*, 830–833, University of New South Wales, Sydney, Australia.

Allen, R. G., Tasumi, M., & Morse, A. (2005a, February 8–10). Satellite-based evapotranspiration by METRIC and Landsat for western states water management. *US Bureau of reclamation evapotranspiration workshop*, Ft. Collins.

Allen, R. G., Tasumi, M., & Trezza, R. (2005b). *METRIC: Mapping evapotranspiration at high resolution–applications manual for Landsat satellite imagery*. Kimberly: University of Idaho.

Allen, R. G., Tasumi, M., & Trezza, R. (2007). Satellite-based energy balance for mapping evapotranspiration with internalized calibration (METRIC)-model. *Journal of Irrigation and Drainage Engineering, 133*(4), 380–394.

Anderson, J. L. (2001). An ensemble adjustment Kalman filter for data assimilation. *Monthly Weather Review, 129*(12), 2884–2903.

Anderson, M. C., Norman, J. M., Diak, G. R., Kustas, W. P., & Mecikalski, J. R. (1997). A two-source time-integrated model for estimating surface fluxes using thermal infrared remote sensing. *Remote Sensing of Environment, 60*(2), 195–216.

Anderson, M. C., Norman, J. M., Kustas, W. P., Li, F., Prueger, J. H., & Mecikalski, J. R. (2005). Effects of vegetation clumping on two-source model estimates of surface energy fluxes from an agricultural landscape during SMACEX. *Journal of Hydrometeorology, 6*(6), 892–909.

Angus, D. E., & Watts, P. J. (1984). Evapotranspiration- How good is the Bowen Ratio Method. *Agricultural Water Management, 8*(1–3), 133–150.

Bastiaanssen, W. G. M. (1995). *Regionalization of surface flux densities and moisture indicators in composite terrain: A remote sensing approach under clear skies in Mediterranean climates* (p. 273). Wageningen: Wageningen Agricultural University. ISBN 90-5485-465-0.

Bastiaanssen, W. G. M. (2000). SEBAL-based sensible and latent heat fluxes in the irrigated Gediz Basin, Turkey. *Journal of Hydrology, 229*(1), 87–100.

Bastiaanssen, W. G. M., Menenti, M., Feddes, R. A., & Holtslag, A. A. M. (1998a). A remote sensing surface energy balance algorithm for land (SEBAL). 1. Formulation. *Journal of Hydrology, 212*, 198–212.

Bastiaanssen, W. G. M., Pelgrum, H., Wang, J., Ma, Y., Moreno, J. F., Roerink, G. J., & Van der Wal, T. (1998b). A remote sensing surface energy balance algorithm for land (SEBAL). Part 2: Validation. *Journal of Hydrology, 212*, 213–229.

Bastiaanssen, W. G. M., Noordman, E. J. M., Pelgrum, H., Davids, G., Thoreson, B. P., & Allen, R. G. (2005). SEBAL model with remotely sensed data to improve water-resources management under actual field conditions. *Journal of Irrigation and Drainage Engineering, 131*(1), 85–93.

Bastiaanssen, W., Thoreson, B., Clark, B., & Davids, G. (2010). Discussion of "Application of SEBAL model for mapping evapotranspiration and estimating surface energy fluxes in south-central Nebraska" by Ramesh K. Singh, Ayse Irmak, Suat Irmak, and Derrel L. Martin. *Journal of Irrigation and Drainage Engineering, 136*(4), 282–283.

Batra, N., Islam, S., Venturini, V., Bisht, G., & Jiang, L. E. (2006). Estimation and comparison of evapotranspiration from MODIS and AVHRR sensors for clear sky days over the Southern Great Plains. *Remote Sensing of Environment, 103*(1), 1–15.

Blad, B. L., & Rosenberg, N. J. (1974). Lysimetric calibration of the Bowen ratio-energy balance method for evapotranspiration estimation in the Central Great Plains. *Journal of Applied Meteorology, 13*, 227–236.

Blad, B. L., & Rosenberg, N. J. (1976). Measurement of crop temperature by leaf thermocouple, infrared thermometry and remotely sensed thermal imagery. *Agronomy Journal, 68*(4), 635–641.

Boni, G., Entekhabi, D., & Castelli, F. (2001). Land data assimilation with satellite measurements for the estimation of surface energy balance components and surface control on evaporation. *Water Resources Research, 37*(6), 1713–1722.

Bouttier, F., & Courtier, P. (2002). *Data assimilation concepts and methods March 1999*. ECMWF lecture notes European Centre for Medium-Range Weather Forecasts, Reading, England, 59 pp.

Brutsaert, W. (1975). On a derivable formula for long-wave radiation from clear skies. *Water Resources Research, 11*(5), 742–744.

Brutsaert, W. (1982). *Evaporation into the atmosphere: Theory, history, and applications* (p. 299). Dordrecht: Reidel Publishing Co.

Brutsaert, W. (1986). Catchment-scale evaporation and the atmospheric boundary layer. *Water Resources Research, 22*(9S), 39S–45S.

Brutsaert, W., & Sugita, M. (1992). Application of self-preservation in the diurnal evolution of the surface energy budget to determine daily evaporation. *Journal of Geophysical Research, 97*, 18377–18382.

Burba, G., & Anderson, D. (2010). *A brief practical guide to eddy covariance flux measurements: Principles and workflow examples for scientific and industrial applications*. Lincoln: LI-COR Biosciences. 212 pp.

Businger, J., Wyngaard, J. C., Izumi, Y., & Bradley, E. F. (1971). Flux-profile relationships in the atmospheric surface layer. *Journal of the Atmospheric Sciences, 28*(2), 181–189.

Cammalleri, C., Anderson, M. C., Ciraolo, G., D'Urso, G., Kustas, W. P., La Loggia, G., & Minacapilli, M. (2012). Applications of a remote sensing-based two-source energy balance algorithm for mapping surface fluxes without in situ air temperature observations. *Remote Sensing of Environment, 124*, 502–515.

Campbell, G. S., & Norman, J. M. (1998). *An introduction to environmental biophysics*. New York: Springer.

Caparrini, F., Castelli, F., & Entekhabi, D. (2003). Mapping of land-atmosphere heat fluxes and surface parameters with remote sensing data. *Boundary-Layer Meteorology, 107*(3), 605–633.

Caparrini, F., Castelli, F., & Entekhabi, D. (2004). Variational estimation of soil and vegetation turbulent transfer and heat flux parameters from sequences of multisensor imagery. *Water Resources Research, 40*(12).

Carlson, T. (2007). An overview of the "triangle method" for estimating surface evapotranspiration and soil moisture from satellite imagery. *Sensors, 7*(8), 1612–1629.

Carlson, T. N., Capehart, W. J., & Gillies, R. R. (1995). A new look at the simplified method for remote sensing of daily evapotranspiration. *Remote Sensing of Environment, 54*(2), 161–167.

Choudhury, B. J. (1989). Estimating evaporation and carbon assimilation using infrared temperature data: Vistas in modeling. In G. Asrar (Ed.), *Theory and applications of optical remote sensing* (pp. 628–690). New York: Wiley.

Choudhury, B. J. (1994). Synergism of multispectral satellite observations for estimating regional land surface evaporation. *Remote Sensing of Environment, 49*(3), 264–274.

Choudhury, B. J., Reginato, R. J., & Idso, S. B. (1986). An analysis of infrared temperature observations over wheat and calculation of latent heat flux. *Agricultural and Forest Meteorology, 37*(1), 75–88.

Cleugh, H. A., Leuning, R., Mu, Q., & Running, S. W. (2007). Regional evaporation estimates from flux tower and MODIS satellite data. *Remote Sensing of Environment, 106*(3), 285–304.

Compaoré, H., Hendrickx, J. M., Hong, S. H., Friesen, J., van de Giesen, N. C., Rodgers, C., Szarzynski, J., & Vlek, P. L. (2008). Evaporation mapping at two scales using optical imagery in the White Volta Basin, Upper East Ghana. *Physics and Chemistry of the Earth, Parts A/B/C, 33*(1), 127–140.

Courtier, P., Andersson, E., Heckley, W., Vasiljevic, D., Hamrud, M., Hollingsworth, A., Rabier, F., Fisher, M., & Pailleux, J. (1998). The ECMWF implementation of three-dimensional variational assimilation (3D-Var). I: Formulation. *Quarterly Journal of the Royal Meteorological Society, 124*(550), 1783–1807.

Crow, W. T., & Kustas, W. P. (2005). Utility of assimilating surface radiometric temperature observations for evaporative fraction and heat transfer coefficient retrieval. *Boundary-Layer Meteorology, 115*(1), 105–130.

Daughtry, C. S. T., Kustas, W. P., Moran, M. S., Pinter, P. J., Jackson, R. D., Brown, P. W., Nichols, W. D., & Gay, L. W. (1990). Spectral estimates of net radiation and soil heat flux. *Remote Sensing of Environment, 32*(2), 111–124.

Diak, G. R., Rabin, R. M., Gallo, K. P., & Neale, C. M. (1995). Regional-scale comparisons of NDVI, soil moisture indices from surface and microwave data and surface energy budgets evaluated from satellite and in-situ data. *Remote Sensing Reviews, 12*, 355–382.

Engman, E. T., & Gurney, R. J. (1991). *Remote sensing in hydrology*. London: Chapman and Hall Ltd.

Farahani, H. J., Howell, T. A., Shuttleworth, W. J., & Bausch, W. C. (2007). Evapotranspiration: Progress in measurement and modeling in agriculture. *Transactions of the ASABE, 50*(5), 1627–1638.

Fisher, J. B., Tu, K. P., & Baldocchi, D. D. (2008). Global estimates of the land–atmosphere water flux based on monthly AVHRR and ISLSCP-II data, validated at 16 FLUXNET sites. *Remote Sensing of Environment, 112*(3), 901–919.

Gao, W., Coulter, R. L., Lesht, B. M., Qiu, J., & Wesely, M. L. (1998). Estimating clear-sky regional surface fluxes in the southern Great Plains Atmospheric Radiation Measurement site with ground measurements and satellite observations. *Journal of Applied Meteorology, 37*(1), 5–22.

Garratt, J. R., & Hicks, B. B. (1973). Momentum, heat and water vapour transfer to and from natural and artificial surfaces. *Quarterly Journal of the Royal Meteorological Society, 99*(422), 680–687.

Gebremichael, M., Wang, J., & Sammis, T. W. (2010). Dependence of remote sensing evapotranspiration algorithm on spatial resolution. *Atmospheric Research, 96*(4), 489–495.

Gentine, P., Entekhabi, D., Chehbouni, A., Boulet, G., & Duchemin, B. (2007). Analysis of evaporative fraction diurnal behavior. *Agricultural and Forest Meteorology, 143*(1–2), 13–29.

Gillies, R. R., & Carlson, T. N. (1995). Thermal remote sensing of surface soil water content with partial vegetation cover for incorporation into climate models. *Journal of Applied Meteorology, 34*(4), 745–756.

Gokmen, M., Vekerdy, Z., Verhoef, A., Verhoef, W., Batelaan, O., & van der Tol, C. (2012). Integration of soil moisture in SEBS for improving evapotranspiration estimation under water stress conditions. *Remote Sensing of Environment, 121*, 261–274.

Goward, S. N., Cruickshanks, G. D., & Hope, A. S. (1985). Observed relation between thermal emission and reflected spectral radiance of a complex vegetated landscape. *Remote Sensing of Environment, 18*(2), 137–146.

Gowda, P. H., Chavez, J. L., Colaizzi, P. D., Evett, S. R., Howell, T. A., & Tolk, J. A. (2008). ET mapping for agricultural water management: Present status and challenges. *Irrigation Science, 26*(3), 223–237.

Gurney, R. J., & Camillo, P. J. (1984). Modelling daily evapotranspiration using remotely sensed data. *Journal of Hydrology, 69*(1), 305–324.

Hatfield, J. L. (1983). Evapotranspiration obtained from remote sensing methods. *Advances in Irrigation, 2*, 396–416.

Hatfield, J. L., Perrier, A., & Jackson, R. D. (1983). Estimation of evapotranspiration at one time-of-day using remotely sensed surface temperatures. *Agricultural Water Management, 7*(1), 341–350.

Hemakumara, H. M., Chandrapala, L., & Moene, A. F. (2003). Evapotranspiration fluxes over mixed vegetation areas measured from large aperture scintillometer. *Agricultural Water Management, 58*(2), 109–122.

Hoedjes, J. C. B., Zuurbier, R. M., & Watts, C. J. (2002). Large aperture scintillometer used over a homogeneous irrigated area, partly affected by regional advection. *Boundary-Layer Meteorology, 105*(1), 99–117.

Hoedjes, J. C. B., Chehbouni, A., Ezzahar, J., Escadafal, R., & De Bruin, H. A. R. (2007). Comparison of large aperture scintillometer and eddy covariance measurements: Can thermal infrared data be used to capture footprint-induced differences? *Journal of Hydrometeorology, 8*(2), 144–159.

Hong, S. H., Hendrickx, J. M., & Borchers, B. (2009). Up-scaling of SEBAL derived evapotranspiration maps from Landsat (30 m) to MODIS (250 m) scale. *Journal of Hydrology, 370*(1), 122–138.

Hope, A. S. (1988). Estimation of wheat canopy resistance using combined remotely sensed spectral reflectance and thermal observations. *Remote Sensing of Environment, 24*(2), 369–383.

Huang, C., Li, X., Lu, L., & Gu, J. (2008). Experiments of one-dimensional soil moisture assimilation system based on ensemble Kalman filter. *Remote Sensing of Environment, 112*(3), 888–900.

Idso, S. B., Jackson, R. D., & Reginato, R. J. (1975). Estimating evaporation: A technique adaptable to remote sensing. *Science, 189*(4207), 991–992.

Idso, S. B., Jackson, R. D., Pinter, P. J., Reginato, R. J., & Hatfield, J. L. (1981). Normalizing the stress-degree-day parameter for environmental variability. *Agricultural Meteorology, 24*, 45–55.

Jackson, R. D. (1985). Evaluating evapotranspiration at local and regional scales. *Proceedings of the IEEE, 73*(6), 1086–1096.

Jackson, R. D., Reginato, R. J., & Idso, S. B. (1977). Wheat canopy temperature: a practical tool for evaluating water requirements. *Water Resources Research, 13*(3), 651–656.

Jackson, R. D., Idso, S. B., Reginato, R. J., & Pinter, P. J., Jr. (1981). Canopy temperature as a crop water stress indicator. *Water Resources Research, 17*(4), 1133–1138.

Jackson, R. D., Kustas, W. P., & Choudhury, B. J. (1988). A reexamination of the crop water stress index. *Irrigation Science, 9*(4), 309–317.

Jarvis, P. G. (1976). The interpretation of the variations in leaf water potential and stomatal conductance found in canopies in the field. *Philosophical Transactions of the Royal Society of London. B, Biological Sciences, 273*(927), 593–610.

Jia, L., Su, Z., van den Hurk, B., Menenti, M., Moene, A., De Bruin, H. A., Javier Baselga Yrisarry, J., Ibanez, M., & Cuesta, A. (2003). Estimation of sensible heat flux using the Surface Energy Balance System (SEBS) and ATSR measurements. *Physics and Chemistry of the Earth, Parts A/B/C, 28*(1), 75–88.

Jiang, L., & Islam, S. (1999). A methodology for estimation of surface evapotranspiration over large areas using remote sensing observations. *Geophysical Research Letters, 26*(17), 2773–2776.

Jiang, L., & Islam, S. (2001). Estimation of surface evaporation map over southern Great Plains using remote sensing data. *Water Resources Research, 37*(2), 329–340.

Jiang, L., & Islam, S. (2003). An intercomparison of regional latent heat flux estimation using remote sensing data. *International Journal of Remote Sensing, 24*(11), 2221–2236.

Kalma, J. D., & Jupp, D. L. B. (1990). Estimating evaporation from pasture using infrared thermometry: Evaluation of a one-layer resistance model. *Agricultural and Forest Meteorology, 51*(3), 223–246.

Kim, H. W., Hwang, K., Mu, Q., Lee, S. O., & Choi, M. (2012). Validation of MODIS 16 global terrestrial evapotranspiration products in various climates and land cover types in Asia. *KSCE Journal of Civil Engineering, 16*(2), 229–238.

Kohsiek, W., Meijninger, W. M. L., Moene, A. F., Heusinkveld, B. G., Hartogensis, O. K., Hillen, W. C. A. M., & De Bruin, H. A. R. (2002). An extra large aperture scintillometer for long range applications. *Boundary-Layer Meteorology, 105*(1), 119–127.

Kohsiek, W., Meijninger, W. M. L., DeBruin, H. A. R., & Beyrich, F. (2006). Saturation of the large aperture scintillometer. *Boundary-Layer Meteorology, 121*(1), 111–126.

Kumar, S. V., Reichle, R. H., Peters-Lidard, C. D., Koster, R. D., Zhan, X., Crow, W. T., Eylander, J. B., & Houser, P. R. (2008). A land surface data assimilation framework using the land information system: Description and applications. *Advances in Water Resources, 31*(11), 1419–1432.

Kustas, W. P., & Daughtry, C. S. (1990). Estimation of the soil heat flux/net radiation ratio from spectral data. *Agricultural and Forest Meteorology, 49*(3), 205–223.

Kustas, W. P., & Norman, J. M. (1996). Use of remote sensing for evapotranspiration monitoring over land surfaces. *Hydrological Sciences Journal, 41*(4), 495–516.

Kustas, W. P., & Norman, J. M. (1997). A two-source approach for estimating turbulent fluxes using multiple angle thermal infrared observations. *Water Resources Research, 33*(6), 1495–1508.

Kustas, W. P., & Norman, J. M. (1999). Evaluation of soil and vegetation heat flux predictions using a simple two-source model with radiometric temperatures for partial canopy cover. *Agricultural and Forest Meteorology, 94*(1), 13–29.

Kustas, W. P., Choudhury, B. J., Moran, M. S., Reginato, R. J., Jackson, R. D., Gay, L. W., & Weaver, H. L. (1989). Determination of sensible heat flux over sparse canopy using thermal infrared data. *Agricultural and Forest Meteorology, 44*(3), 197–216.

Kustas, W. P., Jackson, T. J., Schmugge, T. J., Parry, R., Goodrich, D. C., Amer, S. A., Bach, L. B., Keefer, T. O., Weltz, M. A., Moran, M. S., Clarke, T. R., Pinter, P. J., Jr., Claassen, H., Riggs, A. C., Stannard, D. I., Hipps, L. E., Swiatek, E., Blanford, J. H., Shutko, A. M., Doraiswamy, P. C., Daughtry, C. S. T., Perry, E. M., Nichols, W. D., Pinker, R. T., Huete, A. R., Qi, J., Van Leeuwen, J. D., Humes, K. S., Washburne, J., Chehbouni, A., Vidal, A., Gellman, D. I., Clements, W. E., Grant, T. A., Dubois, P., & van Zyl, J. (1991). An interdisciplinary field study of the energy and water fluxes in the atmosphere–biosphere system over semiarid rangelands: Description and some preliminary results. *Bulletin of the American Meteorological Society, 72*(11), 1683–1705.

Kustas, W. P., Norman, J. M., Anderson, M. C., & French, A. N. (2003). Estimating subpixel surface temperatures and energy fluxes from the vegetation index–radiometric temperature relationship. *Remote Sensing of Environment, 85*(4), 429–440.

Lhomme, J.-P., & Elguero, E. (1999). Examination of evaporative fraction diurnal behavior using a soil-vegetation model coupled with a mixed-layer model. *Hydrology and Earth System Sciences, 3*(2), 259–270.

Li, F., & Lyons, T. J. (1999). Estimation of regional evapotranspiration through remote sensing. *Journal of Applied Meteorology, 38*(11), 1644–1654.

Li, F., Kustas, W. P., Prueger, J. H., Neale, C. M., & Jackson, T. J. (2005). Utility of remote sensing-based two-source energy balance model under low-and high-vegetation cover conditions. *Journal of Hydrometeorology, 6*(6), 878–891.

Li, Z.-L., Tang, R., Wan, Z., Bi, Y., Zhou, C., Tang, B., Yan, G., & Zhang, X. (2009). A review of current methodologies for regional evapotranspiration estimation from remotely sensed data. *Sensors, 9*(5), 3801–3853.

Liang, S. (2005). *Quantitative remote sensing of land surfaces* (Vol. 23). Hoboken: Wiley-Interscience. 560 pp.

Long, D., & Singh, V. P. (2012). A modified surface energy balance algorithm for land (M-SEBAL) based on a trapezoidal framework. *Water Resources Research, 48*(2).

Long, D., & Singh, V. P. (2013). Assessing the impact of end-member selection on the accuracy of satellite-based spatial variability models for actual evapotranspiration estimation. *Water Resources Research, 49*(5), 2601–2618. doi:10.1002/wrcr. 20208.

Long, D., Singh, V. P., & Li, Z.-L. (2011). How sensitive is SEBAL to changes in input variables, domain size and satellite sensor? *Journal of Geophysical Research, 116*(D21), D21107.

Lundberg, A., & Halldin, S. (2001). Snow interception evaporation. Review of measurement techniques, processes, and models. *Theoretical and Applied Climatology, 70*(1–4), 117–133.
Luquet, D., Vidal, A., Dauzat, J., Begue, A., Olioso, A., & Clouvel, P. (2004). Using directional TIR measurements and 3D simulations to assess the limitations and opportunities of water stress indices. *Remote Sensing of Environment, 90*(1), 53–62.
Margulis, S. A., & Entekhabi, D. (2003). Variational assimilation of radiometric surface temperature and reference-level micrometeorology into a model of the atmospheric boundary layer and land surface. *Monthly Weather Review, 131*(7), 1272–1288.
Margulis, S. A., McLaughlin, D., Entekhabi, D., & Dunne, S. (2002). Land data assimilation and estimation of soil moisture using measurements from the Southern Great Plains 1997 Field Experiment. *Water Resources Research, 38*(12), 35-1.
Margulis, S. A., Kim, J., & Hogue, T. (2005). A comparison of the triangle retrieval and variational data assimilation methods for surface turbulent flux estimation. *Journal of Hydrometeorology, 6*(6), 1063–1072.
Massman, W. J. (1999). A model study of $kB_H{}^{-1}$ for vegetated surfaces using 'localized near-field' Lagrangian theory. *Journal of Hydrology, 223*, 27–43.
Mauser, W., & Schädlich, S. (1998). Modelling the spatial distribution of evapotranspiration on different scales using remote sensing data. *Journal of Hydrology, 212*, 250–267.
McCabe, M. F., & Wood, E. F. (2006). Scale influences on the remote estimation of evapotranspiration using multiple satellite sensors. *Remote Sensing of Environment, 105*(4), 271–285.
McLaughlin, D. (1995). Recent developments in hydrologic data assimilation. *Reviews of Geophysics-Richmond Virginia Then Washington, 33*, 977–984.
McLaughlin, D., Zhou, Y., Entekhabi, D., & Chatdarong, V. (2006). Computational issues for large-scale land surface data assimilation problems. *Journal of Hydrometeorology, 7*(3), 494–510.
Mecikalski, J. R., Diak, G. R., Anderson, M. C., & Norman, J. M. (1999). Estimating fluxes on continental scales using remotely sensed data in an atmospheric-land exchange model. *Journal of Applied Meteorology, 38*(9), 1352–1369.
Meijninger, W. M. L., Hartogensis, O. K., Kohsiek, W., Hoedjes, J. C. B., Zuurbier, R. M., & De Bruin, H. A. R. (2002). Determination of area-averaged sensible heat fluxes with a large aperture scintillometer over a heterogeneous surface–Flevoland field experiment. *Boundary-Layer Meteorology, 105*(1), 37–62.
Menenti, M., & Choudhury, B. J. (1993). Parameterization of land surface evaporation by means of location dependent potential evaporation and surface temperature range. In H. J. Bolle, R. A. Feddes, & J. D. Kalma (Eds.), *Exchange processes at the land surface for a range of space and time scales*. Yokohama: IAHS.
Monin, A. S., & Obukhov, A. (1954). Basic laws of turbulent mixing in the surface layer of the atmosphere. *Trudy Geofizicheskogo Instituta, Akademiya Nauk SSSR, 24*, 163–187.
Monteith, J. L. (1965). Evaporation and environment. In G. E. Fogg (Ed.), *The state and movement of water in living organisms* (pp. 205–234). Cambridge: Cambridge University Press.
Monteith, J., & Unsworth, M. (2007). *Principles of environmental physics* (3rd ed.). London: Academic. 418 pp.
Moran, M. S., & Jackson, R. D. (1991). Assessing the spatial distribution of evapotranspiration using remotely sensed inputs. *Journal of Environmental Quality, 20*(4), 725–737.
Moran, M. S., Jackson, R. D., Raymond, L. H., Gay, L. W., & Slater, P. N. (1989). Mapping surface energy balance components by combining Landsat Thematic Mapper and ground-based meteorological data. *Remote Sensing of Environment, 30*(1), 77–87.
Moran, M. S., Clarke, T. R., Inoue, Y., & Vidal, A. (1994). Estimating crop water deficit using the relation between surface-air temperature and spectral vegetation index. *Remote Sensing of Environment, 49*(3), 246–263.
Moran, M. S., Rahman, A. F., Washburne, J. C., Goodrich, D. C., Weltz, M. A., & Kustas, W. P. (1996). Combining the Penman-Monteith equation with measurements of surface temperature and reflectance to estimate evaporation rates of semiarid grassland. *Agricultural and Forest Meteorology, 80*(2), 87–109.

Mu, Q., Heinsch, F. A., Zhao, M., & Running, S. W. (2007). Development of a global evapotranspiration algorithm based on MODIS and global meteorology data. *Remote Sensing of Environment, 111*(4), 519–536.

Mu, Q., Zhao, M., & Running, S. W. (2011). Improvements to a MODIS global terrestrial evapotranspiration algorithm. *Remote Sensing of Environment, 115*(8), 1781–1800.

Nemani, R. R., & Running, S. W. (1989). Estimation of regional surface resistance to evapotranspiration from NDVI and thermal-IR AVHRR data. *Journal of Applied Meteorology, 28*(4), 276–284.

Nemani, R., Pierce, L., Running, S., & Goward, S. (1993). Developing satellite-derived estimates of surface moisture status. *Journal of Applied Meteorology, 32*(3), 548–557.

Norman, J. M., & Becker, F. (1995). Terminology in thermal infrared remote sensing of natural surfaces. *Agricultural and Forest Meteorology, 77*(3), 153–166.

Norman, J. M., Kustas, W. P., & Humes, K. S. (1995). A two-source approach for estimating soil and vegetation energy fluxes from observations of directional radiometric surface temperature. *Agricultural and Forest Meteorology, 77*, 263–293.

Norman, J. M., Kustas, W. P., Prueger, J. H., & Diak, G. R. (2000). Surface flux estimation using radiometric temperature: A dual-temperature-difference method to minimize measurement errors. *Water Resources Research, 36*(8), 2263–2274.

Norman, J. M., Anderson, M. C., & Kustas, W. P. (2006, August). Are single-source, remote-sensing surface-flux models too simple? In G. D'Urso, M. A. Osann Jochum, & J. Moreno (Eds.), *Proceedings of the international conference on earth observation for vegetation monitoring and water management* (Vol. 852, pp. 170–177). American Institute of Physics, Melville, New York, USA.

Novick, K. A., Oren, R., Stoy, P. C., Siqueira, M. B. S., & Katul, G. G. (2009). Nocturnal evapotranspiration in eddy-covariance records from three co-located ecosystems in the Southeastern US: Implications for annual fluxes. *Agricultural and Forest Meteorology, 149*(9), 1491–1504.

Oki, T., & Kanae, S. (2006). Global hydrological cycles and world water resources. *Science, 313*(5790), 1068–1072.

Paulson, C. A. (1970). The mathematical representation of wind speed and temperature profiles in the unstable atmospheric surface layer. *Journal of Applied Meteorology, 9*(6), 857–861.

Price, J. C. (1980). The potential of remotely sensed thermal infrared data to infer surface soil moisture and evaporation. *Water Resources Research, 16*(4), 787–795.

Price, J. C. (1982). On the use of satellite data to infer surface fluxes at meteorological scales. *Journal of Applied Meteorology, 21*, 1111–1122.

Price, J. C. (1990). Using spatial context in satellite data to infer regional scale evapotranspiration. *IEEE Transactions on Geoscience and Remote Sensing, 28*(5), 940–948.

Priestley, C. H. B., & Taylor, R. J. (1972). On the assessment of surface heat flux and evaporation using large-scale parameters. *Monthly Weather Review, 100*(2), 81–92.

Rango, A. (1994). Application of remote sensing methods to hydrology and water resources. *Hydrological Sciences Journal, 39*(4), 309–320.

Reginato, R. J., Jackson, R. D., & Pinter, P. J. (1985). Evapotranspiration calculated from remote multispectral and ground station meteorological data. *Remote Sensing of Environment, 18*(1), 75–89.

Reichle, R. H. (2008). Data assimilation methods in the Earth sciences. *Advances in Water Resources, 31*(11), 1411–1418.

Reichle, R. H., McLaughlin, D. B., & Entekhabi, D. (2002a). Hydrologic data assimilation with the ensemble Kalman filter. *Monthly Weather Review, 130*(1), 103–114.

Reichle, R. H., Walker, J. P., Koster, R. D., & Houser, P. R. (2002b). Extended versus ensemble Kalman filtering for land data assimilation. *Journal of Hydrometeorology, 3*(6), 728–740.

Robinson, A. R., & Lermusiaux, P. F. (2000). Overview of data assimilation. *Harvard Reports in Physical/Interdisciplinary Ocean Science, 62*, 1–13.

Roerink, G. J., Su, Z., & Menenti, M. (2000). S-SEBI: A simple remote sensing algorithm to estimate the surface energy balance. *Physics and Chemistry of the Earth, Part B: Hydrology, Oceans and Atmosphere, 25*(2), 147–157.

Ryu, Y., Baldocchi, D. D., Black, T. A., Detto, M., Law, B. E., Leuning, R., Miyata, A., Reichstein, M., Vargas, R., Ammann, C., Beringer, J., Flanagan, L. B., Gu, L., Hutley, L. B., Kim, J., McCaughey, H., Moors, E. J., Rambal, S., & Vesala, T. (2012). On the temporal upscaling of evapotranspiration from instantaneous remote sensing measurements to 8-day mean daily-sums. *Agricultural and Forest Meteorology, 152*, 212–222.

Sánchez, J. M., Kustas, W. P., Caselles, V., & Anderson, M. C. (2008). Modelling surface energy fluxes over maize using a two-source patch model and radiometric soil and canopy temperature observations. *Remote Sensing of Environment, 112*(3), 1130–1143.

Seguin, B. (1984). Estimation de I'evapotranspiration a partir de I'Infra-rouge thermIque. In *Proceedings of 2nd International Conference on Spectral Signatures of Objects in Remote Sensing* (pp. 427–446). Bordeaux.

Seguin, B., & Itier, B. (1983). Using midday surface temperature to estimate daily evaporation from satellite thermal IR data. *International Journal of Remote Sensing, 4*(2), 371–383.

Seguin, B., Baelz, S., Monget, J. M., & Petit, V. (1982). Utilisation de la thermographie IR pour l'estimation de l'évaporation régionale I. Mise au point méthodologique sur le site de la Crau. *Agronomie, 2*(1), 7–16.

Seguin, B., Courault, D., & Guerif, M. (1994). Surface temperature and evapotranspiration: Application of local scale methods to regional scales using satellite data. *Remote Sensing of Environment, 49*(3), 287–295.

Sellers, P. J., Heiser, M. D., & Hall, F. G. (1992). Relations between surface conductance and spectral vegetation indices at intermediate (100 m^2 to 15 km^2) length scales. *Journal of Geophysical Research, [Atmospheres], 97*(D17), 19033–19059.

Seo, D. J., Koren, V., & Cajina, N. (2003). Real-time variational assimilation of hydrologic and hydrometeorological data into operational hydrologic forecasting. *Journal of Hydrometeorology, 4*(3), 627–641.

Shuttleworth, W. J. (2007). Putting the 'vap' into evaporation. *Hydrology and Earth System Sciences Discussions, 11*(1), 210–244.

Shuttleworth, W. J., & Wallace, J. S. (1985). Evaporation from sparse crops-an energy combination theory. *Quarterly Journal of the Royal Meteorological Society, 111*(469), 839–855.

Shuttleworth, W. J., Gurney, R. J., Hsu, A. Y., & Ormsby, J. P. (1989). FIFE: The variation in energy partition at surface flux sites. *International Association of Hydrological Sciences Publication, 186*, 67–74.

Singh, R. K., Irmak, A., Irmak, S., & Martin, D. L. (2008). Application of SEBAL model for mapping evapotranspiration and estimating surface energy fluxes in south-central Nebraska. *Journal of Irrigation and Drainage Engineering, 134*(3), 273–285.

Soer, G. J. R. (1980). Estimation of regional evapotranspiration and soil moisture conditions using remotely sensed crop surface temperatures. *Remote Sensing of Environment, 9*(1), 27–45.

Stisen, S., Sandholt, I., Nørgaard, A., Fensholt, R., & Jensen, K. H. (2008). Combining the triangle method with thermal inertia to estimate regional evapotranspiration—Applied to MSG-SEVIRI data in the Senegal River basin. *Remote Sensing of Environment, 112*(3), 1242–1255.

Su, Z. (2002). The Surface Energy Balance System (SEBS) for estimation of turbulent heat fluxes. *Hydrology and Earth System Sciences Discussions, 6*(1), 85–100.

Su, Z. (2005). Hydrological applications of remote sensing. Surface fluxes and other derived variables–surface energy balance. In *Encyclopedia of hydrological sciences.* Hoboken: Wiley.

Su, Z., Li, X., Zhou, Y., Wan, L., Wen, J., & Sintonen, K. (2003, July 21–25). Estimating areal evaporation from remote sensing. *IEEE Transactions on Geoscience and Remote Sensing*. Toulouse.

Suzuki, K., & Nakai, Y. (2008). Canopy snow influence on water and energy balances in a coniferous forest plantation in northern Japan. *Journal of Hydrology, 352*(1), 126–138.

Talagrand, O. (1997). Assimilation of observations, an introduction. *Journal-Meteorological Society of Japan Series, 2*(75), 81–99.

Tang, R. (2011). *Retrieval of land surface evapotranspiration from remotely sensed surface temperature-fractional vegetation cover characteristic space*. Ph. D. thesis, Institute of Geographic Sciences and Natural Resources Research, Chinese Academy of Sciences (in Chinese).

Tang, R., Li, Z.-L., & Tang, B. (2010). An application of the Ts–VI triangle method with enhanced edges determination for evapotranspiration estimation from MODIS data in arid and semi-arid regions: Implementation and validation. *Remote Sensing of Environment, 114*(3), 540–551.

Tang, R., Li, Z.-L., & Chen, K. S. (2011a). Validating MODIS-derived land surface evapotranspiration with in situ measurements at two AmeriFlux sites in a semiarid region. *Journal of Geophysical Research: Atmospheres (1984–2012), 116*(D4), D04106.

Tang, R., Li, Z.-L., Jia, Y., Li, C., Sun, X., Kustas, W. P., & Anderson, M. C. (2011b). An intercomparison of three remote sensing-based energy balance models using Large Aperture Scintillometer measurements over a wheat–corn production region. *Remote Sensing of Environment, 115*(12), 3187–3202.

Tang, R., Li, Z.-L., Chen, K. S., Zhu, Y., & Liu, W. (2012). Verification of land surface evapotranspiration estimation from remote sensing spatial contextual information. *Hydrological Processes, 26*(15), 2283–2293.

Tang, R., Li, Z.-L., Chen, K. S., Jia, Y., Li, C., & Sun, X. (2013a). Spatial-scale effect on the SEBAL model for evapotranspiration estimation using remote sensing data. *Agricultural and Forest Meteorology, 174*, 28–42.

Tang, R., Li, Z.-L., & Sun, X. (2013b). Temporal upscaling of instantaneous evapotranspiration: an intercomparison of four methods using eddy covariance measurements and MODIS data. *Remote Sensing of Environment, 138*(4), 102–118. doi:10.1016/j.rse.2013.07.001.

Tang, R., Li, Z.-L., Jia, Y., Li, C., Chen, K.-S., Sun, X., & Lou, J. (2013c). Evaluating one- and two-source energy balance models in estimating surface evapotranspiration from Landsat-derived surface temperature and field measurements. *International Journal of Remote Sensing, 34*(9–10), 3299–3313.

Teixeira, A. D. C., Bastiaanssen, W. G. M., Ahmad, M. D., & Bos, M. G. (2009). Reviewing SEBAL input parameters for assessing evapotranspiration and water productivity for the Low-Middle Sao Francisco River basin, Brazil: Part A: Calibration and validation. *Agricultural and Forest Meteorology, 149*(3), 462–476.

Timmermans, W. J., Kustas, W. P., Anderson, M. C., & French, A. N. (2007). An intercomparison of the surface energy balance algorithm for land (SEBAL) and the two-source energy balance (TSEB) modeling schemes. *Remote Sensing of Environment, 108*(4), 369–384.

Todd, R. W., Evett, S. R., & Howell, T. A. (2000). The Bowen ratio-energy balance method for estimating latent heat flux of irrigated alfalfa evaluated in a semi-arid, advective environment. *Agricultural and Forest Meteorology, 103*(4), 335–348.

Trezza, R. (2002). *Evapotranspiration using a satellite-based surface energy balance with standardized ground control*. Ph.D. dissertation, USU, Logan, UT, 339 pp.

Tuzet, A., Perrier, A., & Leuning, R. (2003). Stomatal control of photosynthesis and transpiration: Results from a soil–plant–atmosphere continuum model. *Plant, Cell and Environment, 26*, 1097–1116.

Twine, T. E., Kustas, W. P., Norman, J. M., Cook, D. R., Houser, P., Meyers, T. P., Prueger, J. H., Starks, P. J., & Wesely, M. L. (2000). Correcting eddy-covariance flux underestimates over a grassland. *Agricultural and Forest Meteorology, 103*(3), 279–300.

Van den Hurk, B. J. J. M. (2001). Energy balance based surface flux estimation from satellite data, and its application for surface moisture assimilation. *Meteorology and Atmospheric Physics, 76*(1–2), 43–52.

Venturini, V., Bisht, G., Islam, S., & Jiang, L. E. (2004). Comparison of evaporative fractions estimated from AVHRR and MODIS sensors over South Florida. *Remote Sensing of Environment, 93*(1), 77–86.

Verhoef, A., De Bruin, H. A. R., & Van Den Hurk, B. J. J. M. (1997). Some practical notes on the parameter kB-1 for sparse vegetation. *Journal of Applied Meteorology, 36*(5), 560–572.

Vinukollu, R. K., Wood, E. F., Ferguson, C. R., & Fisher, J. B. (2011). Global estimates of evapotranspiration for climate studies using multi-sensor remote sensing data: Evaluation of three process-based approaches. *Remote Sensing of Environment, 115*(3), 801–823.

Wang, K., Li, Z., & Cribb, M. (2006). Estimation of evaporative fraction from a combination of day and night land surface temperatures and NDVI: A new method to determine the Priestley–Taylor parameter. *Remote Sensing of Environment, 102*(3), 293–305.

Wang, K., Wang, P., Li, Z., Cribb, M., & Sparrow, M. (2007). A simple method to estimate actual evapotranspiration from a combination of net radiation, vegetation index, and temperature. *Journal of Geophysical Research, [Atmospheres], 112*(D15), D15107.

Webb, E. (1970). Profile relationships: The log-linear range, and extension to strong stability. *Quarterly Journal of the Royal Meteorological Society, 96*(407), 67–90.

Webb, E. K., Pearman, G. I., & Leuning, R. (1980). Correction of flux measurements for density effects due to heat and water vapour transfer. *Quarterly Journal of the Royal Meteorological Society, 106*(447), 85–100.

Wood, E. F., Su, H., McCabe, M., & Su, B. (2003). Estimating evaporation from satellite remote sensing. *IEEE Transactions on Geoscience and Remote Sensing, 2*, 1163–1165.

Yao, Y., Liang, S., Cheng, J., Liu, S., Fisher, J. B., Zhang, X., Jia, K., Zhao, X., Qin, Q., Zhao, B., Han, S., Zhou, G., Zhou, G., Li, Y., Zhao, S., & Zhao, S. (2013). MODIS-driven estimation of terrestrial latent heat flux in China based on a modified Priestley-Taylor algorithm. *Agricultural and Forest Meteorology, 171–172*, 187–202.

Yuan, W., Liu, S., Yu, G., Bonnefond, J. M., Chen, J., Davis, K., Desal, A. R., Goldstein, A. H., Gianelle, D., Rossi, F., Suyker, A. E., & Verma, S. B. (2010). Global estimates of evapotranspiration and gross primary production based on MODIS and global meteorology data. *Remote Sensing of Environment, 114*(7), 1416–1431.

Zhan, X., Kustas, W. P., & Humes, K. S. (1996). An intercomparison study on models of sensible heat flux over partial canopy surfaces with remotely sensed surface temperature. *Remote Sensing of Environment, 58*(3), 242–256.

Zhang, K., Kimball, J. S., Nemani, R. R., & Running, S. W. (2010). A continuous satellite-derived global record of land surface evapotranspiration from 1983 to 2006. *Water Resources Research, 46*(9), W09522. doi:10.1029/2009WR008800.

Chapter 7
Applications of Thermal Remote Sensing in Agriculture Drought Monitoring and Thermal Anomaly Detection

7.1 Agricultural Drought Monitoring

Drought is a natural hazard and complex phenomenon. Drought is considered to be one of the most costly hazards because it can lead to reduced water supply and consequently have substantial effects on agriculture and socioeconomic activities (Riebsame et al. 1991). The severity of drought depends on its duration, intensity, spatial extent, and local socioeconomic conditions. Climate change due to global warming has altered precipitation patterns (Milly et al. 2002) and triggered more intensive droughts in recent years (Mishra and Singh 2009). The detrimental impacts on crop production affect large areas and therefore have high socioeconomic costs. These impacts have led to increasing attention to drought monitoring. Monitoring droughts at the regional scale is of primary importance for water planning and management to reduce possible impacts on crop production.

The evolution and characterization of the droughts from one type to the other with the continued deficiency of water over extended time duration is commonly defined based on the conditions caused by the aridity and reduction in water supplies. The droughts are most commonly classified as (Wilhite and Glantz 1985):

1. A meteorological drought is defined as a deviation from normal precipitation conditions due to climate variability or an increase in temperature over a period of time for a specific region, further causing higher evaporation and transpiration.
2. An agricultural drought occurs after a meteorological drought, and characteristic lack of adequate soil moisture causes a certain crop or plant water stress and obstructs its ability to grow and thrive during a particular time.
3. A hydrological drought occurs when precipitation has been reduced for an extended period of time and water supplies found in streams, lakes, rivers, and reservoirs are deficient with demand exceeding supply.
4. A socioeconomic drought is a condition when the physical water supplies are so low that they negatively affect the community where the drought is occurring.

H. Tang and Z.-L. Li, *Quantitative Remote Sensing in Thermal Infrared: Theory and Applications*, Springer Remote Sensing/Photogrammetry,
DOI 10.1007/978-3-642-42027-6_7,

Table 7.1 Thermal inertia values for some typical materials (Sobrino and Cuenca 1999)

Material	Thermal inertia ($Ws^{1/2}/(m^2 \cdot K)$)
Water and clouds	5,000
Ice	2,000
Snow	150
Dry sand	590
Wet sand	2,500
Dry clay	550
Wet clay	2,200
Shale	1,900
Granite	2,200
Bush	2,000
Grass	2,100
Corn	2,700
Alfalfa	2,900
Oats	2,500
Wood land	4,200

Although there are further classifications of drought, this section will concentrate on agricultural drought as agriculture is the economic basis of most of the countries in the world and agricultural drought has important consequences on food security and social stability. Satellite-based remote sensing is very suitable for monitoring drought conditions because instruments and sensors onboard satellites allow for both regional- and global-scale observations of land surfaces, providing spatially continuous measurements across the landscape, which can give detailed information about land surface patterns and conditions (Kogan 1997). Satellite observations can also have a high repeat visit frequency over the same geographic area, allowing for comparisons of data collected over time. Thermal infrared data which are sensitive to soil moisture are suitable for drought monitoring.

7.1.1 Apparent Thermal Inertia

The thermal inertia (TI, in $W \cdot s^{1/2}/(m^2 \cdot K)$) of a material or surface determines its resistance to temperature variations and is function of the material's bulk density, specific heat capacity, and thermal conductivity (Pohn et al. 1974). Its mathematical formulation is

$$\mathrm{TI} = \sqrt{\rho \cdot K \cdot c} \tag{7.1}$$

where K is the thermal conductivity [J/(m · s · K)], ρ is the bulk density [kg/m^3], and c is the specific heat capacity of the material, [J/(kg · K)]. As these three properties are material specific, different materials can be distinguished by their TI, offering numerous applications in terrestrial and planetary geology. Table 7.1 shows typical TI values for different surface types (Sobrino and Cuenca 1999). Water bodies have a higher TI than dry soils and rocks and exhibit a lower diurnal temperature fluctuation. When soil water content increases, TI proportionally increases as well. Furthermore, since the thermal conductivity of soils changes with fluctuating moisture level, TI can be used as a soil moisture indicator, given all other soil properties unchanged over space or time (Kahle 1977). Unfortunately, bulk density, specific heat capacity, and thermal conductivity cannot be derived from remotely sensed data, thus mapping of the TI through remote sensing technique requires alternative approaches.

A much more simple formulation of TI is the apparent thermal inertia (ATI) which is derived directly from multispectral remote sensing imagery (Mitra and Majumdar 2004; Verstraeten et al. 2006):

$$\mathrm{ATI} = C \cdot \frac{1-\alpha}{\Delta T_{\mathrm{s}}} \tag{7.2}$$

where C is the solar correction factor, α is the surface albedo, and ΔT_{s} is the difference between the afternoon and midnight surface temperatures. ATI represents the temporal and spatial variability of soil and canopy moisture. The strategy to derive soil water content is based on the rationale that high (low) ATI values correspond to maximum (minimum) soil moisture contents. Soil moisture content can be estimated using a linear model, a logarithmic model, or an exponential model, established by regression methods (Verstraeten et al. 2008).

ATI was initially found to be of limited use in areas with strong evapotranspiration (Price 1985). Wet surfaces allow considerable evaporation and transpiration during the daytime, thus reducing daytime surface temperatures through evaporative cooling. Nevertheless, ATI recently received renewed interest for mapping of both geology (Mitra and Majumdar 2004) and soil moisture (Verstraeten et al. 2006) because of its simple formulation which requires only remote sensing data. The coupling between ATI and soil moisture is not straightforward. True TI can be converted to soil moisture when soil properties are known (Lu et al. 2009). Figure 7.1 shows the ATI image derived from MODIS data at 0.05° spatial resolution on March 26, 2009, for the southern part of the African continent and the corresponding AMSR-E soil moisture image at 0.25°spatial resolution on the same day. The comparison of both products shows that the ATI largely reflects the regional soil moisture pattern. The ATI value is below 0.04 in the southwest area and along the Atlantic coastline. The same location with low soil moisture (below 0.1 m^3/m^3) is shown in the AMSR-E image. The trend of higher ATI values toward the east and north is also reflected on the soil moisture image. Some specific dry or wet features can also be recognized in both images. There is a clear relationship between AMSR-E-derived soil moisture and MODIS-derived ATI.

7.1.2 Temperature Condition Index

Rising leaf temperatures are good indicators of plant moisture stress and precede the onset of drought. This thermal response can occur even when the plants are green, as

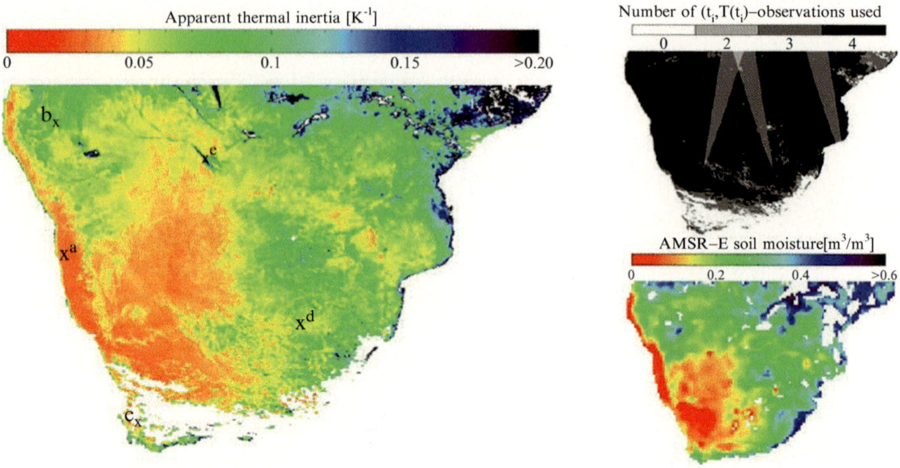

Fig. 7.1 *Left panel*: the apparent thermal inertia derived from MODIS data at 0.05° spatial resolution on March 26, 2009, for the southern part of the African continent. *Right panel, top*: number of MODIS observations used to derive apparent thermal inertia on March 26, 2009. *Right panel, bottom*: VUA-NASA AMSR-E soil moisture at 0.25° spatial resolution on March 26, 2009, interpolated using a 7-day temporal average (Reproduced from Van doninck et al. (2011), with permission from Elsevier)

stomatal closure to minimize water loss by transpiration results in a decreased latent heat flux. At the same time, due to the requirement that the energy flux must balance, there will be an increase in the sensible heat flux which may result in increased leaf temperatures. Plant moisture stress occurs when the demand for water exceeds the available soil moisture level. The potential for stress exists when the water in the soil storage is not enough to sustain the current growth. This is especially important for crops. Daytime thermal data have been used frequently to determine vegetation condition.

Kogan (1995) developed the temperature condition index (TCI), a proxy for thermal condition, which can be calculated only from a long time series of daytime satellite-derived land surface temperatures. The TCI is defined as

$$\mathrm{TCI} = \frac{T_{\mathrm{s\,max}} - T_{\mathrm{s}}}{T_{\mathrm{s\,max}} - T_{\mathrm{s\,min}}} \times 100 \qquad (7.3)$$

where TCI is the temperature condition index [–], T_{s} is the land surface temperature [K], T_{smax} is the maximum land surface temperature from all images within the data set [K], and T_{smin} is the minimum land surface temperature from all images within the data set [K]. The perceived major advantage of this approach is that only a suitable time series of daytime thermal remotely sensed data are required.

Figure 7.2 shows the 7-day composite TCI at a 4-km spatial resolution, which is calculated using the brightness temperatures measured by channel 4 (central wavelength is about 11 μm) of AVHRR. There are above-normal precipitations

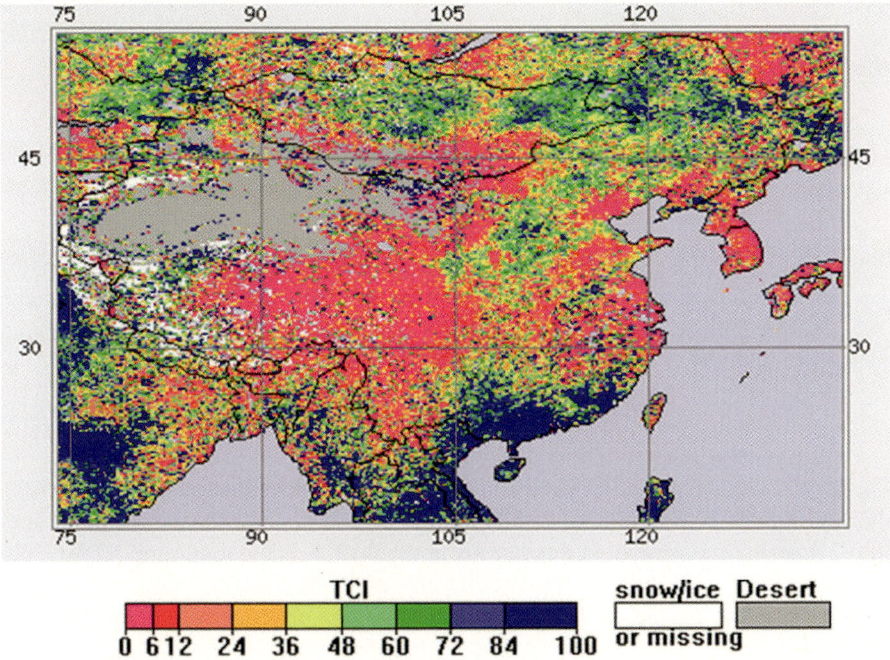

Fig. 7.2 A 7-day composite NOAA AVHRR temperature condition index on August 26, 2013. The lower value represents the drier conditions (http://www.star.nesdis.noaa.gov/smcd/emb/vci/VH/index.php)

in the northeast area of China, so it is under wet condition in some regions corresponding to a TCI value larger than 80. The precipitation is below normal in the North Plain of China, resulting in droughts in some regions with TCI values below 12. The surface temperature is high in East and West China areas corresponding to TCIs lower than 12. South China is wet with a TCI value of 80–100.

7.1.3 Vegetation Health Index

Kogan (2001, 2002) developed another index, the vegetation health index (VHI), which is an additive combination of vegetation condition index (VCI) and TCI, to monitor vegetation health, moisture, and thermal conditions as well as to determine drought-affected areas. The VCI, TCI, and VHI have been used as tools for drought detection and vegetation stress mostly in the context of agricultural production in different parts of the world.

Higher VCI values correspond to favorable moisture condition and represent unstressed vegetation. Subtle changes in vegetation health due to thermal stress in specific could be monitored through analysis of TCI data (Kogan 2001, 2002).

Table 7.2 Vegetation Health Index classification scheme (Kogan 2001)

Drought classes	Vegetation Health Index range
Extreme drought	<10
Severe drought	<20
Moderate drought	<30
Mild drought	<40
No drought	>40

While VCI and TCI characterize the varying moisture and thermal conditions of vegetation, VHI represents overall vegetation health, which was used by Kogan (2001) who gave five classes (Table 7.2) of VHI that were used for drought mapping. VHI is computed and expressed as

$$\mathrm{VHI} = 0.5\mathrm{VCI} + 0.5\mathrm{TCI} \tag{7.4}$$

where VCI is calculated from Normalized Difference Vegetation Index (NDVI). $\mathrm{VCI} = 100 \times (\mathrm{NDVI} - \mathrm{NDVI}_{\mathrm{min}})/(\mathrm{NDVI}_{\mathrm{max}} - \mathrm{NDVI}_{\mathrm{min}})$, $\mathrm{NDVI}_{\mathrm{max}}$ is the maximum NDVI from all images within the data set and $\mathrm{NDVI}_{\mathrm{min}}$ is the minimum NDVI from all images within the data set. The value range of VCI is 0–100.

Figure 7.3 shows the VHI of Southern Africa in El Niño years and La Niña years (http://www.star.nesdis.noaa.gov/smcd/emb/vci/VH/vh_case_elnino.php). El Niño is one part of a multiyear cycle of the interaction between the ocean and atmosphere. A typical characteristic of La Niña cycle is the emergence of a huge mass of warm water in the central and eastern Pacific during El Niño and is cold during La Niña years. In Southern Africa, during El Niño years, rainfall reduces considerably, leading to drought and severe vegetation stress with low values of VHI (0–20) corresponding to red-color areas in the image. In La Niña years, the effect is opposite and therefore the VHI value is high (70–100) corresponding to blue-color areas. The variation of VHI shows that the typical patterns of vegetation condition are severe drought-related vegetation stress in El Niño and favorable conditions in La Niña years.

7.1.4 Normalized Difference Temperature Index

The normalized difference temperature index (NDTI) is defined as (McVicar and Jupp 1998)

$$\mathrm{NDTI} = \frac{T_{\infty} - T_{\mathrm{s}}}{T_{\infty} - T_0} \tag{7.5}$$

where T_{∞} is the modeled surface temperature if there is an infinite surface resistance, that is, evapotranspiration (ET) is zero; T_{s} is the surface temperature observed from the sensor; and T_0 is the modeled surface temperature if there is

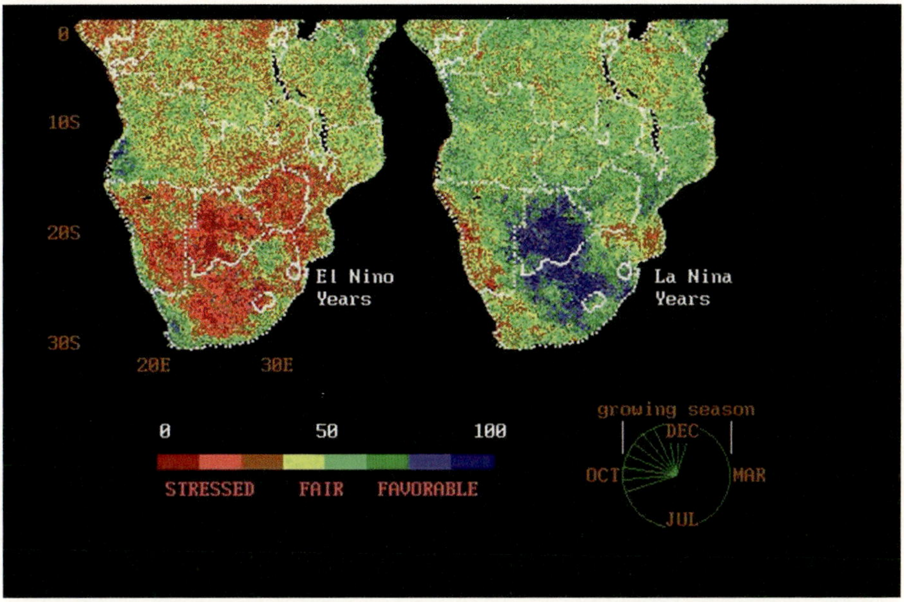

Fig. 7.3 The composite 7-day Vegetation Health Index of the second week in December in El Niño years and La Niña years over Southern Africa. VHI is an average for the two most recent El Niño and La Niña years. The data were acquired from NOAA-9, NOAA-11, and NOAA-14 polar orbiting spacecrafts. The values of VHI for stressful conditions are below 35, 36–65 for fair conditions, and above 65 for favorable conditions (http://www.star.nesdis.noaa.gov/smcd/emb/vci/VH/vh_case_elnino.php)

zero surface resistance; hence actual ET equals potential ET. NDTI is developed to remove seasonal trends from the analysis of daytime land surface temperatures. According to McVicar and Jupp (1998), there are a number of advantages to derive the NDTI:

- NDTI is a very close approximation of moisture availability. NDTI can be developed as a standard product, which is an easily computable surrogate for moisture availability. The required modeling techniques to determine NDTI for the time and date of a satellite overpass have been developed.
- When NDVI is high, NDTI is also high. This is of benefit for the intercomparisons of the remotely sensed products.

McVicar and Jupp (2002) estimated NDTI as shown in Fig. 7.4 to acquire moisture availability in the Murray–Darling Basin in southeast Australia, using the AVHRR remotely sensed data combined with meteorological data. Figure 7.4a shows the NDTI image for spring (September 22, 1987) and Fig. 7.4b shows the NDTI image for summer (December 25, 1987). Changes in the NDTI between the two dates are shown in Fig. 7.4c, where blue is an increase, green fairly constant, and red a decrease between the two dates. Murray River crosses the southwest of

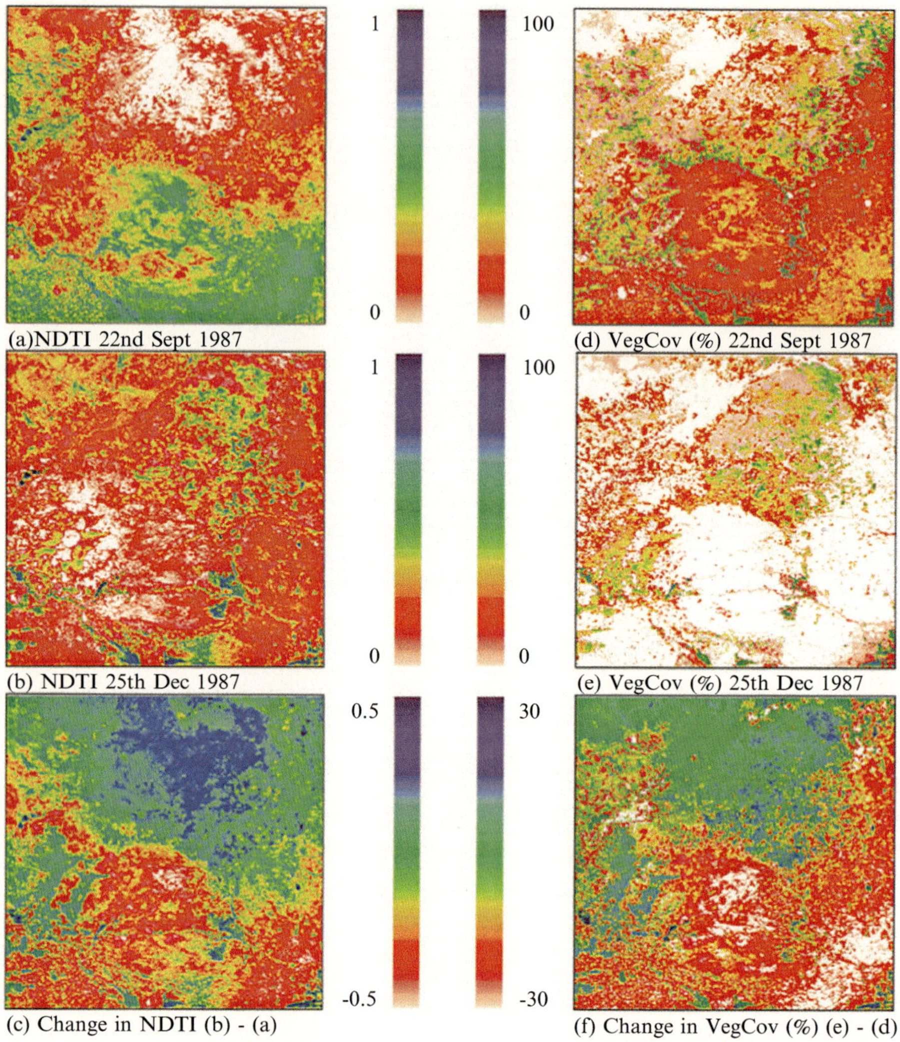

Fig. 7.4 Images of the NDTI in the Murray–Darling Basin in southeast Australia. (**a**) NDTI on September 22, 1987; (**b**) NDTI on December 25, 1987; and (**c**) the difference of NDTI between these two dates, calculated as (**b**) – (**a**), (**d**) to (**f**) is same as (**a**) to (**c**), but for vegetation cover (VegCov) (Reproduced from McVicar and Jupp (2002), with permission from Elsevier)

the study area, and Fig. 7.4a shows that the southwest is the only area where the NDTI > 0.8, with the southern portion of the wheat belt in the southeast having a NDTI of about 0.5. In Fig. 7.4b, some areas with a NDTI > 0.8 are lakes and some are associated with flood irrigation of crops along the Murray River. In the north, there are scattered areas with NDTI values of approximately 0.5, which correspond to areas where the NDTI has increased between the two dates (Fig. 7.4c); these areas have received scattered rainfall. The AVHRR-derived estimates of vegetation cover are shown for spring (Fig. 7.4d) and summer (Fig. 7.4e). In Fig. 7.4d, e, the remnant deep-rooted forests are identified by having a stable vegetation cover between the two dates. The changes between the two dates in NDTI (Fig. 7.4c) and vegetation cover (Fig. 7.4f) are similar.

7.1.5 Vegetation Supply Water Index

When the water supply is inadequate, crops become water stressed by drought and vegetation index will be decreased. Meanwhile, crop stomatal is closed to reduce the leaf surface transpiration, resulting in an increase of the crop canopy temperature. Therefore, vegetation supply water index (VSWI) (Li et al. 1998) combining the vegetation index and canopy temperature reflects the drought conditions during the growing season. VSWI is expressed as

$$\mathrm{VSWI} = K_{\mathrm{C}} \cdot \frac{\mathrm{NDVI}}{T_{\mathrm{s}}} \tag{7.6}$$

where $K_{\mathrm{C}} = 10{,}000$ is an enhance factor.

Figure 7.5 shows the combined 1-km VSWI distribution of China from April 22 to May 6, 2012, using MODIS data. The northwest area is under drought conditions corresponding to a VSWI value of 0–100 due to low vegetation and little precipitation all year round. Cropland in China mainly locates in the southeast area. The precipitation is below normal in the North Plain, resulting in drought in some regions with a VSWI range of 100–150. Large areas in the south are wet with a VSWI value of 250–350. However, water-stressed conditions still exist in some south regions of China, such as Yunnan and Sichuan provinces with a VSWI range of 110–130.

7.1.6 Temperature Vegetation Dryness Index

The temperature-vegetation index (T_{s}-VI) triangle space emerges when the study area is large enough and the variability in the surface moisture and vegetation cover conditions is also large. Figure 7.6 shows a diagram of the T_{s}-VI triangle space. As the VI increases along the x-axis, the maximum T_{s} decreases. The wet edge consists of a group of points forming a horizontal or sloping line for different surface types. For a given VI, the T_{s} increases progressively as a result of water stress in the surface soil from the minimum value at the wet edge to the maximum value at the dry edge, whereas the soil moisture decreases from maximum to minimum values correspondingly. Specifically, the wet edge is close to potential evapotranspiration, whereas evapotranspiration is close to zero at the dry edge because soil moisture

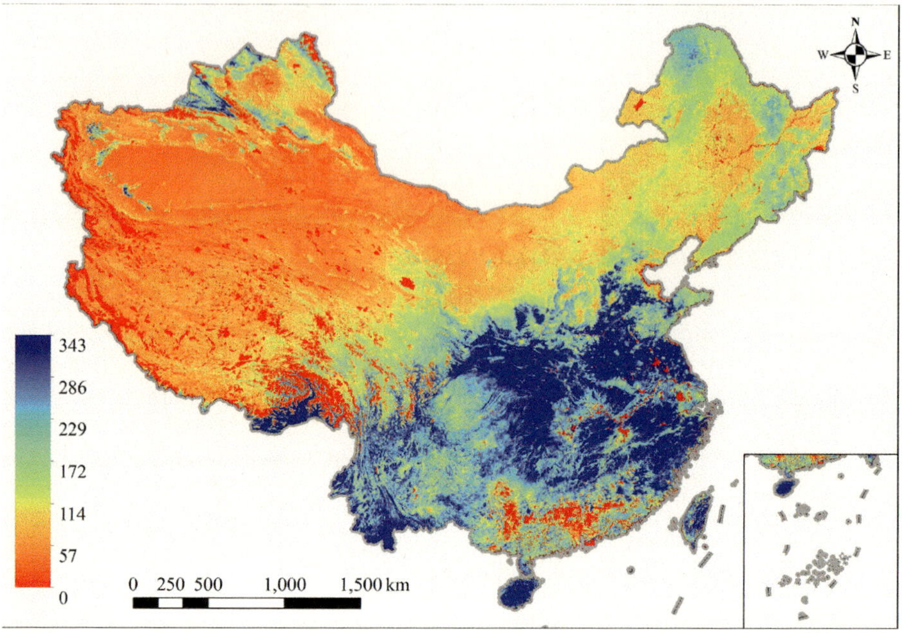

Fig. 7.5 The combined 1-km VSWI distribution of China from April 22 to May 6, 2012, using MODIS data. *Red color* corresponds to water-stressed area; *blue color* corresponds to nonstressed area

is less than the wilting point. Carlson et al. (1995) and Gillies et al. (1997) used a soil-vegetation-atmosphere transfer (SVAT) model to derive the surface soil moisture availability and fractional vegetation cover from the analysis of the T_s-VI space. Sandholt et al. (2002) proposed a simplified land surface dryness index, the temperature-vegetation dryness index (TVDI), which is based on an empirical parameterization of the T_s-VI space. The results indicated that the spatial pattern of the TVDI was closely related to the surface soil moisture simulated with the MIKE SHE model. The TVDI isolines can be regarded as several superimposed lines of the T_s-VI slope. Because the TVDI may easily be estimated for each pixel without any ancillary data, it is widely used to monitor soil moisture (Wang et al. 2004; Mallick et al. 2009; Gao et al. 2011).

TVDI is defined as (Sandholt et al. 2002)

$$\mathrm{TVDI} = \frac{T_s - T_{\min}}{T_{\max} - T_{\min}} \tag{7.7}$$

where $T_{\max}$ is the maximum surface temperature at the dry edge and $T_{\min}$ is the minimum surface temperature at the wet edge. TVDI has the values of 1 at the dry edge and 0 at the wet edge.

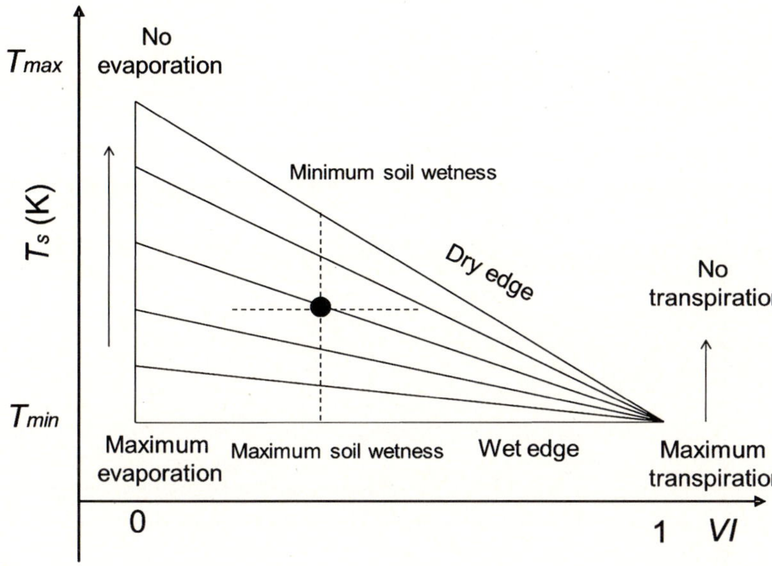

Fig. 7.6 Diagram of the T_s-VI triangle space. Soil wetness is minimal along the dry edge and is maximal along the wet edge. Soil wetness for a given pixel is linearly interpolated between the minimum and maximum values (Reproduced from Sun et al. (2012a), with permission from Elsevier)

In reality, zero evapotranspiration rarely occurs for dense vegetation covers, even in semiarid environments, primarily because of the soil water uptake from the root zone (Stisen et al. 2008). Consequently, the dry edge observed in the remote sensing data is characterized by lower temperatures compared with the theoretical dry edge, and thus, the observed dry edge is not assigned the minimum evapotranspiration. In that way, the observed dry edge determined by scatter plots is usually lower than the theoretical dry edge (Stisen et al. 2008). Zhang et al. (2005, 2008) adopted a trapezoid $T_s - F_v$ (T_s-VI) space in their model as shown in Fig. 7.7. The dashed line represents the observed dry edge, which is determined by the scatterplots. The upper solid line represents the theoretical dry edge, under extreme water stressed conditions, which is estimated from the energy balance principle. The T_{sd} and T_{vd} represent the surface temperatures of the theoretical dry bare soil and full-cover vegetation surface, respectively. The T_{sw} and T_{vw} represent the surface temperatures of the saturated wet, bare soil and full-cover vegetation surface, respectively, which are assumed to be equal and represented by the wet edge temperature (T_w). The dry edge and the wet edge form two physical limits of the $T_s - F_v$ (T_s-VI) trapezoid.

In the $T_s - F_v$ (T_s-VI) trapezoid space (Fig. 7.7), the largest T_s (T_{max}) value along the theoretical dry edge represents the driest conditions when the relative surface soil moisture is zero because T_s reaches a physical maximum value when no evaporative cooling occurs. In this case, complete stomatal closure occurs for the vegetated

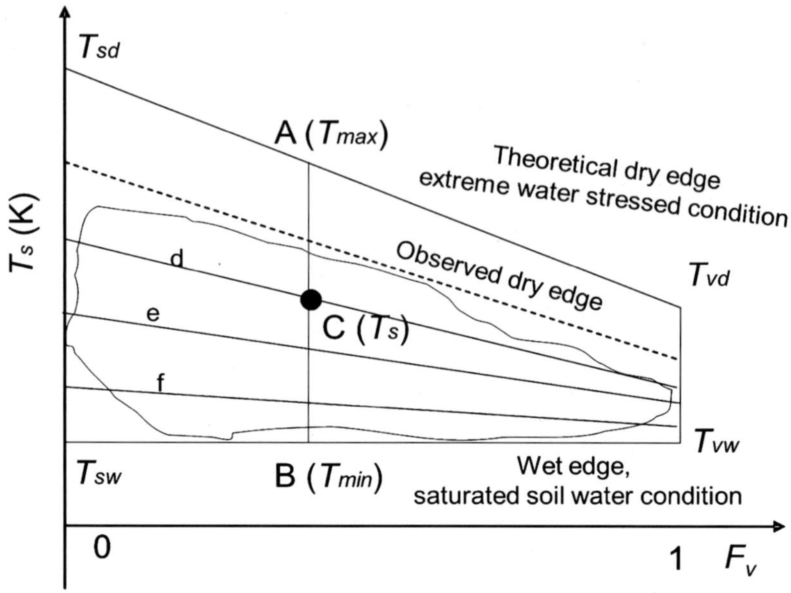

Fig. 7.7 The $T_s - F_v$ (T_s-VI) trapezoid space. The *dashed line* represents the observed dry edge, which is directly determined by the scatterplots from the remote sensing data. The *upper solid line* is the theoretical dry edge, estimated from the energy balance principle, representing the extreme water stressed conditions. The *bottom solid line* is the wet edge, representing the saturated soil water conditions. The *solid lines d, e,* and *f* represent the isolines of the relative soil moisture. The value of the soil wetness index is zero along the theoretical dry edge, and its value is 1 along the wet edge (Reproduced from Sun et al. (2012a), with permission from Elsevier)

part (Moran et al. 1994). The wettest (saturated) soil conditions are represented by the minimum T_s (T_{min}) value along the wet edge, when the surface soil wetness is the greatest (relative surface soil moisture equals 1). The moisture availability is assumed to vary linearly from the dry edge to the wet edge. The pixels on each isoline, such as d, e, and f in Fig. 7.7, represent the same value of relative surface soil moisture. This is in agreement with previous interpretations of the T_s-VI space (Mallick et al. 2009).

Sun et al. (2012a) compared the advanced temperature-vegetation dryness index (ATVDI) and the simple temperature-vegetation dryness index (STVDI), which is calculated from the theoretical dry edge and from the observed dry edge as shown in Fig. 7.7, respectively. Figure 7.8 shows the images of the ATVDI and STVDI produced from MODIS data in the Southern Great Plain in the USA on day of year (DOY) 152 and 213, 2004. On DOY 152, the ATVDI and STVDI distributions exhibit a close resemblance because there has been no precipitation for 2 weeks before DOY 152; therefore, the soil surface is under relatively dry conditions. Both the ATVDI and STVDI are about 0.8 on the southwest of the image. On DOY 213, STVDI is about 0.9 in the southwest, and it is greater than 0.5 in most areas of the image. However, the ATVDI is lower than 0.5 for almost all the pixels. Prior to

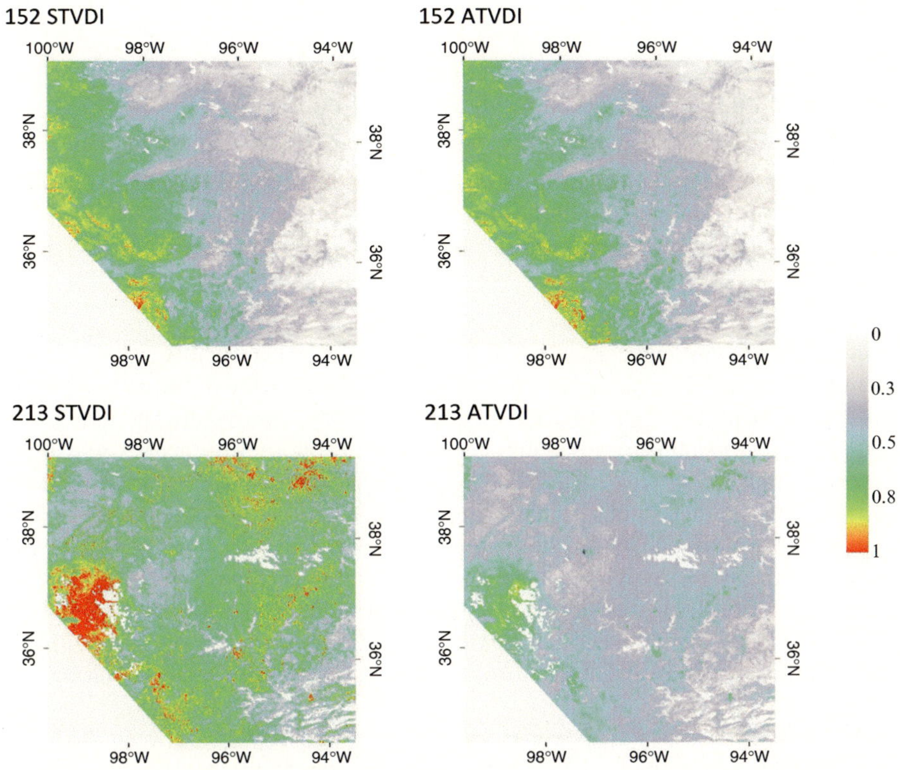

Fig. 7.8 Images of the advanced temperature-vegetation dryness index (*ATVDI*) and the simple temperature-vegetation dryness index (*STVDI*) from MODIS data on DOY 152 and 213, 2004, in the Southern Great Plain in the USA. ATVDI is calculated from the theoretical dry edge and STVDI is calculated from the observed dry edge, as shown in Fig. 7.7. *Red color* represents dry conditions and *blue color* represents wet conditions (Reproduced from Sun et al. (2012a), with permission from Elsevier)

DOY213, rainfall lasts for nearly a week, and consequently, the soil surface is under wet conditions. Therefore, the distribution of the ATVDI on this day apparently reflects the real soil moisture conditions.

7.1.7 Crop Water Stress Index

The crop water stress index (CWSI) is a measure of the relative transpiration rate occurring from a plant at the time of measurement using measurements of plant temperature and the vapor pressure deficit. Jackson et al. (1981) presents the theory behind the energy balance that separates net radiation from the sun into sensible heat flux that heats the air and latent heat flux that is used for transpiration.

7.1.7.1 Empirical CWSI

Idso et al. (1981) and Jackson et al. (1981) pioneered a method using the daytime canopy temperature, called the CWSI, which is now used routinely for assessing crop health and establishing irrigation schedule at the field scale. CWSI is expressed as (Idso et al. 1981)

$$\mathrm{CWSI} = \frac{(T_c - T_a) - D_2}{D_1 - D_2} \tag{7.8}$$

where D_1 is the maximum canopy and air temperature difference for a stressed crop (the maximum stressed baseline), D_2 is the lower limit of the canopy and air temperature difference for a well-watered crop (the nonwater-stressed baseline), T_c is the measured canopy surface temperature, and T_a is the air temperature.

A CWSI of 0 indicates no water stress, and a value of 1 represents maximum water stress. The crop water stress that signals the need for irrigation is crop specific, and its study should consider factors such as yield response to water stress, probable crop price, and water cost. As the crop undergoes water stress, the stomata closes, the transpiration decreases, and the leaf temperature increases. When a plant is transpiring fully, the leaf temperature is 1–4 degrees below air temperature and the CWSI is 0. As the transpiration decreases, the leaf temperature rises and can reach to 4–6 degrees above air temperature. When the plant is no longer transpiring, the CWSI is 1.

The CWSI is calculated by determining the relative distance between the lower baseline representing nonstress conditions and the upper baseline representing no transpiration from a plot of canopy–air temperature difference versus vapor pressure deficit. Several methods are used to determine the upper and lower limits in the CWSI equation. Idso (1982) demonstrated that the lower limit of the CWSI is a linear function of vapor pressure deficit (VPD) for a number of crops and locations. The canopy–air temperature differences for a well-watered crop (lower limit) and a severely stressed crop (upper limit) can be calculated as

$$\begin{aligned} D_1 &= A + B \times \mathrm{VPG} \\ D_2 &= A + B \times \mathrm{VPD} \end{aligned} \tag{7.9}$$

where VPD is the air vapor pressure deficit [Pa], and A (intercept) and B (slope) are the linear regression coefficients. VPG is the difference between the saturation vapor pressure evaluated at air temperature (T_a) and that at a temperature equal to $T_a + A$. Thus, with a measure of relative humidity (RH), air temperature, and canopy temperature, it is now possible to determine CWSI. The slope and intercept parameters are determined from experimental data, so the CWSI calculated using this method is often referred as the "empirical" CWSI. There have been many applications of field crops using the empirical CWSI, some of which are very successful. For example, Reginato and Howe (1985) found that cotton yield showed

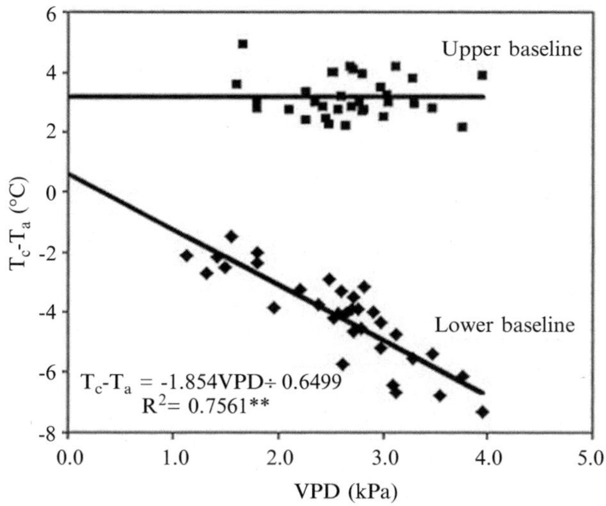

Fig. 7.9 The canopy–air temperature difference ($T_c - T_a$) versus air vapor pressure deficit (VPD) for nonwater stressed and maximally stressed soybeans. R^2, the coefficient of determination; **, significant at the 1 % probability level ($P < 0.01$) (Reproduced from Candogan et al. (2013), with permission from Elsevier)

the first signs of decline when the average of the CWSI during the season was greater than 0.2. For corn, a CWSI value of 0.3–0.4 might be a conservative timing parameter to avoid excess irrigation (Yazar et al. 1999).

Candogan et al. (2013) conducted field experiments for 2 years (2005–2006) to evaluate the use of the CWSI for irrigation schedule of soybeans under a subhumid climate of Bursa, Turkey. Figure 7.9 shows the upper and lower limit baselines. The value of $(T_c - T_a)_{upper}$ was calculated as 3.2 °C using the combined data of the 2 years. Also, the linear equation for the lower limit baseline was defined as $(T_c - T_a)_{lower} = -1.8549\text{VPD} + 0.6499$. Figure 7.10a, b show the seasonal course of CWSI values computed for each irrigation treatment in 2005 and 2006, respectively. The highest CWSI value calculated for T100 treatment was 0.36. During the growing seasons of 2005 and 2006, CWSI values reached 0.47 and 0.37 for T75, 0.50 and 0.57 for T50, and 0.68 and 0.80 for T25, respectively. Throughout both growing seasons, CWSI of T0 (rain fed) fluctuated close to 1. The CWSI values were affected not only by $T_c - T_a$, depending on the transpiration of crop, but also by the VPD of the air, and thus, CWSI was very sensitive to water stress in the soybean crops. Based on these results a maximum CWSI value of nearly 0.36 could be offered as a threshold number for excellent irrigation timing and CWSI over that values could cause water stress and thus yield reduction. Consequently, an irrigation schedule that considers water stress could be employed when the cost of water is high and/or water is scarce.

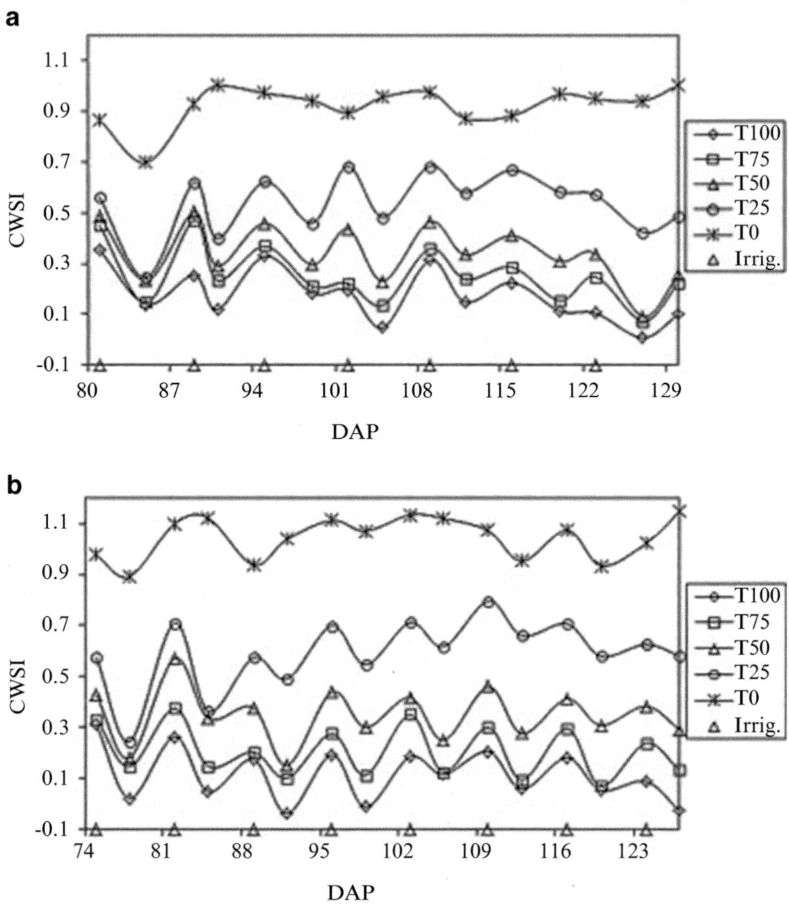

Fig. 7.10 Variations of the crop water stress index (*CWSI*) for each irrigation treatment of soybeans during the (**a**) 2005 and (**b**) 2006 experiments. DAP means day after planting. The canopy temperature (T_c) was measured with a handheld infrared thermometer. Irrigations were scheduled based on the replenishment of 100 % (T100), 75 % (T75), 50 % (T50), 25 % (T25), and 0 % (T0) of soil water depletion from a soil depth of 90 cm using a 7-day irrigation interval (Reproduced from Candogan et al. (2013), with permission from Elsevier)

7.1.7.2 Theoretical CWSI

Another method is the so-called "theoretical" CWSI. It has the same equation as Eq. (7.8), but the coefficients are determined by combining the Penman–Monteith equation with a one-dimensional energy balance equation. By rearranging terms of the surface energy balance equation, Jackson et al. (1981) were able to develop an equation to predict the canopy minus air temperature difference ($T_c - T_a$):

$$T_c - T_a = \frac{r_a (R_n - G)}{\rho C_p} \frac{\gamma (1 + r_c/r_a)}{\Delta + \gamma (1 + r_c/r_a)} - \frac{\text{VPD}}{\Delta + \gamma (1 + r_c/r_a)} \tag{7.10}$$

where R_n is the net radiation [W/m^2], G is the soil heat flux or the energy flux density leaving the lower canopy layer [W/m^2], ρ is the air density [kg/m^3], C_p is the specific heat at constant pressure [J/kg/°C], γ is the psychrometric constant [Pa/°C], Δ is the slope of the saturated vapor pressure versus temperature curve [Pa/°C], r_a is the aerodynamic resistance [s/m], and r_c is the canopy resistance [s/m]. In the case of a nonwater stressed crop, assuming that $r_c = 0$,

$$D_2 = \frac{r_a (R_n - G)}{\rho C_p} \frac{\gamma}{\Delta + \gamma} - \frac{\text{VPD}}{\Delta + \gamma} \tag{7.11}$$

In the case of the upper limit (nontranspiring crop), canopy resistance will approach infinity, and Eq. (7.10) reduces to

$$D_1 = \frac{r_a (R_n - G)}{\rho C_p} \tag{7.12}$$

Equations (7.11) and (7.12) can be used to determine the CWSI as given in Eq. (7.8), which in fact can be expressed as

$$\text{CWSI} = 1 - \frac{\text{ET}}{\text{ET}_p} \tag{7.13}$$

where ET and ET_p are the daily actual evapotranspiration and the daily potential evapotranspiration, respectively. CWSI has been used extensively for such important farm applications as irrigation scheduling, predicting crop yields, and detecting certain plant diseases. Since it requires a measurement of foliage temperature and most remote sensing systems measure the surface temperature of the mixed medium (both canopy and soil), the application has generally been limited to fully vegetated sites, such as agricultural fields (Moran et al. 1996).

7.1.8 Water Deficit Index

The previous discussions of the CWSI have assumed that a measure of canopy temperature is available or that the crop completely covers the soil surface. To consider the situation of mixed pixel, Moran et al. (1994) developed the water deficit index (WDI) that uses both surface–air temperature difference and a vegetation index to estimate the relative water status of a field. As shown in Fig. 7.11, the distribution of surface–air temperature difference is plotted against a spectral

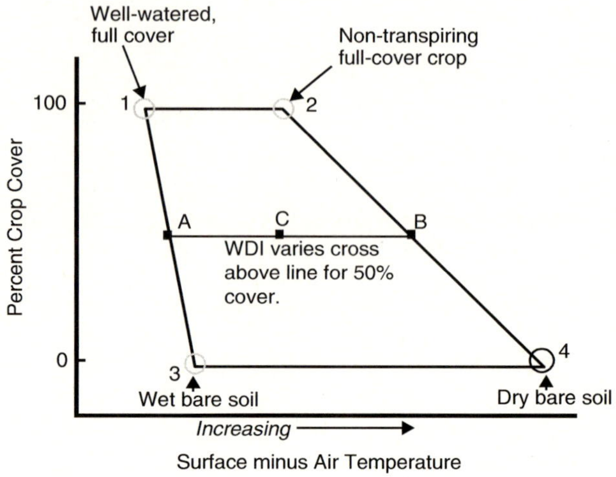

Fig. 7.11 Illustration of the water deficit index trapezoid (Reproduced from Moran et al. (1994), with permission from Elsevier)

vegetation index to form a trapezoidal shape that encompasses all possible values of surface temperature for both full-cover and partially vegetated surfaces. Vegetation indices can be used to calculate percent vegetation cover. The upper left of the trapezoid corresponds to a well-watered crop at 100 % cover and the upper right to a nontranspiring crop at 100 % cover (points 1 and 2, respectively). These two points are the same as those for the upper and lower limits of the standard CWSI. The lower portion of the trapezoid (bare soil) is bounded by a wet and dry soil surface. For full-cover, well-watered vegetation,

$$(T_s - T_a)_1 = \frac{r_a (R_n - G)}{\rho C_p} \frac{\gamma \left(1 + r_{cp}/r_a\right)}{\Delta + \gamma \left(1 + r_{cp}/r_a\right)} - \frac{\text{VPD}}{\Delta + \gamma \left(1 + r_{cp}/r_a\right)} \tag{7.14}$$

where r_{cp} is the canopy resistance at potential evapotranspiration and the subscript "n" of $(T_s - T_a)_n$ refers to vertex n, as shown in Fig. 7.11, and for full-cover vegetation with no available water,

$$(T_s - T_a)_2 = \frac{r_a (R_n - G)}{\rho C_p} \frac{\gamma (1 + r_{cx}/r_a)}{\Delta + \gamma (1 + r_{cx}/r_a)} - \frac{\text{VPD}}{\Delta + \gamma (1 + r_{cx}/r_a)} \tag{7.15}$$

where r_{cx} is the canopy resistance associated with nearly complete stomatal closure. For saturated bare soil, where $r_c = 0$ (the case of a free water surface),

$$(T_s - T_a)_3 = \frac{r_a (R_n - G)}{\rho C_p} \frac{\gamma}{\Delta + \gamma} - \frac{\text{VPD}}{\Delta + \gamma} \tag{7.16}$$

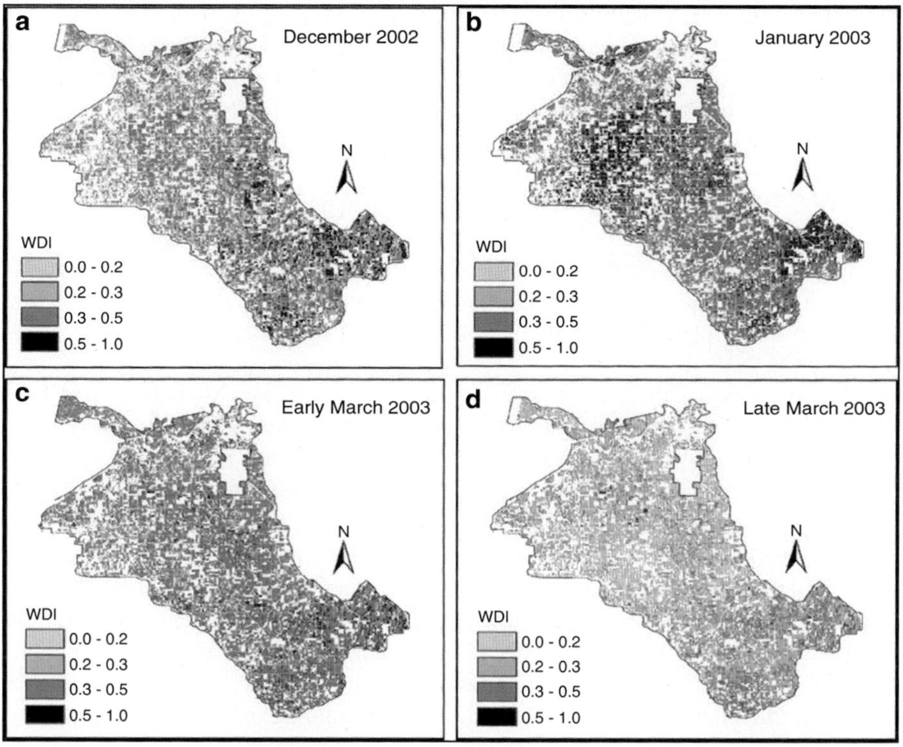

Fig. 7.12 WDI calculated for different crop stages in Yaqui Valley. (**a**) December 2002, (**b**) January 2003, (**c**) early March 2003, and (**d**) late March 2003 (Reproduced from Mendez-Barroso et al. (2008), with permission from Elsevier)

and for dry bare soil, where $r_c = \infty$ (analogous to complete stomatal closure),

$$(T_s - T_a)_4 = \frac{r_a (R_n - G)}{\rho C_p} \tag{7.17}$$

Given corners of the trapezoid, the WDI for a measured percent cover becomes

$$\text{WDI} = \frac{(T_s - T_a) - (T_s - T_a)_{\min}}{(T_s - T_a)_{\max} - (T_s - T_a)_{\min}} \tag{7.18}$$

where $(T_s - T_a)_{\min}$ is the surface minus air temperature determined by the line from points 1 to 3 for the percent cover of interest ("wet" line) and $(T_s - T_a)_{\max}$ is the temperature difference on the line formed between points 2 and 4 ("dry" line). Graphically, WDI is equal to the ratio of distances AC/AB as shown in Fig. 7.11. Thus, WDI = 0.0 for well-watered conditions and WDI = 1.0 for maximum stress conditions.

In Mendez-Barroso et al. (2008), the spatial and temporal distribution of the WDI for wheat were determined through analysis of ETM + data onboard the LANDSAT 7 satellite in the Yaqui Valley, Sonora, Mexico. Figure 7.12a shows the spatial distribution of WDI calculated for December 16, 2002. Areas with very low stress (WDI < 0.2) are observed in the west part of the Yaqui Valley, near the Yaqui River. This behavior may be explained by local soil conditions that favor the retention of soil moisture or parcels that have been just irrigated. On the other hand, the higher WDI values were concentrated mainly in the southeastern part of the Yaqui Valley with values between 0.5 and 0.7. Figure 7.12b shows the WDI values for January 1, 2003. On this day, high stress indices can be observed in the southeastern region. The center-west region also begins to show values of high water stress, with WDI = 0.5–0.7. By this date, all of the wheat has been established and the first auxiliary irrigation had been applied. In this year, the local water management recommended applying the first irrigation around 60 days after planting. WDI values on March 6, 2003, as shown in Fig. 7.12c, show a range of values between 0.3 and 0.5. At this time of the year, the second and third auxiliary irrigations had been applied, since the local water management recommended applying them near to flowering and grain filling. On the other hand, some very high WDI values were present in the southeastern region. This concentration of high WDI values can be caused by local soil texture properties as this area is characterized by coarse soil texture (sandy–stony soil). Even if a heavy irrigation is applied, the soils in this area tend to retain less water compared to other regions. Finally, WDI values corresponding to March 22, 2003, are presented in Fig. 7.12d. Clearly, almost all the entire Yaqui Valley shows WDI values ranging from 0 to 0.3, indicating favorable soil moisture conditions for the crop, mainly in the central and west region. However, the tendency of high WDI values in the southeastern region continues.

7.1.9 Evaporative Stress Index

Spatial and temporal variations in instantaneous ET at the continental scale are primarily due to variability in soil moisture availability (antecedent precipitation), radiative forcing (cloud cover, sun angle), vegetation amount, and local atmospheric conditions such as air temperature, wind speed, and vapor pressure deficit. Potential ET describes the evaporation rate expected when soil moisture is nonlimiting, ideally capturing response to all other forcing variables. To isolate effects due to spatially varying soil moisture availability, an evaporative stress index (ESI) can be developed from model flux estimates (Anderson et al. 2007), given by 1 minus the ratio of actual to potential ET following the formulation of the CWSI and WDI. Evaporative stress indices associated with the canopy (ESI_c), the soil surface (ESI_s), and the combined plant–soil system (ESI) can be derived as

$$\begin{aligned} ESI_c &= 1 - \frac{ET_c}{ET_{pc}} \\ ESI_s &= 1 - \frac{ET_s}{ET_{ps}} \\ ESI &= 1 - \frac{ET}{ET_p} \end{aligned} \tag{7.19}$$

where ET_c, ET_s, and ET are the modeled actual ET fluxes from the canopy, soil, and total system, respectively, and ET_{pc}, ET_{ps}, and ET_p are the potential ET from the canopy, soil, and total system, respectively. These indices have a value of 0 when there is no stress and a value of 1 when evapotranspiration has been cut off because of the stress-induced stomatal closure or complete drying of the soil surface. Among the existing regional thermal infrared remote sensing drought indices, the ESI is unique in that it has a clearly defined physical meaning in terms of the impact on evaporative fluxes from the soil and canopy (Anderson et al. 2007). It therefore has additional utility in terms of direct assimilation into other types of simulation systems, such as numerical weather prediction and surface hydrology models. It also has the advantage of being quantitatively verifiable in comparison with ground measurements.

On the basis of the exponential ET model and Hargreaves equation using MODIS NDVI composite products at 0.05 spatial resolution and 16-day temporal resolution, monthly ESI products for April–September 2004 have been produced in the agricultural crop growing cycle (Yao et al. 2011). Figure 7.13 shows that the values of ESI have decreased from Northwest to Southeast of China from April to September of 2004. During the entire growing season (April–September), the arid and semiarid regions of Northwest China with the highest ESI values have been suffering from water shortages. Specifically, the Xinjiang Uyghur autonomous region is a special desert region characterized by more severe drought. From the beginning of May, Northeast China including the Northeast Plain and Songliao Plain has experienced more droughts. This may stem from the intense soil degradation, due to the agricultural intensification reducing the water-holding capacity of agricultural ecosystems (Tao et al. 2005). In both June and July, ESI shows the extremely dry conditions at Huanghuaihai Plain locating in the midstream and downstream basins of the Yellow River, Huaihe River, and Haihe River where irrigation activities are dependent on the river water. This can be attributed to the fact that abnormal precipitation deficiency has raised the likelihood of increased agricultural droughts and their effect on crop production. In contrast, the major regions of Southeast China remain consistently high ET and low ESI throughout April–September, where annual precipitation is more than 1,200 mm and vegetation is provided water via precipitation. Therefore, ESI reflects the variation of drought over China and provides a good example for assessing the surface drought.

7.1.10 Comparison of Different Drought Indices

Various remote sensing indices of agriculture drought have been proposed at present, but it is hard to decide which one is the most appropriate. Sun et al. (2012b) performed an intercomparison among remote sensed indices for monitoring agricultural drought using MODIS products. The results indicated that these indices were not exactly the same because different indices used different features of land surface to represent agricultural drought. Some indices are suitable for agricultural

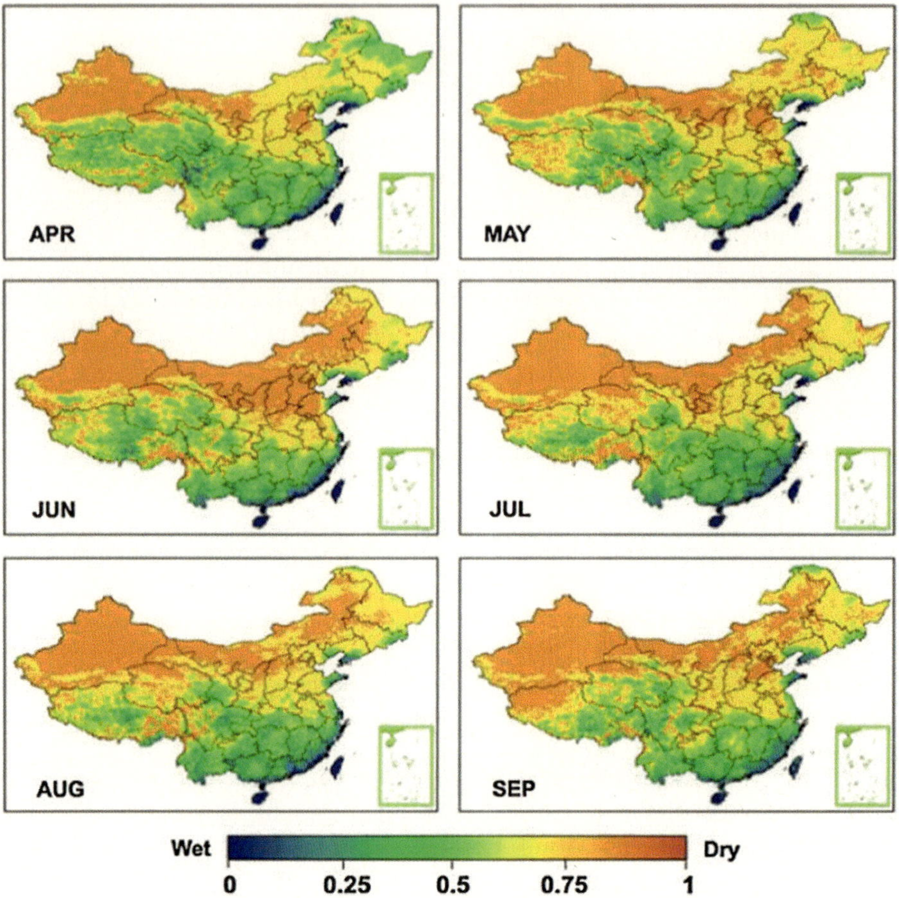

Fig. 7.13 Spatial distribution of monthly ESI for April–September of 2004 over China at 0.25° resolution based on the exponential evapotranspiration (*ET*) model and Hargreaves equation using radiation flux from the Global Energy and Water Cycle Experiment (*GEWEX*) Surface Radiation Budget (*SRB*) products and air temperature from the NCEP/NCAR Reanalysis Project (Reproduced from Yao et al. (2011), with permission from Springer)

drought early warning. Others (e.g., TVDI) are more suitable for drought monitoring. Table 7.3 gives a brief summary of the commonly applied drought indices that are developed based on remote sensing surface temperature.

7.2 Thermal Anomaly Detection

A thermal anomaly is an unexpected increase in the radiant temperature value of a pixel relative to its neighbors. Thermal anomalies can be caused by geothermal, forest fires, straw burning, underground coal fires, earthquakes, and volcanic

Table 7.3 Summary of the commonly applied remote sensing drought indices

Drought index	Input	Advantages	Disadvantages	References
ATI	Day and night T_s, albedo	Good physical foundation	Only suitable for bare soil or sparse vegetated areas	Mitra and Majumdar (2004)
TCI	T_s	Simple; suitable for large areas	Only suitable for homogeneous area	Kogan (1995)
VHI	T_s, NDVI	Simple; suitable for large areas	Only suitable for homogeneous area; not suitable for short-term moisture monitoring	Kogan (2001, 2002)
NDTI	T_s	Has more robust physical foundation than TCI	Complicated to calculate T_o and T_∞	McVicar and Jupp (1998)
VSWI	T_s, NDVI	Simple; suitable for large areas	Sensitive to remote sensing data errors	Li et al. (1998)
TVDI	T_s, NDVI	Suitable for low and high vegetated areas	Uncertainties in the determination of dry edge	Sandholt et al. (2002)
CWSI	T_s, T_a, RH	Good physical foundation; suitable for monitoring crop moisture and irrigation	Only suitable for highly vegetated areas	Idso et al. (1981) and Jackson et al. (1981)
WDI	T_s, T_a, NDVI	Suitable for lowly and highly vegetated areas	Uncertainties in the calculation of theoretical boundaries	Moran et al. (1994)
ESI	T_s, T_a, RH, WS, NDVI	Good physical foundation	Complicated for ET calculation	Anderson et al. (2007)

activities, among others. This section focuses mainly on the active fire detection and characterization and presents a brief overview of the urban heat island effect. Section 7.2.1 introduces the background and theoretical basis for fire detection. Section 7.2.2 describes the characteristics of the AVHRR and the MODIS sensors which are two most widely used sensors for long-term and large-scale fire monitoring. Sections 7.2.3 and 7.2.4 present methods for active fire detection and characterization. Section 7.2.5 presents two examples of active fire detection. Section 7.2.6 presents a brief overview of the urban heat island effect.

7.2.1 Background

Biomass burning is a key earth system process, a major element of the terrestrial carbon cycle, and a major source of atmospheric trace gases and aerosols (Crutzen et al. 1979; Crutzen and Andreae 1990; Andreae and Merlet 2001). Depending on their size, location, and time, fires can significantly modify land surface properties (Levine 1991), influence atmospheric chemistry and air quality, and perturb the radiative budget (Li et al. 1995; Li 1998). Therefore, fire has been identified by the international community as an important variable for the global climate observing system and an essential climate variable for the framework convention on climate change.

Satellite-based fire detection has traditionally relied on channels located near 4-μm to exploit the high levels of blackbody radiation emitted at typical fire temperature in this spectral region. The Planck function $B(\lambda, T)$, which describes the spectral radiance emitted at wavelength λ by a blackbody at temperature T, is given by

$$B(\lambda, T) = c_1 \lambda^{-5} \left[\exp\left(\frac{c_2}{\lambda T} \right) - 1 \right]^{-1} \tag{7.20}$$

where c_1 is the first Planck constant (3.741832×10^{-16} W μm^4 sr^{-1} m^{-2}), and c_2 is the second Planck constant (1.43876×10^{-2} μm K).

Planck radiances for blackbody temperatures from 300 to 800 K versus wavelengths are shown in Fig. 7.14. It is evident that the blackbody radiance increases with increasing temperature. Moreover, the peak of the Planck function shifts toward shorter wavelengths, as temperature increases. This phenomenon is due to the fact that the radiation of middle infrared (MIR) channel (MODIS channel 22) increases more rapidly than that of thermal infrared (TIR) channel (MODIS channel 31) at higher temperatures, which can be seen from Fig. 7.15. The differences in the radiometric response of MIR and TIR channels can be used to detect active fires.

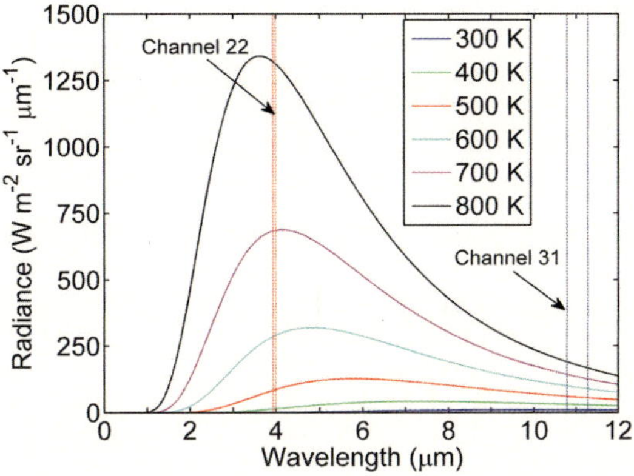

Fig. 7.14 Planck radiances for blackbody temperatures from 300 to 800 K. The *dashed lines* in *red* and *blue* represent the bandwidths of the MODIS channels 22 and 31, respectively (Reproduced from Matson and Dozier (1981), with permission from American Society for Photogrammetry and Remote Sensing)

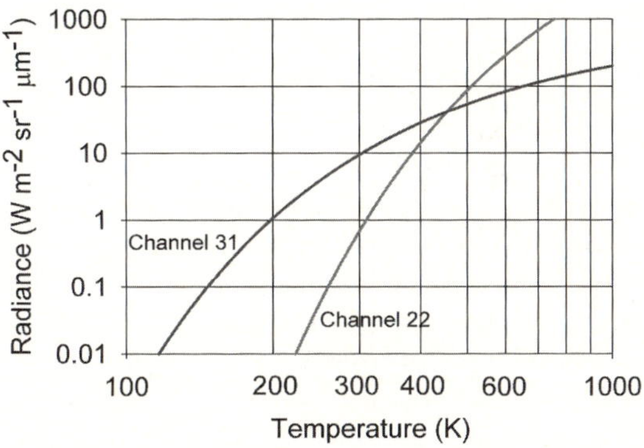

Fig. 7.15 Integrated Planck radiances for the MODIS channels 22 and 31, as calculated by convoluting the Planck function with the spectral response functions of the MODIS instrument (Reproduced from Dozier (1981), with permission from Elsevier)

7.2.2 Characteristics of the AVHRR and MODIS Sensors

Satellite remote sensing of fires has been achieved using a number of sensors. The AVHRR instrument onboard the NOAA polar orbiting satellites and the MODIS instrument onboard the Earth Observation System's (EOS) polar orbiting satellites are two most widely used sensors for long-term and large-scale fire monitoring (Robinson 1991; Kaufman et al. 1998a; Giglio et al. 1999, 2003; Li et al. 2000a, b; Justice et al. 2002).

The AVHRR instrument has two major advantages for fire monitoring. First, it provides daily coverage at a moderate resolution (~1 km), which is critical for operational global fire monitoring. Second, it has a wide spectral coverage ranging from visible to TIR. All channels are sensitive to certain attributes of fire but contain different information (see Table 7.4). Channels 1 and 2 provide data capable of detecting, monitoring, and measuring smoke emissions. Channel 3 is extremely sensitive to fire at the subpixel level. Although it has a lower temperature saturation point, ~321 K, it is the most important channel for fire detection. Channels 4 and 5 are far less sensitive to subpixel hotspots, but they are often helpful to fire

Table 7.4 AVHRR channels used for active fire monitoring

Channel number	Center wavelength (μm)	Denotation	Purpose	References
1	0.63	R_1	Bright surface, sun glint, and coastal false alarm rejection; cloud masking	Franca et al. (1995) and Arino and Melinotte (1998)
2	0.86	R_2	Bright surface, sun glint, and coastal false alarm rejection; cloud masking	Arino and Melinotte (1998) and Li et al. (2000a)
3	3.75	T_3	Active fire detection	Kaufman et al. (1990) and Li et al. (1997)
4	10.8	T_4	Active fire detection, cloud masking	Kaufman et al. (1990) and Li et al. (1997)
5	12.0	T_5	Cloud masking	Franca et al. (1995) and Li et al. (2000a)

detection when combined with the other channels. Nevertheless, AVHRR is also subject to numerous shortcomings, notably insufficient diurnal sampling, cloud contamination, and rapid saturation in channel 3. Some of the shortcomings are overcome by the MODIS instrument.

The MODIS instrument is designed to include characteristics specifically for fire detection and provides a unique capability over existing sensors in terms of fire monitoring. The brightness temperatures from two 3.96-μm channels (T_{21} and T_{22}) and the 11.03-μm channel (T_{31}) are used for fire monitoring. T_{21} and T_{22} saturate at approximately 500 K and 331 K, respectively. Because the low saturation channel (channel 22) is less noisy and has a smaller quantization error, T_{22} is always used for fire detection. Nevertheless, T_{22} is replaced by T_{21} when it saturates or has missing data. T_{31} saturates at approximately 400 K for the Terra-MODIS and 340 K for the Aqua-MODIS. The brightness temperature from the 12.02-μm channel (T_{32}) is used for cloud masking. The reflectances from the 0.65-μm and 0.86-μm channels (R_1 and R_2) are used to reject false alarms and mask clouds. The reflectance from the 2.13-μm channel (R_7) is used to reject water-induced false alarms. A summary of the MODIS channels used for active fire monitoring is shown in Table 7.5.

7.2.3 Fire Detection Theory

There are two types of algorithms to detect active fires (Eva and Flasse 1996; Boles and Verbyla 2000; Li et al. 2001; Ichoku et al. 2003): fixed-threshold algorithms and contextual algorithms. The fixed-threshold algorithms have the best performance for regional and unique landscapes (Kaufman et al. 1990; Kennedy et al. 1994;

Table 7.5 MODIS channels used for active fire monitoring

Channel number	Center wavelength (μm)	Denotation	Purpose	References
1	0.65	R_1	Sun glint and coastal false alarm rejection; cloud masking	Kaufman et al. (1998a) and Justice et al. (2002)
2	0.86	R_2	Bright surface, sun glint, and coastal false alarm rejection; cloud masking	Justice et al. (2002) and Giglio et al. (2003)
7	2.13	R_7	Sun glint and coastal false alarm rejection	Giglio et al. (2003) and He and Li (2011)
21	3.96	T_{21}	High-range channel for active fire detection	Kaufman et al. (1998a) and Giglio et al. (2003)
22	3.96	T_{22}	Low-range channel for active fire detection	Kaufman et al. (1998a) and Giglio et al. (2003)
31	11.03	T_{31}	Active fire detection, cloud masking	Kaufman et al. (1998a) and Justice et al. (2002)
32	12.02	T_{32}	Cloud masking	Giglio et al. (2003) and Wang et al. (2007)

Reproduced from Giglio et al. (2003), with permission from Elsevier

Li et al. 1997; Pozo et al. 1997; Rauste et al. 1997; Li et al. 2000a, b). In contrast, the contextual algorithms use dynamic thresholds and thus are more applicable to variable surface conditions (Lee and Tag 1990; Flasse and Ceccato 1996; Nakayama et al. 1999; Giglio et al. 2003; Lasaponara et al. 2003; Wang et al. 2007). Fixed threshold algorithms generally rely on preset absolute thresholds and consider a single pixel at a time, while contextual algorithms compute relative thresholds based on statistics calculated from neighboring pixels.

7.2.3.1 Fixed Threshold Algorithms

The fixed threshold algorithms apply empirically defined thresholds to discriminate between pixels containing fire and those of the nonburning surroundings and clouds. The majority of the fixed threshold algorithms consist of three basic steps (Kaufman et al. 1990): (1) using MIR channel to identify all potential fires, (2) using TIR channel to eliminate clouds, and (3) using the difference between brightness temperatures in MIR and TIR channels to isolate fires from warm background. The fixed threshold algorithms have been applied to detect fires in various biomes such as tropical forests (Kaufman et al. 1990), savanna fires in West Africa (Kennedy et al. 1994; Franca et al. 1995), and boreal forests (Flannigan and Vonder Haar 1986; Li et al. 1997, 2000a). Seven fixed threshold algorithms are summarized in Table 7.6.

Table 7.6 Summary of seven fixed threshold algorithms

Algorithm	Ecosystem	Data	Fire identification criteria
Flannigan and Vonder Haar (1986)	Boreal forest in Canada	NOAA-7 AVHRR	(1) $T_3 > \overline{T}_{3b}$ (2) $T_4 > \overline{T}_{4b}$ (3) $\Delta T_{34} \geq 10$ K for daytime and $\Delta T_{34} \geq 8$ K for nighttime
Kaufman et al. (1990)	Tropical forest in Brazil	NOAA-9 AVHRR	(1) $T_3 \geq 320$ K (2) $\Delta T_{34} \geq 10$ K (3) $T_4 > 250$ K
Kennedy et al. (1994)	Savanna in West Africa	NOAA-11 AVHRR	(1) $T_3 \geq 320$ K (2) $\Delta T_{34} \geq 15$ K (3) $T_4 > 295$ K (4) $R_2 \leq 0.16$
Franca et al. (1995)	Savanna and tropical forest in West Africa	NOAA-11 AVHRR	(1) $T_3 \geq 320$ K (2) $\Delta T_{34} \geq 15$ K (3) $T_4 > 287$ K (4) $0 \leq \Delta T_{45} \leq 5$ K (5) $R_1 \leq 0.09$
Li et al. (1997)	Boreal forest in Canada	NOAA-11 AVHRR	(1) $T_3 \geq 316$ K (2) $\Delta T_{34} \geq 10$ K (3) $T_4 > 245$ K
Arino and Melinotte (1998)	Tropics in Africa	NOAA AVHRR	(1) $T_3 \geq 320$ K (2) $\Delta T_{34} \geq 15$ K (3) $T_4 > 245$ K (4) $R_1 \leq 0.25$ (5) $\lvert R_1 - R_2 \rvert > 0.01$
Li et al. (2000a, b)	Boreal forest in Canada	NOAA-14 AVHRR	(1) $T_3 > 315$ K (2) $\Delta T_{34} \geq 14$ K (3) $T_4 \geq 260$ K (4) $R_2 \leq 0.22$ (5) $\Delta T_{45} \leq 4.1$ K or $\Delta T_{34} \geq 19$ K

Flannigan and Vonder Haar (1986) first used a threshold algorithm for boreal forest fire monitoring in Canada using NOAA-7 AVHRR data. This algorithm identifies fire pixels by means of the following thresholds: (1) $T_3 > \overline{T}_{3b}$, (2) $T_4 > \overline{T}_{4b}$, and (3) $\Delta T_{34} > 10$ K for daytime and $\Delta T_{34} > 8$ K for nighttime, where $\Delta T_{34} = T_3 - T_4$, T_3 and T_4 are the brightness temperatures in the AVHRR channels 3 and 4, respectively, and $\overline{T}_{3b}$ and $\overline{T}_{4b}$ are the mean of T_3 and of T_4 for the background pixels, respectively.

Kaufman et al. (1990) proposed a threshold-based algorithm to detect tropical forest fires in Brazil using NOAA-9 AVHRR data. They introduced brightness temperature in the AVHRR channel 4 to eliminate highly reflective clouds. A pixel was classified as fire if it met a set of three criteria: (1) $T_3 \geq 320$ K, (2) $\Delta T_{34} \geq 10$ K, and (3) $T_4 > 250$ K. The first criterion is used to identify potential fire pixels. The second criterion is used to eliminate warm background. The third criterion is used to eliminate highly reflective clouds.

Kennedy et al. (1994) improved the algorithm developed by Kaufman et al. (1990) for savanna fire monitoring in West Africa using NOAA-11 AVHRR data. Modified thresholds were tested for fire detecting in the ecosystems of West Africa. Moreover, they introduced a threshold of reflectance in the AVHRR channel 2 (R_2) to eliminate highly reflective surfaces. This algorithm identified fire pixels in terms of four criteria: (1) $T_3 \geq 320$ K, (2) $\Delta T_{34} \geq 15$ K, (3) $T_4 > 295$ K, and (4) $R_2 \leq 0.16$.

Franca et al. (1995) used an algorithm to detect fires of Savanna and tropical forest in West Africa using NOAA-11 AVHRR data. They introduced the difference between brightness temperatures in the AVHRR channels 4 and 5 to eliminate thin cirrus clouds. In addition, reflectance in the AVHRR channel 1 (R_1) was used to eliminate highly reflective surfaces. This algorithm identifies fire pixels using the following threshold criteria: (1) $T_3 \geq 320$ K, (2) $\Delta T_{34} \geq 15$ K, (3) $T_4 > 287$ K, (4) $0 \leq \Delta T_{45} \leq 5$ K, and (5) $R_1 \leq 0.09$, where $\Delta T_{45} = T_4 - T_5$, T_5 is the brightness temperature in the AVHRR channel 5.

Li et al. (1997) developed an algorithm based on the algorithm proposed by Kaufman et al. (1990) for boreal forest fire monitoring in Canada using NOAA-11 AVHRR data. Sun glint was found to be the prominent problem resulting in false alarms, due to specular reflection of solar radiation from the lakes, rivers, and swamps in the study areas. To eliminate this problem, all forward scattering measurements were discarded from further analysis. A pixel was identified as fire in terms of three criteria: (1) $T_3 \geq 316$ K, (2) $\Delta T_{34} \geq 10$ K, and (3) $T_4 > 245$ K.

Arino and Melinotte (1998) used an algorithm for monitoring fires in Africa using NOAA AVHRR data. They introduced the absolute difference between reflectances in the AVHRR channels 1 and 2 ($|R_1 - R_2|$) to eliminate false alarms due to sun glint. The following criteria were used to identify fire pixels: (1) $T_3 \geq 320$ K, (2) $\Delta T_{34} \geq 15$ K, (3) $T_4 > 245$ K, (4) $R_1 \leq 0.25$, and (5) $|R_1 - R_2| > 0.01$.

Li et al. (2000a, b) developed an algorithm to detect boreal forest fires in Canada using NOAA-14 AVHRR data. This algorithm consists of five major steps: (1) pixels with $T_3 > 315$ K are identified as potential fire pixels, (2) pixels with $\Delta T_{34} \leq 14$ K are considered as false fire pixels caused by warm background, (3) pixels with $T_4 \leq 260$ K are eliminated as false fire pixels caused by highly reflective clouds, (4) pixels with $R_2 \geq 0.22$ are considered as false fire pixels caused by highly reflective clouds and surfaces, and (5) pixels with $\Delta T_{45} \geq 4.1$ K and $\Delta T_{34} < 19$ K are removed as false fire pixels caused by thin cirrus clouds with warm background.

7.2.3.2 Contextual Algorithms

The contextual algorithms (Cuomo et al. 2001; Lasaponara et al. 2003; He and Li 2011, 2012) use dynamic thresholds, relying on the contrast between a potential fire pixel and its background pixels to detect fires. These algorithms are more flexible and effective in variable surface conditions than fixed threshold algorithms. The majority of the contextual algorithms are composed of three basic steps: (1) preliminary thresholds to identify potential fire pixels, (2) contextual tests to confirm the potential fire pixels, and (3) thresholds to reject false alarms. For the contextual algorithms, it is vital to characterize the background pixels neighboring a potential fire pixel. Valid background pixels in a window centered on the potential fire pixel are identified as those pixels that (1) contain usable observations, (2) are located on lands, (3) are not cloud-contaminated, and (4) are not background fire pixels. The window starts from 3×3 pixels and increases to a maximum of 21×21 pixels until at least 25 % of the pixels within the window are valid background pixels. If an insufficient number of valid background pixels are identified, the potential pixel is classified as unknown and is not subject to further test.

Justice et al. (1996) developed a contextual algorithm for active fire detection using NOAA AVHRR data. This contextual algorithm consists of two steps: potential fire detection and confirmation. A pixel is flagged as fire using the following criteria: (1) $T_3 \geq 316$ K, (2) $\Delta T_{34} \geq \overline{\Delta T}_{34b} + \max(2\sigma_{34b}, 5\text{ K})$, and (3) $T_4 \geq \overline{T}_{4b}$, where $\overline{\Delta T}_{34b}$ and σ_{34b} are the mean and standard deviation of ΔT_{34} for the background pixels, respectively. The criterion 1 is used to identify potential pixels, and the criteria 2 and 3 are used to confirm the potential pixels.

Based on the principles proposed by Justice et al. (1996), Flasse and Ceccato (1996) developed a contextual active fire detection algorithm using NOAA AVHRR data. This algorithm has been used to generate a Global Fire Product (Stroppiana et al. 2000). A pixel is classified as fire in terms of the following criteria: (1) $T_3 > 311$ K, (2) $\Delta T_{34} > 8$ K, (3) $R_2 < 0.2$, (4) $T_3 > \overline{T}_{3b} + 2\sigma_{3b} + 3$ K, and (5) $\Delta T_{34} > \overline{\Delta T}_{34b} + 2\sigma_{34b}$, where σ_{3b} is the standard deviation of T_3 for the background pixels.

Giglio et al. (1999) compared the algorithms developed by Justice et al. (1996) and Flasse and Ceccato (1996) and proposed a contextual fire detection algorithm based on the algorithm of Justice et al. (1996). Daytime pixels are identified as fires if the following criteria are satisfied: (1) $T_3 > 310$ K, (2) $\Delta T_{34} > 6$ K, (3) $R_2 < 0.25$, (4) $\Delta T_{34} \geq \overline{\Delta T}_{34b} + \max(2.5\sigma_{34b}, 4\text{ K})$, and (5) $T_4 > \overline{T}_{4b} + 2\sigma_{3b} - 3$ K. Nighttime pixels are flagged as fires if the following conditions are satisfied: (1) $T_3 > 308$ K, (2) $\Delta T_{34} > 4$ K, and (3) $\Delta T_{34} \geq \overline{\Delta T}_{34b} + \max(2.5\sigma_{34b}, 4\text{ K})$.

Based on the heritage algorithms developed for AVHRR (Giglio et al. 1999), Kaufman et al. (1998a) developed the MODIS version 3 active fire detection algorithm. This contextual algorithm consists of absolute fire detection and relative fire detection (Justice et al. 2002). For the absolute fire detection, the algorithm requires that at least one of two conditions is satisfied: (1) $T_{22} > 360$ K (330 K at night), and (2) $T_{22} > 330$ K (315 K at night) and $\Delta T > 25$ K (10 K at night), where $\Delta T = T_{22} - T_{31}$, and T_{22} and T_{31} are the brightness temperatures in the MODIS channels 22 and 31, respectively. If either of these absolute criteria is not satisfied, the relative fire detection is performed in terms of the following criteria: (1) $T_{22} > \overline{T}_{22b} + 3\sigma_{22b}$, and (2) $\Delta T > M_{\Delta Tb} + 3\sigma_{\Delta Tb}$, where $\overline{T}_{22b}$ and σ_{22b} are the mean and standard deviation of T_{22} for the background pixels, respectively, and $M_{\Delta Tb}$ and $\sigma_{\Delta Tb}$ are the median and standard deviation of ΔT for the background pixels, respectively. Valid background pixels are identified as those pixels with $T_{22} < 325$ K (315 K at night) and $\Delta T < 20$ K (10 K at night). If either σ_{22b} or $\sigma_{\Delta Tb}$ is below 2 K, a value of 2 K is used instead. Daytime fire pixels are rejected as false alarms due to sun glint if the following criteria are satisfied: (1) $R_1 > 0.3$, (2) $R_2 > 0.3$, and (3) glint angle <40°, where R_1 and R_2 are the reflectances in the MODIS channels 1 and 2, respectively.

Although the MODIS version 3 active fire detection algorithm (Kaufman et al. 1998a) functioned reasonably well, two significant problems limited the overall quality of the product (Giglio et al. 2003). Firstly, persistent false detections occurred in some deserts and sparsely vegetated land surfaces. Secondly, relatively small fires were frequently not detected. Consequently, Giglio et al. (2003) developed an improved contextual fire detection algorithm that offers superior sensitivity to smaller, cooler fires and significantly lowers false alarms. This MODIS version 4 active fire detection algorithm consists of nine steps:

1. Cloud and water masking. Daytime pixels are identified as clouds if one of the following criteria is satisfied: (1) $R_1 + R_2 > 0.9$, (2) $T_{32} < 265$ K, and (3) $R_1 + R_2 > 0.7$ and $T_{32} < 285$ K, where T_{32} are the brightness temperature in the MODIS channel 32. Nighttime pixels are flagged as clouds if the single condition is satisfied: $T_{32} < 265$ K. Water pixels are eliminated using the MODIS land/sea mask product.
2. Identification of potential fire pixels. Daytime pixels with $T_{22} > 310$ K, $\Delta T > 10$ K, and $R_2 < 0.3$ are identified as potential fire pixels. Nighttime pixels with $T_{22} > 305$ K and $\Delta T > 10$ K are flagged as potential fire pixels.
3. Absolute threshold test. Absolute threshold criterion with $T_{22} > 360$ K for daytime or $T_{22} > 320$ K for nighttime is used to identify fire pixels.
4. Background characterization. Valid background pixels in a window centered on a potential fire pixel are identified as those that (1) contain usable observations, (2) are located on land, (3) are not cloud-contaminated, and (4) are not background fire pixels. Background fire pixels are defined as those with $T_{22} > 325$ K and $\Delta T > 20$ K for daytime or with $T_{22} > 310$ K and $\Delta T > 10$ K for nighttime, where $\Delta T = T_{22} - T_{31}$.
5. Contextual threshold tests. If the background characterization is successful, a series of contextual threshold tests are used for relative fire detection. These tests are (1) $\Delta T > \overline{\Delta T}_b + 3.5\delta_{\Delta Tb}$, (2) $\Delta T > \overline{\Delta T}_b + 6$ K, (3) $T_{22} > \overline{T}_{22b} + 3\delta_{22b}$, (4) $T_{31} > \overline{T}_{31b} + \delta_{31b} - 4$ K, and (5) $\delta_{22bf} > 5$ K, where $\overline{\Delta T}_b$ and $\delta_{\Delta Tb}$ are the mean and mean absolute deviation of ΔT for the background pixels, $\overline{T}_{22b}$ and δ_{22b} are the mean and mean absolute deviation of T_{22} for the background pixels, $\overline{T}_{31b}$ and δ_{31b} are the mean and mean absolute deviation of T_{31} for the background pixels, and δ_{22bf} is the mean absolute deviation of T_{22} for the background fire pixels.
6. Tentative fire detection. A daytime pixel is tentatively classified as fire if one of the following criteria is satisfied: (1) step 2 is true, and (2) tests 1–3 of step 5 are

true and test 4 or 5 of step 5 is true. A nighttime pixel is definitively classified as fire if one of the following criteria is satisfied: (1) step 2 is true, and (2) tests 1–3 of step 5 are true.

7. Sun glint rejection. A tentative fire pixel during daytime is rejected as sun glint and classified as a nonfire pixel if one or more of the following criteria are satisfied: (1) $\theta_g < 2°$; (2) $\theta_g < 8°$, $R_1 > 0.1$, $R_2 > 0.2$, and $R_7 > 0.12$; and (3) $\theta_g < 12°$ and $N_{aw} + N_w > 0$, where θ_g is the angle between vectors pointing in the surface-to-satellite and specular reflection directions (i.e., $\theta_g = \cos\theta_v \cos\theta_s - \sin\theta_v \sin\theta_s \cos\varphi$); θ_v and θ_s are the view and solar zenith angles, respectively; φ is the relative azimuth angle; R_7 is the reflectance in the MODIS channel 7; N_w is the number of water pixels surrounding a potential fire pixel; and N_{aw} is the number of water pixels within eight pixels surrounding a tentative fire pixel.
8. Desert boundary rejection. A tentative fire pixel during daytime is rejected as a hot desert boundary surface and classified as a nonfire pixel if all of the following conditions are satisfied: (1) $N_f > 0.1N_v$, (2) $N_f \geq 4$, (3) $R_2 > 0.15$, (4) $\overline{T}_{22bf} < 345$ K, (5) $\delta_{22bf} < 3$ K, and (6) $T_{22} < \overline{T}_{22bf} + 6\delta_{22bf}$, where N_f is the number of background fire pixels surrounding a potential fire pixel, N_v is the number of valid background pixels surrounding a potential fire pixel, and $\overline{T}_{22bf}$ is the mean of T_{22} for the background fire pixels.
9. Coastal false alarm rejection. A tentative fire pixel during daytime is rejected as a coastal false alarm and classified as a nonfire pixel if step 2 is not satisfied and $N_{uw} > 0$, where N_{uw} is the number of unmasked water pixels surrounding a potential fire pixel. Valid background pixels are considered to be unmasked water pixels if all conditions are satisfied: (1) $R_2 < 0.15$, (2) $R_7 < 0.05$, and (3) NDVI < 0, where NDVI $= (R_2 - R_1)/(R_2 + R_1)$.

Based on the MODIS version 4 contextual algorithm (Giglio et al. 2003) and a smoke detection algorithm (Xie et al. 2005), Wang et al. (2007) developed a contextual algorithm for small and cool fire detection using MODIS data. Four main differences between this algorithm and the MODIS version 4 contextual algorithm are identified as follows:

1. A threshold of $T_{28} < 255$ K is used to reject cloud edge, where T_{28} is the brightness temperature in the MODIS channel 28.
2. Four criteria are used to identify smoke pixels: (1) $0.15 \leq (R_8 - R_{19})/(R_8 + R_{19}) \leq 0.5$, (2) $(R_9 - R_7)/(R_9 + R_7) \geq 0.3$, (3) $(R_8 - R_3)/(R_8 + R_3) \leq 0.09$, and (4) $R_8 \geq 0.09$, where R_3, R_8, R_9, and R_{19} are the reflectances in the MODIS channels 3, 8, 9, and 19, respectively. Potential fire pixels are identified as the pixels within a 14×14 pixels window centered on a smoke pixel.
3. A threshold of $T_{22} > 293$ K is used to replace the threshold of $T_{22} > 310$ K in the MODIS version 4 contextual algorithm to detect small and cool fires.
4. Steps 7–9 in the MODIS version 4 contextual algorithm are not used in this algorithm.

7.2.4 *Fire Characterization*

In addition to active fire detection, it is important to characterize active fire properties in terms of subpixel fire temperature and area and fire radiative power (FRP).

7.2.4.1 Subpixel Fire Temperature and Area

Dozier (1981) first proposed a bi-spectral method to retrieve the temperature and area of a subpixel fire. This method uses the difference in the response of the 3.8-μm MIR and 10.8-μm TIR channels to a subpixel high temperature target. Although originally developed for the AVHRR, this method can be applied to any sensor having similar MIR and TIR channels. For a target at temperature T_f occupying a fractional area p, where $0 < p < 1$, located within a uniform background at temperature T_b, the observed radiances in the AVHRR channels 3 and 4 are

$$R_3 = pB\left(\lambda_3, T_f\right) + (1 - p)\, B\left(\lambda_3, T_b\right) \tag{7.21}$$

$$R_4 = pB\left(\lambda_4, T_f\right) + (1 - p)\, B\left(\lambda_4, T_b\right) \tag{7.22}$$

where $B(\lambda, T)$ is the Plank function and R_3 and R_4 are the observed radiances in the AVHRR channels 3 and 4, respectively.

The bi-spectral method has been applied to derive the subpixel temperature and area of hot targets using different sensors (Matson and Dozier 1981; Flannigan and Vonder Haar 1986; Matson and Holben 1987; Prins and Menzel 1992; Langaas 1993; Oertel et al. 2004; Qian and Kong 2012). Applications of this method generally require the following assumptions: (1) the pixel consists of only two components (the hot target and the background) with different temperatures, (2) the hot target and the background are assumed to be blackbodies with unit emissivity in both channels, (3) the background temperature can be estimated from surrounding pixels, and (4) all atmospheric effects are neglected.

The assumptions used in the original bi-spectral method are unrealistic. To create a more realistic retrieval, Giglio and Kendall (2001) pointed out that the atmospheric effects must be taken into account and the assumption of unit background emissivity must be removed. With these modifications, the observed radiances in the AVHRR channels 3 and 4 are

$$R_3 = \tau_3 pB\left(\lambda_3, T_f\right) + (1 - p)\, R_{b,3} \tag{7.23}$$

$$R_4 = \tau_4 pB\left(\lambda_4, T_f\right) + (1 - p)\, R_{b,4} \tag{7.24}$$

where τ_3 and τ_4 are the atmospheric transmittances in the AVHRR channels 3 and 4, respectively, and $R_{b,3}$ and $R_{b,4}$ are the background radiances in the AVHRR channels 3 and 4, respectively, which are estimated from the average radiances of several neighboring nonfire pixels.

Despite the improved calculations, there are numerous well-documented limitations of the bi-spectral method for fire characterization, including improper selection of TIR background temperature, band-to-band co-registration errors, and point spread function (PSF) differences between the MIR and TIR channels (Langaas 1993; Giglio and Kendall 2001; Giglio and Justice 2003; Shephard and Kennelly 2003; Calle et al. 2009). To eliminate the effects of the band-to-band co-registration errors and PSF differences on fire property retrievals (Shephard and Kennelly 2003; Calle et al. 2009), Zhukov et al. (2006) applied the bi-spectral method on a prepixel-cluster basis rather than on a per-pixel basis.

Dennison et al. (2006) developed a method to retrieve subpixel properties using Multiple Endmember Spectral Mixture Analysis (MESMA). The MESMA assumes that the radiative signature of each pixel is a result of a linear combination of subpixel end-members (Roberts et al. 1998):

$$R_i = \sum_{k=1}^{N} f_k \times R_{ik} + \varepsilon_i \quad \text{and} \quad \sum_{k=1}^{N} f_k = 1 \tag{7.25}$$

where R_i is the observed radiance of the pixel in channel i, R_{ik} is the observed radiance of end-member k in channel i, f_k is the fraction area of end-member k, N is the number of end-members, and ε_i is the residual error in channel i.

Model fit is assessed using root-mean-squared error (RMSE):

$$\text{RMSE} = \left[\frac{\sum_{i=1}^{M} (\varepsilon_i)^2}{M} \right]^{1/2} \tag{7.26}$$

where M is the number of channels.

The MESMA method has been applied to retrieve the temperature and area of subpixel fires using the Airborne Visible Infrared Imaging Spectrometer (AVIRIS, Dennison et al. 2006), Advanced Spaceborne Thermal Emission and Reflection Radiometer (ASTER, Eckmann et al. 2009), and MODIS (Eckmann et al. 2008, 2010) data.

7.2.4.2 Fire Radiative Power

Fire radiative power (FRP) is a measure of the radiant energy liberated per unit time from burning vegetation via the rapid oxidation of fuel carbon. FRP is therefore related to the rate of fuel combustion and carbon volatization (Roberts et al. 2005). Temporal integration of FRP over a fire's lifetime provides a measure of the total fire radiative energy (FRE), which is proportional to the fuel mass combusted and carbon volatized (Roberts et al. 2005). For a fire pixel, the true FRP from all fire components is given by the Stefan–Boltzmann law (Giglio et al. 2008; Roberts and Wooster 2008):

$$\mathrm{FRP}_{\mathrm{TRUE}} = A_{\mathrm{pixel}} \varepsilon \sigma \sum_{i=1}^{n} p_i T_i^4 \tag{7.27}$$

where $\mathrm{FRP}_{\mathrm{TRUE}}$ is the fire radiative power (W), σ is the Stefan–Boltzmann constant (5.67×10^{-8} W m^{-2} K^{-4}), ε is the effective mean emissivity of all fire components over all emitting wavelengths, p_i is the fractional area of the ith fire component, T_i is the temperature of the ith fire component, and A_{pixel} is the area of the pixel.

Equation (7.27) is inappropriate to calculate the $\mathrm{FRP}_{\mathrm{TRUE}}$ from remotely sensed measurements because the spatial resolution of satellite sensors is inadequate for resolving the temperature and area of each fire component (Roberts et al. 2005). Nevertheless, assuming a single, homogeneous subpixel fire within a pixel, the temperature and area of the subpixel fire derived from the bi-spectral method (Dozier 1981) can be used to estimate FRP via Eq. (7.27) (Wooster et al. 2003; Zhukov et al. 2006),

$$\mathrm{FRP}_{\mathrm{BS}} = \sigma \left(T_{\mathrm{f}}^4 - T_{\mathrm{b}}^4 \right) A_{\mathrm{f}} \tag{7.28}$$

where T_{f} is the temperature of the subpixel fire, A_{f} is the area of the subpixel fire, and T_{b} is the brightness temperature of the background in the TIR channel.

Two major disadvantages are identified for the bi-spectral method to estimate FRP. Firstly, the bi-spectral method is sensitive to the co-registration errors and PSF differences between the MIR and TIR channels (Shephard and Kennelly 2003). This can cause the radiant power contribution from a fire to vary between the spectral bands of the sensor, with the results that the fire may contribute proportionally more to the MIR channel signal of a fire pixel than to the TIR channel (Wooster et al. 2005). Secondly, the bi-spectral method relies on an estimate of the TIR background signal. This can be problematic because the background contribution to the pixel-averaged TIR radiance of the fire pixels is likely to be significantly larger than the fire contribution, and thus inaccuracy in the TIR background radiance estimation can introduce significant errors into the temperature and area of the subpixel fire derived from the bi-spectral method (Wooster et al. 2003).

An empirical method was proposed to relate the FRP of a fire pixel to the brightness temperature in the MODIS MIR channel (Kaufman et al. 1998a, b):

$$\mathrm{FRP}_{\mathrm{MODIS}} = 4.34 \times 10^{-19} \left(T_{\mathrm{MIR}}^8 - T_{\mathrm{MIR,b}}^8 \right) A_{\mathrm{pixel}} \tag{7.29}$$

where T_{MIR} and $T_{\mathrm{MIR,b}}$ are the brightness temperatures of the fire pixel and the background in the MODIS MIR channel.

A major advantage of the MODIS method is that it relies only on quantification of the fire pixel in a single MIR channel, removing problems related to interchannel spatial co-registration (Wooster et al. 2005). Furthermore, because it uses the MIR channel, where the signal from the fire is at a maximum, the effect of uncertainty in the background signal is lessened when compared to situations where quantification of the TIR signal is also required (Wooster et al. 2005). The major disadvantage of the MODIS method is that, unlike the bi-spectral method, variations in fire temperature are not explicitly accounted for (Wooster et al. 2003).

Based on a fourth-order power-law approximation to the Plank function, Wooster et al. (2003) developed a method to estimate FRP from the MIR radiance of a fire pixel:

$$\mathrm{FRP_{MIR}} = \frac{A_{\mathrm{pixel}}\sigma\varepsilon}{a\varepsilon_{\mathrm{MIR}}}\left(\mathrm{R_{MIR}} - \mathrm{R_{MIR,b}}\right) \tag{7.30}$$

where R_{MIR} and $R_{\mathrm{MIR,b}}$ are the radiances of the fire pixel and the background in the MIR channel; $\varepsilon_{\mathrm{MIR}}$ is the emissivity in the MIR channel, which is assumed to radiate as a gray body ($\varepsilon = \varepsilon_{\mathrm{MIR}}$); and a is a constant based on the empirical best-fit between emitter temperature and MIR radiance using a power-law approximation to the Plank function.

A major advantage of the MIR method is that it avoids the need to quantify the fire signal in the TIR channel, where variance in the background radiance can be of a similar magnitude to the fire pixel radiance itself (Wooster et al. 2005). However, a major assumption of the MIR method is the necessity to use a fourth-order power-law approximation to the Plank function, because the soundness of this approximation varies somewhat over the temperature range considered appropriate to active fires (i.e., ~600 K < Tf <1,600 K) and breaks down markedly outside this range (Wooster et al. 2005).

The three methods have been used to estimate FRP using different sensors, such as MODIS (Kaufman et al. 1998a, b; Peterson et al. 2013; Peterson and Wang 2013), GOES (Schroeder et al. 2010; Xu et al. 2010), Spinning Enhanced Visible and Infrared Imager (SEVIRI, Roberts et al. 2005; Roberts and Wooster 2008), and Bi-spectral Infrared Detection (BIRD, Wooster et al. 2003; Zhukov et al. 2006).

7.2.5 Fire Detection Application

In this section, we introduce the applications of a fixed threshold algorithm and a contextual algorithm for active fire detection based on the studies of Li et al. (2000a, b) and Wang et al. (2007), respectively.

7.2.5.1 Fire Detection Using a Fixed Threshold Algorithm

Based on the algorithm proposed by Kaufman et al. (1990), Li et al. (2000a) developed a fixed threshold algorithm to detect boreal forest fires in Canada using

Fig. 7.16 Flowchart of the fixed threshold fire detection algorithm (Reproduced from Li et al. (2000a), with permission from Taylor & Francis)

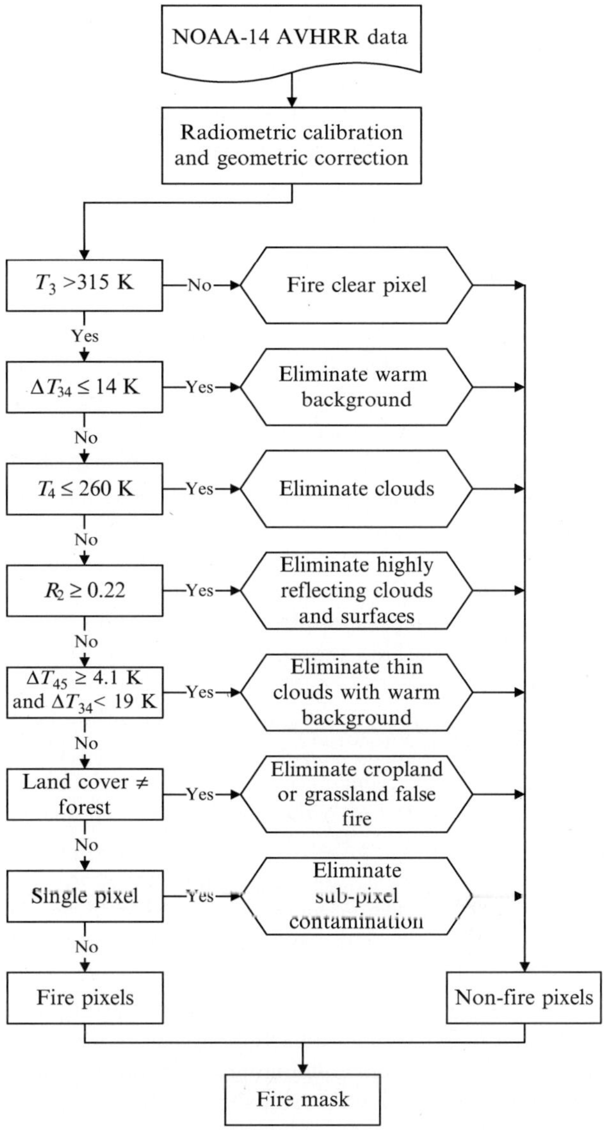

NOAA-14 AVHRR data. This algorithm consists of two major steps: marking potential fires and removing false fires. The threshold values are selected following a trial-and-error approach based on a fire training dataset. Histogram analyses of reflectances and brightness temperatures corresponding to burning and nonburning pixels prove to be an effective mean of obtaining optimized threshold values. The flowchart of the algorithm is shown in Fig. 7.16.

The study area is located in northern Manitoba and Saskatchewan, which is representative of the boreal biome. The efficiency of each threshold test of the algorithm is computed in terms of 24 training scenes of 1,200 km × 1,200 km over the study area during the 1995 fire season. For the training data set, fires are identified by visual inspection of hot spots in the AVHRR channel 3 and smoke plumes in the AVHRR channel 1 in the absence of thick clouds. The results of the tests are given by means of the number of retained true fires (TF) and removed false fires (FF) in each test. As shown in Table 7.7, the algorithm omits 11 % of the TF pixels and eliminates 99 % of the FF pixels after applying all of the threshold tests. The most effective threshold test for eliminating FF pixels appears to be the one of $\Delta T_{34} \leq 14$ K, followed by the one of $R_2 \geq 0.22$. Although 14 % of the "fire pixels" that passed all the tests remain FF pixels, further reduction of these would lead to a larger number of TF pixels being omitted.

A fire mask derived from AVHRR images on June 25, 1995 (see Fig. 7.17), is taken as an example to demonstrate the performance of this algorithm. Fire pixels are marked as red spots and the background is a false color composite denoting vegetation (green), thin cloud over land (blue), cloud (white), and smoke (orange). There are three major fire episodes, one located in Northwest Territories, one in a border region between Saskatchewan and Manitoba, and one in Quebec. Almost all of the detected fires are accompanied by smoke plumes, confirming their occurrence. Cloud cover is the primary limitation of the AVHRR-based fire detection technique, especially for real-time fire surveillance. It poses a less serious problem for obtaining seasonal fire statistics, since cloud cover changes quickly from day to day.

To further evaluate the fire detection algorithm, Li et al. (2000b) uses the detailed and complete records of forest fires gathered by the provincial and territorial fire management agencies of Canada. These fire data in Northwest Territories and Saskatchewan during the 1995 fire season are obtained. Figure 7.18 presents comparisons of these data to the annual composites of fire pixels detected by satellite over the two regions. The blue polygons outline the boundaries of burnt area reported by fire agencies while red dots are locations of active fire pixels detected during the entire fire season. In Saskatchewan, 47 major fire events are reported by fire agencies and all of them are detected by this algorithm. In the Northwest Territories, there are 24 large fires reported by fire agencies and six of them are omitted by this algorithm. However, the omitted fires cover small areas on the order of 1,000 ha. Combining data from the two regions, 7.3 % of the pixels showing fire activity fall outside the burnt area boundaries reported by fire agencies. Li et al. (2000b) visually inspect satellite images for all the cases in which satellite detects fire activities that are not reported by fire agencies. It is determined that 20 % of these fires have smoke associated with them. Most of the large fire clusters detected by satellite show some smoke phenomena. These seemingly real fires are not reported by fire agencies. Only 5.8 % of the pixels that are detected by satellite as fires are not certain. Their sizes are very small (smaller than 300 ha). Therefore, the probability that pixels detected by satellites as fires are true fires (>300 ha) is as high as 94 %.

Table 7.7 Summary of the fire detection tests and the statistics of their efficiency

Test no.	Description	Threshold	Number of true fires	Number of false fires
1	Initial test	$T_3 \geq 315$ K	12,569	168,168
2	Eliminate warm background	$\Delta T_{34} \leq 14$ K	12,569	48,855
3	Eliminate nonforest fires	Land cover	12,569	30,511
4	Eliminate highly reflecting clouds and surfaces	$R_2 \geq 0.22$	12,442	5,665
5	Eliminate thin clouds within warm background	$\Delta T_{45} \geq 4.1$ K and $\Delta T_{34} < 19$ K	11,307	2,673
6	Eliminate clouds	$T_4 \leq 260$ K	11,307	2,673
7	Eliminate single fire pixels		11,160	1,828

Reproduced from Li et al. (2000a), with permission from Taylor & Francis

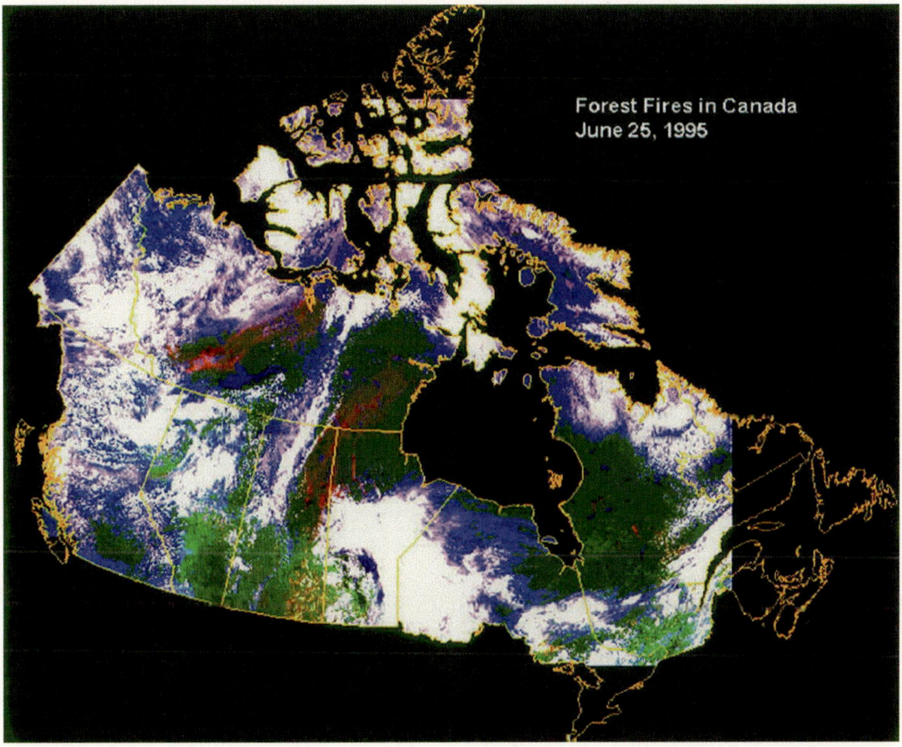

Fig. 7.17 Forest fires detected in June 1995 across Canada. The *red dots* are fire hot spots, superimposed on an AVHRR false color composite image (*green* for vegetation, *orange* for smoke and *white* for clouds, and *blue* for thin cloud over land) (Reproduced from Li et al. (2000a), with permission from Taylor & Francis)

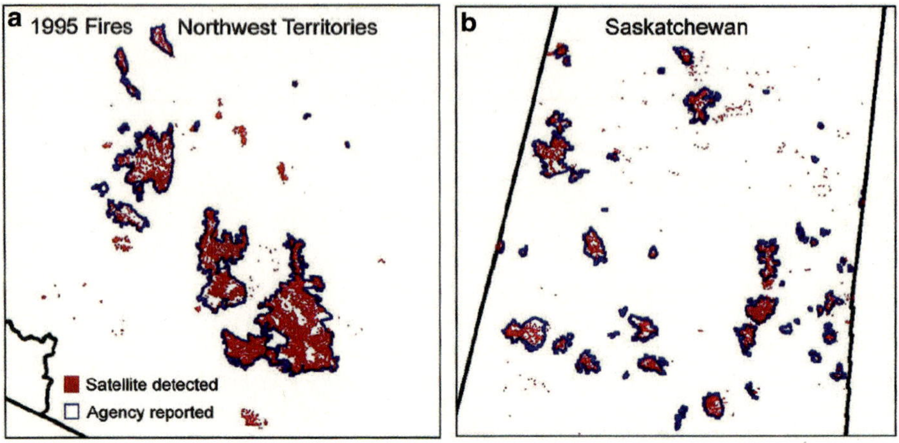

Fig. 7.18 Comparison of the annual composite of fire spots with data obtained by fire agencies for two regions in Canada in 1995. The *blue polygons* outline the boundaries of burnt area reported by fire agencies using conventional means of monitoring while *red dots* are locations of active fire pixels observed by satellite during an entire fire season (Reproduced from Li et al. (2000b), with permission from Taylor & Francis)

7.2.5.2 Fire Detection Using a Contextual Algorithm

Based on the MODIS version 4 contextual fire detection algorithm (Giglio et al. 2003) and a smoke detection algorithm (Xie et al. 2005), Wang et al. (2007) developed a contextual algorithm to detect small and cool fires (referred to as SCF algorithm hereafter) using the reflectances of seven solar reflective channels (R_1, R_2, R_3, R_7, R_8, R_9, and R_{19}) and the brightness temperatures of four infrared channels (T_{22}, T_{28}, T_{31}, and T_{32}). The flowchart of the SCF algorithm is shown in Fig. 7.19.

Wang et al. (2007) selected two wild land fire cases in the southeastern USA to illustrate the performance of the SCF algorithm using MODIS data. One case is the fire event on December 20 and 21, 2004, at the border region between Georgia and Florida along the Atlantic coast and the border region between Mississippi and Alabama along the Gulf coast. The other case is the fire event on September 29, 2003, in the Red River Basin in Mississippi.

In each case, fire detection is performed using both the SCF algorithm and the MODIS version 4 contextual algorithm (referred to as MODIS algorithm hereafter). Fire events detected by the MODIS algorithm are considered as true fires because the MODIS algorithm has been validated systematically and offers a significantly lower false alarm rate (Morisette et al. 2005a, b; Csiszar et al. 2006; Schroeder et al. 2008). Comparative analysis is conducted between earlier and later observations of fire events which are detected by the SCF algorithm but not detected by the MODIS algorithm. If a previously undetected fire event is subsequently detected by the MODIS algorithm and/or by the SCF algorithm and the event is accompanied by obvious smoke plumes, this fire event is believed to be a true fire event which is

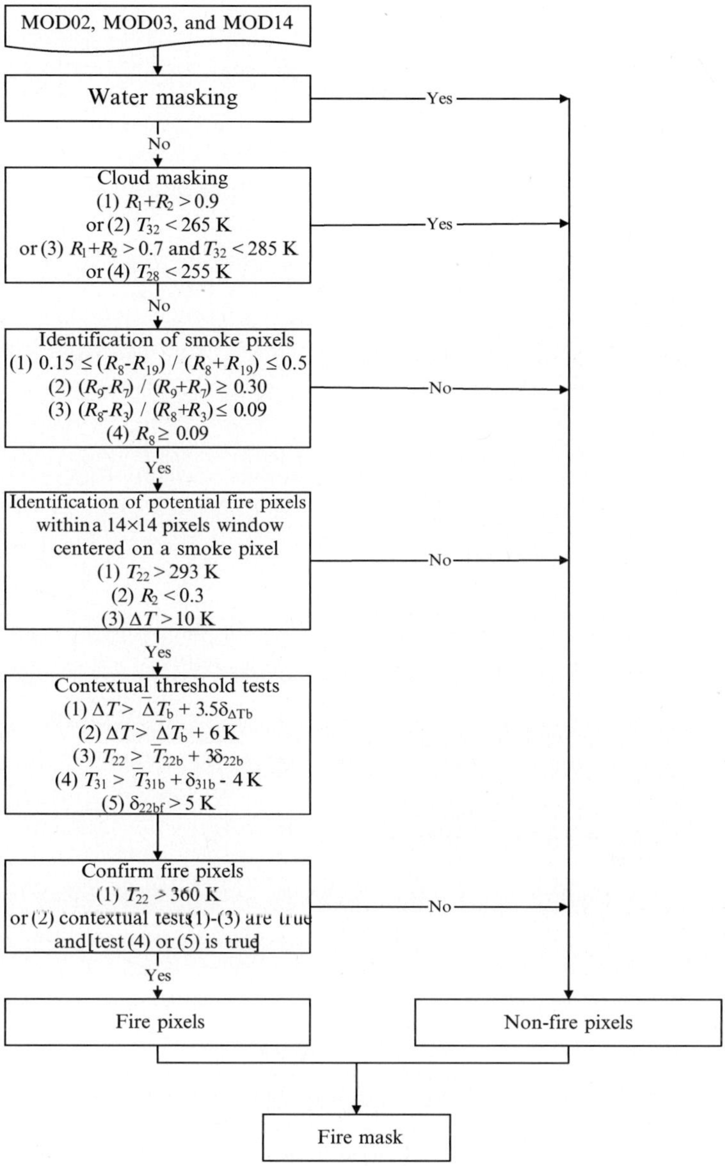

Fig. 7.19 Flowchart of the SCF algorithm

previously omitted by the MODIS algorithm. Fire events, which are only detectable using the SCF algorithm at earlier time, but not detectable using both algorithms in subsequent observations, are considered uncertain spots. Uncertain spots are further inspected using MODIS 250-m true colorimages.

The characteristics of fire events on September 29, 2003, are shown in Fig. 7.20. Eight fire spots are detected by the SCF algorithm but not detected by the MODIS

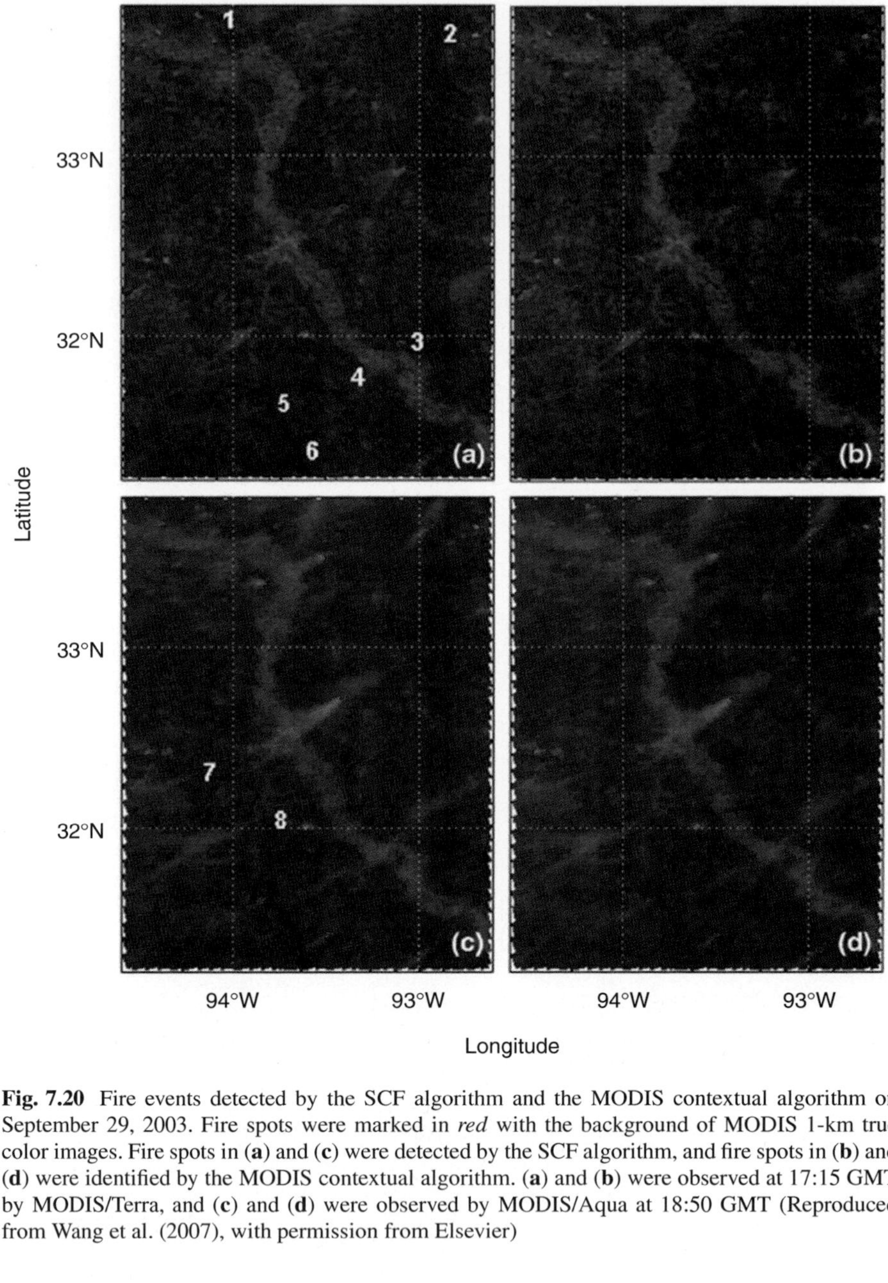

Fig. 7.20 Fire events detected by the SCF algorithm and the MODIS contextual algorithm on September 29, 2003. Fire spots were marked in *red* with the background of MODIS 1-km true color images. Fire spots in (**a**) and (**c**) were detected by the SCF algorithm, and fire spots in (**b**) and (**d**) were identified by the MODIS contextual algorithm. (**a**) and (**b**) were observed at 17:15 GMT by MODIS/Terra, and (**c**) and (**d**) were observed by MODIS/Aqua at 18:50 GMT (Reproduced from Wang et al. (2007), with permission from Elsevier)

algorithm in the fire events of the Red River Basin (see Fig. 7.20a, c). Taking fire spot 2 as an example, both the SCF and MODIS algorithms identify this spot as an active fire with an obvious smoke plume at 18:50 GMT (see Fig. 7.20c, d), but only the SCF algorithm detects this spot as an active fire at 17:15 GMT (see

Table 7.8 Characteristics of fire events on September 29, 2003

Fire spot	Time (GMT)	T_{22} (K)	ΔT (K)	R_2	Scan angle (degree)
1	17:15[a]	309.0	16.2	0.171	19.5
	18:50[b]	310.3	18.2	0.195	45.4
2	17:15[a]	308.1	15.9	0.156	26.5
	18:50[b]	328.7	36.7	0.149	41.8
3	17:15[a]	320.0	25.5	0.157	28.6
	18:50[b]	232.9	28.8	0.157	44.4
4	17:15[a]	308.7	14.4	0.192	27.3
	18:50[b]	315.2	21.5	0.182	45.6
5	17:15[a]	317.2	22.3	0.084	25.1
	18:50[b]	305.2	12.0	0.102	47.0
6	17:15[a]	309.4	16.0	0.161	27.5
	18:50[b]	311.0	17.8	0.163	46.2
7	17:15[a]	327.9	31.6	0.166	21.3
	18:50[b]	305.3	12.1	0.171	47.4
8	17:15[a]	327.3	30.7	0.187	24.0
	18:50[b]	304.3	10.6	0.185	46.6

Reproduced from Wang et al. (2007), with permission from Elsevier
[a]Fire spot is only detected by the SCF algorithm
[b]Fire spot is detected by both the SCF and MODIS algorithms

Fig. 7.20a). In Table 7.8, both T_{22} and ΔT ($\Delta T = T_{22} - T_{31}$) for fire spot 2 increase by approximately 20 K during a period of 1.5 h (i.e., from 17:15 to 18:50 GMT). These results prove that the fire spot at 17:15 GMT is an active fire. Furthermore, fire spot 3 at 17:15 and 18:50 GMT and fire spot 5 at 17:15 GMT satisfy the fire identification threshold of $T_{22} > 310$ K (Table 7.8) and contextual thresholds in the MODIS algorithm, but they are not detected by the MODIS algorithm. These results indicate that the false alarm rejection thresholds in the MODIS algorithm result in omission errors.

The characteristics of fire events on December 20 and 21, 2004, are shown in Fig. 7.21 and Table 7.9. Twelve fire spots are omitted by the MODIS algorithm (see Fig. 7.21a, c, e). Although fire spots 1, 5, and 6 are not accompanied by obvious smoke plumes in the MODIS 1-km true color images, these three spots are accompanied by smoke plumes in the MODIS 500-m true color images. Spots 7–12 are identified as fire spots because they are accompanied by obvious smoke plumes in the MODIS 1-km true colorimages.

Figures 7.20c and 7.21e show that the SCF algorithm is more sensitive to small and cool fires, especially for observations at large scan angles (Tables 7.8 and 7.9). A total of 22 fire spots are omitted by the MODIS algorithm, but detected by the SCF algorithm in the two fire events. Nevertheless, the SCF algorithm fails to detect small fires lacking a visible smoke plume unless they are within the potential fire area of other fires (Wang et al. 2007).

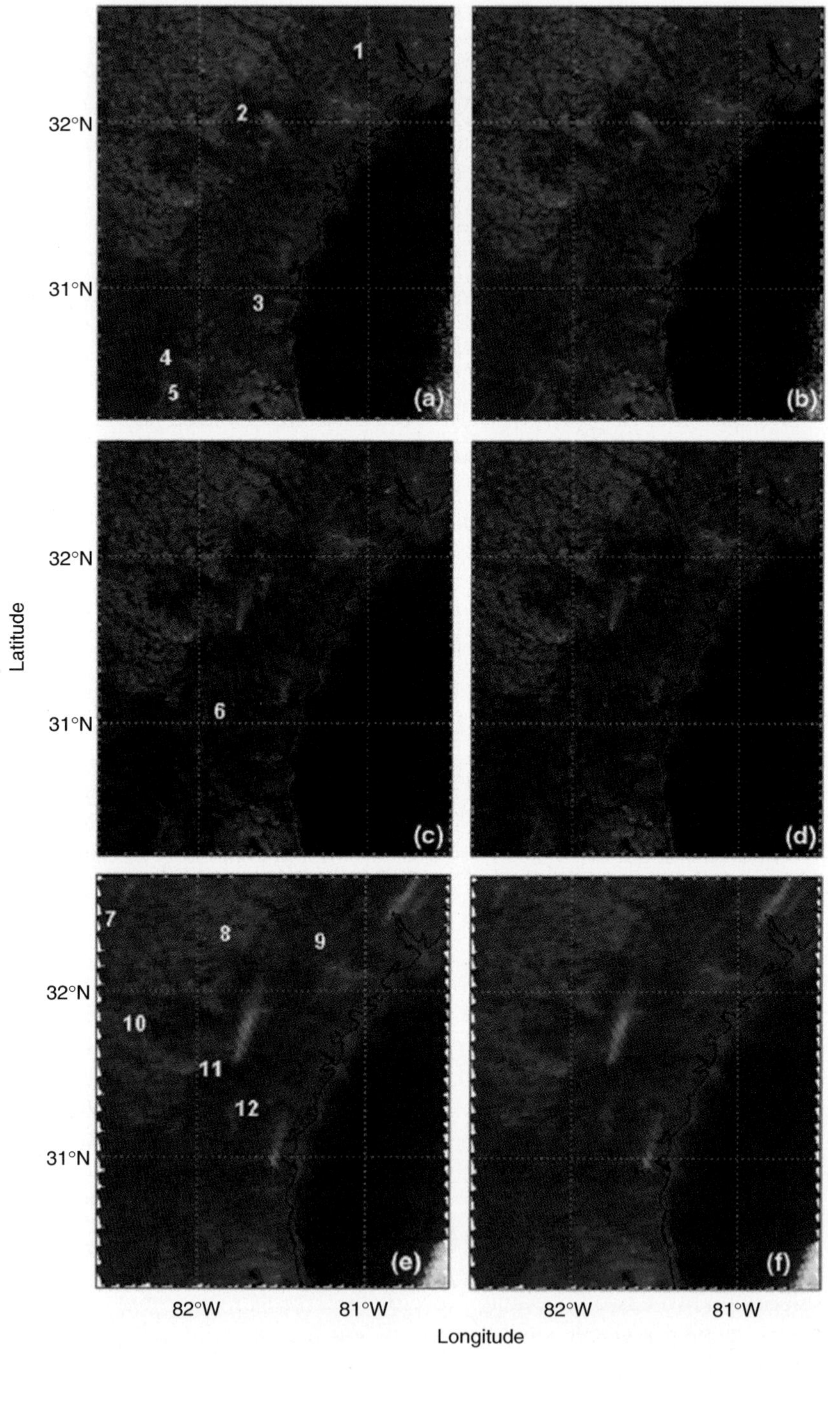
1
2
32°N
31°N
3
4
5
(a)
(b)
32°N
31°N
6
(c)
(d)
Latitude
7
8
9
32°N
10
11
12
31°N
(e)
(f)
82°W
81°W
82°W
81°W
Longitude

Table 7.9 Characteristics of fire events on December 20 and 21, 2004

Fire spot	Time (GMT)	T_{22} (K)	ΔT (K)	R_2	Scan angle (degree)
1	Dec. 20, 18:50[a]	296.4	17.8	0.108	29.2
	Dec. 21, 16:20[b]	32.94	43.3	0.115	17.8
2	Dec. 20, 18:50[a]	302.8	23.1	0.105	24.2
3	Dec. 20, 18:50[a]	293.0	13.2	0.123	23.1
	Dec. 21, 17:55[b]	313.3	26.7	0.141	51.4
4	Dec. 20, 18:50[a]	300.9	19.9	0.137	19.0
5	Dec. 20, 18:50[a]	296.8	16.1	0.124	18.9
6	Dec. 21, 16:20[a]	302.2	16.4	0.107	14.3
7	Dec. 21, 17:55[a]	297.7	11.7	0.118	52.0
8	Dec. 21, 17:55[a]	301.8	14.2	0.125	50.6
9	Dec. 21, 17:55[a]	300.9	15.6	0.090	49.3
10	Dec. 21, 17:55[a]	305.3	18.7	0.131	52.2
11	Dec. 21, 16:20[b]	315.2	26.7	0.110	13.5
	Dec. 21, 17:55[a]	306.1	18.4	0.132	51.4
12	Dec. 21, 17:55[a]	298.9	11.7	0.120	51.2

Reproduced from Wang et al. (2007), with permission from Elsevier
[a]Fire spot is only detected by the SCF algorithm
[b]Fire spot is detected by both the SCF and MODIS algorithms

7.2.6 *Urban Heat Island Effect*

An urban heat island (UHI) is a phenomenon where an urban area is significantly warmer than its surrounding rural areas (Oke 1982). For example, the typical temperature for the center of the city is several degrees higher than that for the surrounding nonurbanized areas. This phenomenon was first investigated and described by Howard in the 1810s. The resulted UHI effect will increase air conditioning and raise pollution levels. Therefore, the study of UHI has attracted more attentions since then.

UHIs are found at both summer and winter and are generally most apparent at night. There are several causes of an UHI. The main reason is that the urbanization modifies the distribution of land cover and land use characteristics, which finally affects the air temperature of the lowest layers of the urban atmosphere. The commonly used materials in urban areas, such as concrete and asphalt, have

Fig. 7.21 Fire events detected by the SCF algorithm and the MODIS contextual algorithm on December 20 and 21, 2004. Fire spots were marked in *red* with the background of MODIS 1-km true color image. Fire spots in (**a**), (**c**), and (**e**) were detected by the SCF algorithm, and fire spots in (**b**), (**d**), and (**f**) were identified by the MODIS contextual algorithm. (**a**) and (**b**), (**c**) and (**d**), and (**e**) and (**f**) were observations on December 20, 18:50 GMT by MODIS/Aqua; December 21, 16:20 GMT by MODIS/Terra; and December 21, 17:55 GMT by MODIS/Aqua, in series (Reproduced from Wang et al. (2007), with permission from Elsevier) significantly different thermal bulk and surface radiative properties, such as heat inertia, thermal conductivity, albedo, and emissivity compared with those in the surrounding rural areas. The changes in the thermal properties of surface materials and lack of evapotranspiration cause a corresponding change in the energy balance of the urban area and may lead to higher temperatures than surrounding rural areas (Oke 1982). Others are due to the geometric effects. The buildings that provide multiple surfaces for the reflection and absorption of solar radiation would result in urban canyon effect. Those buildings may block surface heat into the sky and increase the efficiency of heating for urban areas. Furthermore, the blocking of wind inhibits cooling by convection. Waste heat from automobiles, air conditioning, industry, and other sources also contributes to the UHI (Chen et al. 2011; Sailor 2011; Li and Zhao 2012).

The monitoring of land surface temperature (LST) is of prime importance to the study of UHI. LST is directly linked to surface radiation and energy exchange, the internal climate of buildings, and human comfort in the cities (Weng 2009). Traditionally, UHIs were studied by ground-based observations taken from fixed thermometer networks or by traverses with thermometers mounted on vehicles (Voogt and Oke 2003). With the advent of satellite and aircraft platforms, thermal infrared remote sensing has provided new developments for the study of UHIs.

UHI islands can be defined for different layers of the urban atmosphere, for various surfaces, and even for the subsurface (Oke 1995; Voogt and Oke 1997). Because of the different mechanisms, those definitions are required to be clearly identified. Generally, atmospheric heat islands (AHIs) are the traditional expression of UHIs. An AHI here refers to the excess warmth of the urban atmosphere compared to that in the rural areas. AHIs are defined either for the urban canopy layer from the surface to approximately mean building height or for the urban boundary layer influenced by the underlying urban surface (Voogt and Oke 2003). AHIs are typically detected by in situ sensors at standard meteorological height or by more specialized sensor platforms at tall towers and aircraft, among others. Surface urban heat island (SUHI) is another expression of UHI. SUHIs are observed by thermal remote sensors and are the description of the spatial patterns of upwelling thermal radiance (Voogt and Oke 2003). Compared with AHI, SUHI requires taking the intervening atmosphere and the surface radiative properties into account.

The first SUHI observations from remotely sensed data were reported by Rao (1972). Subsequently, various studies were carried out to make remote observations of SUHI. Those studies can be categorized into three main themes (Voogt and Oke 2003). The first theme is the examination of the spatial structure of urban thermal patters and their relations to urban surface characteristics by using thermal remote sensing. A series of satellite sensors have been developed to retrieve LSTs from space, such as Landsat TM/ETM+, ASTER, AVHRR, and MODIS. Those retrieved LSTs are used with land cover and land use characteristics to assess the spatial patterns of SUHI (Balling and Brazel 1988; Carnahan and Larson 1990; Lougeay et al. 1996). The changes of urban thermal climates with respect to surface characteristics, such as NDVI, are deeply investigated (Eliasson 1992; Quattrochi and Ridd 1994; Lo et al. 1997). Weng et al. (2004) assessed the LST–vegetation abundance relationship at urban areas over a range of scales and pointed out the importance of scale of measurement and modeling when examining those relations. In addition, the surface heat island model proposed by Johnson et al. (1991) is used to analyze the genesis of SUHI (Oke et al. 1991). The surface geometry and surface thermal properties are believed to be the most important factors (Voogt and Oke 2003).

The second theme is the study of urban surface energy balances. Knowledge of urban surface energy balance is important in understanding UHIs (Oke 1982). The urban climate models and remotely sensed observations are combined to study those balances (Carlson et al. 1981; Hafner and Kidder 1999; Voogt and Oke 2003). Bulk heat transfer equation successfully used over agricultural and vegetated surfaces has also been applied to urban areas by taking the anthropogenic heat flux into account (Voogt and Grimmond 2000; Weng 2009). A comprehensive review of the energy balance approach and its development in urban areas was given by Oke (1988). However, Weng (2009) pointed out that it is extremely difficult and expensive to investigate the detailed spatial pattern of energy fluxes in the urban areas when cost, time, instrument, and data calibrations are all considered together.

The last theme is the application of thermal remote sensing to study the relation between AHIs and SUHIs. Empirical models and urban atmosphere models with coincident remote and ground-based observations are used to study the relations between surface and atmospheric air temperature (Caselles et al. 1991; Stoll and Brazel 1992; Ben-Dor and Saaroni 1997). However, there has been no simple general relation except for empirical relations to date (Voogt and Oke 2003). The differing source areas for AHIs and SUHIs should be taken into account when directly comparing LST with atmospheric air temperature. Horizontal transport of heat by wind (Advection) and shadow are other factors determining those relations. Stoll and Brazel (1992) found the relations between LST and atmospheric air temperature at small scale could be explained largely by the atmospheric mixing, mean wind velocity, and thermal properties of surface materials. Those relations will become poorer when the mixture of surfaces is present within the sensor IFOV due to the mixture of subgrid scale advection (Voogt and Oke 2003). When subgrid advection is reduced (such as at night) or some geophysical parameters (such as NDVI) are introduced, the relations are improved (Dousset 1989). Voogt and Oke (2003) concluded that the explanation of those relations remains rooted in detailed study of the surface micrometeorology and geography and those relations are only likely to be predicted by the application of detailed, fully coupled surface–atmosphere models.

The study of urban climate and environment will be difficult without remote sensing (Weng 2009). Voogt and Oke (2003) and Weng (2009) summarized that the developments of thermal remote sensing over urban areas focus on those three aspects: (1) determine appropriate surface radiative and structural parameters from remote sensing, which accurately describe the urban surface and are appropriate for the use of urban atmospheric models; (2) combine various models together, such as sensor geometry models, urban radiative transfer models, and surface energy balance models, to better study the urban thermal environment in and above the urban canopy layer; (3) define an "urban surface" to fully consider the effect of sensor viewing geometry and surface structure on the plan-view of remote sensor measurements; (4) pay more attention to the issue of measurement and modeling scale; the differences between the observational scale of a remote sensing pixel and the operational scale are required to be studied, because the operational scale would vary from city to city with the change of biophysical and anthropogenic settings; and (5) understand the differences of source area between modeled and measured fluxes to well-link LST, the surface energy balance, and atmospheric air temperature in and above the urban canopy layer.

References

Anderson, M. C., Norman, J. M., Mecikalski, J. R., Otkin, J. A., & Kustas, W. P. (2007). A climatological study of evapotranspiration and moisture stress across the continental United States based on thermal remote sensing: 1. Model formulation. *Journal of Geophysical Research, 112*, D10117.

Andreae, M. O., & Merlet, P. (2001). Emission of trace gases and aerosols from biomass burning. *Global Biogeochemical Cycles, 15*, 955–966.

Arino, O., & Melinotte, J. M. (1998). The 1993 Africa fire map. *International Journal of Remote Sensing, 19*, 2019–2023.

Balling, R. C., & Brazel, S. W. (1988). High resolution surface temperature patterns in a complex urban terrain. *Photogrammetric Engineering and Remote Sensing, 54*(9), 1289–1293.

Ben-Dor, E., & Saaroni, H. (1997). Airborne video thermal radiometry as a tool for monitoring microscale structures of the urban heat island. *International Journal of Remote Sensing, 18*, 3039–3053.

Boles, S. H., & Verbyla, D. L. (2000). Comparison of three AVHRR-based fire detection algorithms for interior Alaska. *Remote Sensing of Environment, 72*, 1–16.

Calle, A., Casanova, J.-L., & González-alonso, F. (2009). Impact of point spread function of MSG-SEVIRI on active fire detection. *International Journal of Remote Sensing, 30*, 4567–4579.

Candogan, B. N., Sincik, M., Buyukcangaz, H., Demirtas, C., Goksoy, A. T., & Yazgan, S. (2013). Yield, quality and crop water stress index relationships for deficit-irrigated soybean [Glycine max (L.) Merr.] in sub-humid climatic conditions. *Agricultural Water Management, 118*, 113–121.

Carlson, T. N., Dodd, J. K., Benjamin, S. G., & Cooper, J. N. (1981). Satellite estimation of the surface energy balance, moisture availability and thermal inertia. *Journal of Applied Meteorology, 20*(1), 67–87.

Carlson, T. N., Gillies, R. R., & Schmugge, T. J. (1995). An interpretation of methodologies for indirect measurement of soil water content. *Agricultural and Forest Meteorology, 77*, 191–205.

Carnahan, W. H., & Larson, R. C. (1990). An analysis of an urban heat sink. *Remote Sensing of Environment, 33*(1), 65–71.

Caselles, V., Lopez Garcia, M. J., Melia, J., & Perez Cueva, A. J. (1991). Analysis of the heat-island effect of the city of Valencia, Spain through air temperature transects and NOAA satellite data. *Theoretical and Applied Climatology, 43*, 195–203.

Chen, F., Kusaka, H., Bornstein, R., Ching, J., Grimmond, C. S. B., Grossman-Clarke, S., Loridan, T., Manning, K. W., Martilli, A., Miao, S., Sailor, D., Salamanca, F. P., Taha, H., Tewari, M., Wang, X., Wyszogrodzki, A. A., & Zhang, C. (2011). The integrated WRF/urban modelling system: Development, evaluation, and applications to urban environmental problems. *International Journal of Climatology, 31*(2), 273–288.

Crutzen, P. J., & Andreae, M. O. (1990). Biomass burning in the tropics: Impact on atmospheric chemistry and biogeochemical cycles. *Science, 250*, 1669–1678.

Crutzen, P. J., Heidt, L. E., Krasnec, J. P., Pollock, W. H., & Seiler, W. (1979). Biomass burning as a source of atmospheric gases CO, H_2, N_2O, NO, CH_3Cl and COS. *Nature, 282*, 253–256.

Csiszar, I. A., Morisette, J. T., & Giglio, L. (2006). Validation of active fire detection from moderate-resolution satellite sensors: The MODIS example in Northern Eurasia. *IEEE Transactions on Geoscience and Remote Sensing, 44*, 1757–1764.

Cuomo, V., Lasaponara, R., & Tramutoli, V. (2001). Evaluation of a new satellite-based method for forest fire detection. *International Journal of Remote Sensing, 22*, 1799–1826.

Dennison, P. E., Charoensiri, K., Roberts, D. A., Peterson, S. H., & Green, R. O. (2006). Wildfire temperature and land cover modeling using hyperspectral data. *Remote Sensing of Environment, 100*, 212–222.

Dousset, B. (1989). AVHRR-derived cloudiness and surface temperature patterns over the Los Angeles area and their relationship to land use. In *Proceedings of IGARSS-89* (pp. 2132–2137). New York: IEEE.

Dozier, J. (1981). A method for satellite identification of surface temperature fields of subpixel resolution. *Remote Sensing of Environment, 11*, 221–229.

Eckmann, T. C., Roberts, D. A., & Still, C. J. (2008). Using multiple endmember spectral mixture analysis to retrieve subpixel fire properties from MODIS. *Remote Sensing of Environment, 112*, 3773–3783.

Eckmann, T. C., Roberts, D. A., & Still, C. J. (2009). Estimating subpixel fire sizes and temperatures from ASTER using multiple endmember spectral mixture analysis. *International Journal of Remote Sensing, 30*, 5851–5864.

Eckmann, T. C., Still, C. J., Roberts, D. A., & Michaelsen, J. C. (2010). Variations in subpixel fire properties with season and land cover in southern Africa. *Earth Interactions, 14*, 1–29.

Eliasson, I. (1992). Infrared thermography and urban temperature patterns. *International Journal of Remote Sensing, 13*, 869–879.

Eva, H., & Flasse, S. (1996). Contextual and multiple-threshold algorithms for regional active fire detection with AVHRR data. *Remote Sensing Reviews, 14*, 333–351.

Flannigan, M. D., & Vonder Haar, T. H. (1986). Forest fire monitoring using NOAA satellite AVHRR. *Canadian Journal of Forest Research, 16*, 975–982.

Flasse, S. P., & Ceccato, P. (1996). A contextual algorithm for AVHRR fire detection. *International Journal of Remote Sensing, 17*, 419–424.

Franca, J. R. A., Brustet, J. M., & Fontan, J. (1995). Multispectral remote sensing of biomass burning in West Africa. *Journal of Atmospheric Chemistry, 22*, 81–110.

Gao, Z., Gao, W., & Chang, N.-B. (2011). Integrating temperature vegetation dryness index (TVDI) and regional water stress index (RWSI) for drought assessment with the aid of LANDSAT TM/ETM + images. *International Journal of Applied Earth Observation and Geoinformation, 13*, 495–503.

Giglio, L., & Justice, C. O. (2003). Effect of wavelength selection on characterization of fire size and temperature. *International Journal of Remote Sensing, 24*, 3515–3520.

Giglio, L., & Kendall, J. D. (2001). Application of the Dozier retrieval to wildfire characterization: A sensitivity analysis. *Remote Sensing of Environment, 77*, 34–49.

Giglio, L., Kendall, J. D., & Justice, C. O. (1999). Evaluation of global fire detection algorithm using simulate AVHRR infrared data. *International Journal of Remote Sensing, 20*, 1947–1985.

Giglio, L., Descloitres, J., Justice, C. O., & Kaufman, Y. J. (2003). An enhanced contextual fire detection algorithm for MODIS. *Remote Sensing of Environment, 87*, 273–282.

Giglio, L., Csiszar, I., Restás, Á., Morisette, J., Schroeder, W., Motron, D., & Justice, C. O. (2008). Active fire detection and characterization with the advanced spaceborne thermal emission and reflection radiometer (ASTER). *Remote Sensing of Environment, 112*, 3055–3063.

Gillies, R. R., Kustas, W. P., & Humes, K. S. (1997). A verification of the 'triangle' method for obtaining surface soil water content and energy fluxes from remote measurements of the Normalized Difference Vegetation Index (NDVI) and surface radiant temperature. *International Journal of Remote Sensing, 18*, 3145–3166.

Hafner, J., & Kidder, S. Q. (1999). Urban heat island modeling in conjunction with satellite-derived surface/soil parameters. *Journal of Applied Meteorology, 38*(4), 448–465.

He, L., & Li, Z. (2011). Enhancement of a fire-detection algorithm by eliminating solar contamination effects and atmospheric path radiance: Application to MODIS data. *International Journal of Remote Sensing, 32*, 6273–6293.

He, L., & Li, Z. (2012). Enhancement of a fire detection algorithm by eliminating solar reflection in the mid-IR band: Application to AVHRR data. *International Journal of Remote Sensing, 33*, 7047–7059.

Ichoku, C., Kaufman, Y. J., Giglio, L., Li, Z., Fraser, R. H., Jin, J.-Z., & Park, W. M. (2003). Comparative analysis of daytime fire detection algorithms using AVHRR data for the 1995 fire season in Canada: Perspective for MODIS. *International Journal of Remote Sensing, 24*, 1669–1690.

Idso, S. B. (1982). Non-water-stressed baselines: A key to measuring and interpreting plant water stress. *Agricultural Meteorology, 27*, 59–70.

Idso, S. B., Jackson, R. D., Pinter, P. J., Jr., Reginato, R. J., & Hatfield, J. L. (1981). Normalizing the stress-degree-day parameter for environmental variability. *Agricultural Meteorology, 24*, 45–55.

Jackson, R. D., Idso, S. B., Reginato, R. J., & Pinter, P. J. (1981). Canopy temperature as a crop water stress indicator. *Water Resources Research, 17*, 1133–1138.

Johnson, G. T., Oke, T. R., Lyons, T. J., Steyn, D. G., Watson, I. D., & Voogt, J. A. (1991). Simulation of surface urban heat islands under 'ideal' conditions at night Part 1: Theory and tests against field data. *Boundary-Layer Meteorology, 56*(3), 275–294.

Justice, C. O., Kendall, J. D., Dowty, P. R., & Scholes, R. J. (1996). Satellite remote sensing of fires during the SAFARI campaign using NOAA advanced very high resolution radiometer data. *Journal of Geophysical Research, 101*, 23851–23863.

Justice, C. O., Giglio, L., Korontzi, S., Owens, J., Morisette, J. T., Roy, D., Descloitres, J., Alleaume, S., Petitcolin, F., & Kaufman, Y. (2002). The MODIS fire products. *Remote Sensing of Environment, 83*, 244–262.

Kahle, A. B. (1977). A simple thermal model of the Earth's surface for geologic mapping by remote sensing. *Journal of Geophysical Research, 82*, 1673–1680.

Kaufman, Y. J., Tucker, C. J., & Fung, I. (1990). Remote sensing of biomass burning in the tropics. *Journal of Geophysical Research, 95*, 9927–9939.

Kaufman, Y. J., Justice, C. O., Flynn, L. P., Kendall, J. D., Prins, E. M., Giglio, L., Ward, D. E., Menzel, W. P., & Setzer, A. W. (1998a). Potential global fire monitoring from EOS-MODIS. *Journal of Geophysical Research, 103*(D24), 32215–32238.

Kaufman, Y. J., Kleidman, R. G., & King, M. D. (1998b). SCAR-B fires in the tropics: Properties and remote sensing from EOS-MODIS. *Journal of Geophysical Research, 103*, 31955–31968.

Kennedy, P. J., Belward, A. S., & Grégoire, J.-M. (1994). An improved approach to fire monitoring in West Africa using AVHRR data. *International Journal of Remote Sensing, 15*, 2235–2255.

Kogan, F. N. (1995). Application of vegetation index and brightness temperature for drought detection. *Advances in Space Research, 15*, 91–100.

Kogan, F. N. (1997). Global drought watch from space. *Bulletin of the American Meteorological Society, 78*, 621–636.

Kogan, F. N. (2001). Operational space technology for global vegetation assessment. *Bulletin of the American Meteorological Society, 82*, 1949–1964.

Kogan, F. N. (2002). World droughts in the new millennium from AVHRR-based vegetation health indices. *Eos, Transactions American Geophysical Union, 83*, 557–563.

Langaas, S. (1993). A parameterised bispectral model for savanna fire detection using AVHRR night images. *International Journal of Remote Sensing, 14*, 2245–2262.

Lasaponara, R., Cuomo, V., Macchiato, M. F., & Simoniello, T. (2003). A self-adaptive algorithm based on AVHRR multitemporal data analysis for small active fire detection. *International Journal of Remote Sensing, 24*, 1723–1749.

Lee, T. F., & Tag, P. M. (1990). Improved detection of hotspots using the AVHRR 3.7-um channel. *American Meteorological Society, 71*, 1722–1730.

Levine, J. S. (Ed.). (1991). *Global biomass burning: Atmospheric, climatic and biospheric implications*. Cambridge, MA: The MIT Press.

Li, Z. (1998). Influence of absorbing aerosols on the solar surface radiation budget. *Journal of Climate, 11*, 15–17.

Li, Y., & Zhao, X. (2012). An empirical study of the impact of human activity on long-term temperature change in China: A perspective from energy consumption. *Journal of Geophysical Research: Atmospheres (1984–2012), 117*(D17), D17117.

Li, Z., Barker, H., & Moreau, L. (1995). The variable effect of clouds on atmospheric absorption of solar radiation. *Nature, 376*, 486–490.

Li, Z., Cihlar, J., Moreau, L., Huang, F., & Lee, B. (1997). Monitoring fire activities in the boreal ecosystem. *Journal of Geophysical Research, 102*, 29611–29624.

Li, L., Ying, Z., Feng, Y., & Guoqiang, S. (1998). Monitoring drought of Guizhou using remote sensing vegetation index. *Journal of Guizhou Meteorology, 22*, 50–54.

Li, Z., Nadon, S., & Cihlar, J. (2000a). Satellite-based detection of Canadian boreal forest fires: Development and application of the algorithm. *International Journal of Remote Sensing, 21*, 3057–3069.

Li, Z., Nadon, S., Cihlar, J., & Stocks, B. (2000b). Satellite-based mapping of Canadian boreal forest fires: Evaluation and comparison of algorithms. *International Journal of Remote Sensing, 21*, 3071–3082.

Li, Z., Kaufman, Y. J., Ichoku, C., Fraser, R., Trishchenko, A., Giglio, L., Jin, J.-Z., & Yu, X. (2001). A review of AVHRR-based active fire detection algorithms: Principles, limitations, and recommendations. In F. Ahern, J. G. Goldammer, & C. Justice (Eds.), *Global and regional vegetation fire monitoring from space: Planning and coordinated international effort* (pp. 199–225). Hague: SPB Academic Publishing.

Lo, C. P., Quattrochi, D. A., & Luvall, J. C. (1997). Application of high-resolution thermal infrared remote sensing and GIS to assess the urban heat island effect. *International Journal of Remote Sensing, 18*, 287–304.

Lougeay, R., Brazel, A., & Hubble, M. (1996). Monitoring intra-urban temperature patterns and associated land cover in Phoenix, Arizona using Landsat thermal data. *Geocarto International, 11*, 79–89.

Lu, S., Ju, Z., Ren, T., & Horton, R. (2009). A general approach to estimate soil water content from thermal inertia. *Agricultural and Forest Meteorology, 149*, 1693–1698.

Mallick, K., Bhattacharya, B. K., & Patel, N. K. (2009). Estimating volumetric surface moisture content for cropped soils using a soil wetness index based on surface temperature and NDVI. *Agricultural and Forest Meteorology, 149*, 1327–1342.

Matson, M., & Dozier, J. (1981). Identification of subresolution high temperature sources using a thermal IR sensor. *Photogrammetric Engineering and Remote Sensing, 47*, 1311–1318.

Matson, M., & Holben, B. (1987). Satellite detection of tropical burning in Brazil. *International Journal of Remote Sensing, 8*, 509–516.

McVicar, T. R., & Jupp, D. L. B. (1998). The current and potential operational uses of remote sensing to aid decisions on drought exceptional circumstances in Australia: A review. *Agricultural Systems, 57*, 399–468.

McVicar, T. R., & Jupp, D. L. B. (2002). Using covariates to spatially interpolate moisture availability in the Murray–Darling Basin: A novel use of remotely sensed data. *Remote Sensing of Environment, 79*, 199–212.

Méndez-Barroso, L. A., Garatuza-Payán, J., & Vivoni, E. R. (2008). Quantifying water stress on wheat using remote sensing in the Yaqui Valley, Sonora, Mexico. *Agricultural Water Management, 95*, 725–736.

Milly, P. C. D., Wetherald, R. T., Dunne, K. A., & Delworth, T. L. (2002). Increasing risk of great floods in a changing climate. *Nature, 415*, 514–517.

Mishra, A. K., & Singh, V. P. (2009). Analysis of drought severity-area-frequency curves using a general circulation model and scenario uncertainty. *Journal of Geophysical Research, [Atmospheres], 114*, D06120.

Mitra, D. S., & Majumdar, T. J. (2004). Thermal inertia mapping over the Brahmaputra basin, India using NOAA-AVHRR data and its possible geological applications. *International Journal of Remote Sensing, 25*, 3245–3260.

Moran, M. S., Clarke, T. R., Inoue, Y., & Vidal, A. (1994). Estimating crop water deficit using the relation between surface-air temperature and spectral vegetation index. *Remote Sensing of Environment, 49*, 246–263.

Moran, M. S., Rahman, A. F., Washburne, J. C., Goodrich, D. C., Weltz, M. A., & Kustas, W. P. (1996). Combining the Penman-Monteith equation with measurements of surface temperature and reflectance to estimate evaporation rates of semiarid grassland. *Agricultural and Forest Meteorology, 80*, 87–109.

Morisette, J. T., Giglio, L., Csiszar, I., & Justice, C. O. (2005a). Validation of the MODIS active fire product over Southern Africa with ASTER data. *International Journal of Remote Sensing, 26*, 4239–4264.

Morisette, J. T., Giglio, L., Csiszar, I., Setzer, A., Schroeder, W., Morton, D., & Justice, C. O. (2005b). Validation of MODIS active fire detection products derived from two algorithms. *Earth Interactions, 9*, 1–25.

Nakayama, M., Maki, M., Elvidge, C. D., & Liew, S. C. (1999). Contextual algorithm adapted for NOAA-AVHRR fire detection in Indonesia. *International Journal of Remote Sensing, 20*, 3415–3421.

Oertel, D., Zhukov, B., Thamm, H.-P., Roehrig, J., & Orthmann, B. (2004). Space-borne high resolution fire remote sensing in Benin, West Africa. *International Journal of Remote Sensing, 25*, 2209–2216.

Oke, T. R. (1982). The energetic basis of the urban heat island. *Quarterly Journal of Royal Meteorology Society, 108*, 1–24.

Oke, T. R. (1988). The urban energy balance. *Progress in Physical Geography, 12*(4), 471–508.

Oke, T. R. (1995). The heat island of the urban boundary layer: Characteristics, causes and effects. In J. E. Cermak, A. G. Davenport, E. J. Plate, & D. X. Viegas (Eds.), *Wind climate in cities* (pp. 81–107). Dordrecht: Kluwer.

Oke, T. R., Johnson, G. T., Steyn, D. G., & Watson, I. D. (1991). Simulation of surface urban heat islands under 'ideal' conditions at night Part 2: Diagnosis of causation. *Boundary-Layer Meteorology, 56*, 339–358.

Peterson, D., & Wang, J. (2013). A sub-pixel-based calculation of fire radiative power from MODIS observations: 2. Sensitivity analysis and potential fire weather application. *Remote Sensing of Environment, 129*, 231–249.

Peterson, D., Wang, J., Ichoku, C., Hyer, E., & Ambrosia, V. (2013). A sub-pixel-based calculation of fire radiative power from MODIS observations: 1. Algorithm development and initial assessment. *Remote Sensing of Environment, 129*, 262–279.

Pohn, H. A., Offield, T. W., & Watson, K. (1974). Thermal inertia mapping from satellites – Discrimination of geologic units in Oman. *Journal of Research of the U. S. Geological Survey, 2*, 147–158.

Pozo, D., Olmo, F. J., & Alados-Arboledas, L. (1997). Fire detection and growth monitoring using a multitemporal technique on AVHRR mid-infrared and thermal channels. *Remote Sensing of Environment, 60*, 111–120.

Price, J. C. (1985). On the analysis of thermal infrared imagery: The limited utility of apparent thermal inertia. *Remote Sensing of Environment, 18*, 59–73.

Prins, E. M., & Menzel, W. P. (1992). Geostationary satellite detection of biomass burning in South America. *International Journal of Remote Sensing, 13*, 2783–2799.

Qian, Y., & Kong, X. (2012). A method to retrieve subpixel fire temperature and fire area using MODIS data. *International Journal of Remote Sensing, 33*, 5009–5025.

Quattrochi, D. A., & Ridd, M. K. (1994). Measurement and analysis of thermal energy responses from discrete urban surfaces using remote sensing data. *International Journal of Remote Sensing, 15*, 1991–2022.

Rao, P. K. (1972). Remote sensing of urban heat islands from an environmental satellite. *Bulletin of the American Meteorological Society, 53*, 647–648.

Rauste, Y., Herland, E., Frelander, H., Soini, K., Kuoremaki, T., & Ruokari, A. (1997). Satellite-based forest fire detection for fire control in boreal forests. *International Journal of Remote Sensing, 18*, 2641–2656.

Reginato, R., & Howe, J. (1985). Irrigation scheduling using crop indicators. *Journal of Irrigation and Drainage Engineering, 111*, 125–133.

Riebsame, W. E., Travis, W. R., Changnon, S. A., & Karl, T. (1991). *Drought and natural resources management in the United States: Impacts and implications of the 1987–89 drought*. Boulder: Westview Press.

Roberts, G. J., & Wooster, M. J. (2008). Fire detection and fire characterization over Africa using Meteosat SEVIRI. *IEEE Transactions on Geoscience and Remote Sensing, 46*, 1200–1218.

Roberts, D. A., Gardner, M., Church, R., Ustin, S., Scheer, G., & Green, R. O. (1998). Mapping Chaparral in the Santa Monica Mountains using multiple endmember spectral mixture models. *Remote Sensing of Environment, 65*, 267–279.

Roberts, G., Wooster, M. J., Perry, G. L. W., Drake, N., Rebelo, L.-M., & Dipotso, F. (2005). Retrieval of biomass combustion rates and totals from fire radiative power observations: Application to southern Africa using geostationary SEVIRI imagery. *Journal of Geophysical Research, 110*, D21111. doi:10.1029/2005JD006018.

Robinson, J. M. (1991). Fire from space: Global fire evaluation using infrared remote sensing. *International Journal of Remote Sensing, 12*, 3–24.

Sailor, D. (2011). A review of methods for estimating anthropogenic heat and moisture emissions in the urban environment. *International Journal of Climatology, 31*(2), 189–199.

Sandholt, I., Rasmussen, K., & Andersen, J. (2002). A simple interpretation of the surface temperature/vegetation index space for assessment of surface moisture status. *Remote Sensing of Environment, 79*, 213–224.

Schroeder, W., Prins, E., Giglio, L., Csiszar, I., Schmidt, C., Morisette, J., & Morton, D. (2008). Validation of GOES and MODIS active fire detection products using ASTER and ETM + data. *Remote Sensing of Environment, 112*, 2711–2726.

Schroeder, W., Csiszar, I., Giglio, L., & Schmidt, C. C. (2010). On the use of fire radiative power, area, and temperature estimates to characterize biomass burning via moderate to coarse spatial resolution remote sensing data in the Brazilian Amazon. *Journal of Geophysical Research, 115*, D21121. doi:10.1029/2009JD013769.

Shephard, M. W., & Kennelly, E. J. (2003). Effect of band-to-band coregistration on fire property retrievals. *IEEE Transactions on Geoscience and Remote Sensing, 41*, 2648–2661.

Sobrino, J. A., & Cuenca, J. (1999). Angular variation of thermal infrared emissivity for some natural surfaces from experimental measurements. *Applied Optics, 38*, 3931–3936.

Stisen, S., Sandholt, I., Nørgaard, A., Fensholt, R., & Jensen, K. H. (2008). Combining the triangle method with thermal inertia to estimate regional evapotranspiration – Applied to MSG-SEVIRI data in the Senegal River basin. *Remote Sensing of Environment, 112*, 1242–1255.

Stoll, M. J., & Brazel, A. J. (1992). Surface-air temperature relationships in the urban environment of Phoenix. *Arizona Physical Geography, 13*, 160–179.

Stroppiana, D., Pinnock, S., & Grégoire, J.-M. (2000). The Global Fire Product: Daily fire occurrence from April 1992 to December 1993 derived from NOAA AVHRR data. *International Journal of Remote Sensing, 21*, 1279–1288.

Sun, L., Sun, R., Li, X., Liang, S., & Zhang, R. (2012a). Monitoring surface soil moisture status based on remotely sensed surface temperature and vegetation index information. *Agricultural and Forest Meteorology, 166–167*, 175–187.

Sun, H., Chen, Y., & Sun, H. (2012b). Comparisons and classification system of typical remote sensing indexes for agricultural drought. *Transactions of the Chinese Society of Agricultural Engineering, 28*, 147–154.

Tao, F., Yokozawa, M., Hayashi, Y., & Lin, E. (2005). A perspective on water resources in China: Interactions between climate change and soil degradation. *Climatic Change, 68*, 169–197.

Van Doninck, J., Peters, J., De Baets, B., De Clercq, E. M., Ducheyne, E., & Verhoest, N. E. C. (2011). The potential of multitemporal Aqua and Terra MODIS apparent thermal inertia as a soil moisture indicator. *International Journal of Applied Earth Observation and Geoinformation, 13*, 934–941.

Verstraeten, W. W., Veroustraete, F., van der Sande, C. J., Grootaers, I., & Feyen, J. (2006). Soil moisture retrieval using thermal inertia, determined with visible and thermal spaceborne data, validated for European forests. *Remote Sensing of Environment, 101*, 299–314.

Verstraeten, W., Veroustraete, F., & Feyen, J. (2008). Assessment of evapotranspiration and soil moisture content across different scales of observation. *Sensors, 8*, 70–117.

Voogt, J. A., & Grimmond, C. S. B. (2000). Modeling surface sensible heat flux using surface radiative temperatures in a simple urban area. *Journal of Applied Meteorology, 39*, 1679–1699.

Voogt, J. A., & Oke, T. R. (1997). Complete urban surface temperatures. *Journal of Applied Meteorology, 36*, 1117–1132.

Voogt, J. A., & Oke, T. R. (2003). Thermal remote sensing of urban climate. *Remote Sensing of Environment, 86*(3), 370–384.

Wang, C., Qi, S., Niu, Z., & Wang, J. (2004). Evaluating soil moisture status in China using the temperature–vegetation dryness index (TVDI). *Canadian Journal of Remote Sensing, 30*, 671–679.

Wang, W., Qu, J. J., Hao, X., Liu, Y., & Sommers, W. T. (2007). An improved algorithm for small and cool fire detection using MODIS data: A preliminary study in the southeastern United States. *Remote Sensing of Environment, 108*, 163–170.

Weng, Q. (2009). Thermal infrared remote sensing for urban climate and environmental studies: Methods, applications, and trends. *ISPRS Journal of Photogrammetry and Remote Sensing, 64*(4), 335–344.

Weng, Q., Lu, D., & Schubring, J. (2004). Estimation of land surface temperature vegetation abundance relationship for urban heat island studies. *Remote Sensing of Environment, 89*(4), 467–483.

Wilhite, D. A., & Glantz, M. H. (1985). Understanding: The drought phenomenon: The role of definitions. *Water International, 10*, 111–120.

Wooster, M. J., Zhukov, B., & Oertel, D. (2003). Fire radiative energy for quantitative study of biomass burning: Derivation from the BIRD experimental satellite and comparison to MODIS fire products. *Remote Sensing of Environment, 86*, 83–107.

Wooster, M. J., Roberts, G., Perry, G. L. W., & Kaufman, Y. J. (2005). Retrieval of biomass combustion rates and totals from fire radiative power observations: FRP derivation and calibration relationships between biomass consumption and fire radiative energy release. *Journal of Geophysical Research, 110*, D24311. doi:10.1029/2005JD006318.

Xie, Y., Qu, J., Hao, X., Xiong, J., & Che, N. (2005, May 11–13). *Smoke plume detecting using MODIS measurements in eastern United States*. EastFIRE Conference Proceedings, Fairfax, VA.

Xu, W., Wooster, M. J., Roberts, G., & Freeborn, P. (2010). New GOES imager algorithms for cloud and active fire detection and fire radiative power assessment across North, South and Central America. *Remote Sensing of Environment, 114*, 1876–1895.

Yao, Y., Qin, Q., Fadhil, A., Li, Y., Zhao, S., Liu, S., Sui, X., & Dong, H. (2011). Evaluation of EDI derived from the exponential evapotranspiration model for monitoring China's surface drought. *Environmental Earth Sciences, 63*, 425–436.

Yazar, A., Howell, T. A., Dusek, D. A., & Copeland, K. S. (1999). Evaluation of crop water stress index for LEPA irrigated corn. *Irrigation Science, 18*, 171–180.

Zhang, R., Sun, X., Wang, W., Xu, J., Zhu, Z., & Tian, J. (2005). An operational two-layer remote sensing model to estimate surface flux in regional scale: Physical background. *Science in China (Earth Sciences), 48*, 225–244.

Zhang, R., Tian, J., Su, H., Sun, X., Chen, S., & Xia, J. (2008). Two improvements of an operational two-layer model for terrestrial surface heat flux retrieval. *Sensors, 8*, 6165–6187.

Zhukov, B., Lorenz, E., Oertel, D., Wooster, M., & Roberts, G. (2006). Spaceborne detection and characterization of fires during the bi-spectral infrared detection (BIRD) experimental small satellite mission (2001–2004). *Remote Sensing of Environment, 100*, 29–51.

Chapter 8
Future Development and Perspectives

The overviews in Chaps. 4, 5, and 6 have clearly presented the current status of the retrieval methods of land surface emissivity, temperature, and evapotranspiration from thermal infrared remote sensing data (e.g., methodological development and limitations). Several exercises were also briefed on validating these remote sensing retrievals in these chapters. This chapter provides a perspective of what should be focused on in the future in the remote sensing retrievals and validations of land surface emissivity, temperature, and evapotranspiration.

8.1 Retrieval of Land Surface Emissivity

As an intrinsic property of natural materials, land surface emissivity (LSE) is an important surface parameter in the identification of various surface types and in the determination of land surface temperature by radiometry. Although various methods have already been developed, there is still no best method to retrieve LSE from space. All of the methods either rely on statistics relationships or on assumptions and constraints to solve the inherent ill-posed retrieval problem. Therefore, they might not hold true under some circumstances, and it is necessary to choose the optimum approach to estimate LSEs from space for a particular case by taking the sensor characteristics, the required accuracy, the computation time, as well as the availability of atmospheric temperature and water vapor profiles into account (Li et al. 2013). From the previous discussion in Chap. 4, there are several difficulties in the estimates of LSEs from remotely sensed data. First of all, the presence of the atmosphere between the land surface and the sensors at satellite level disturbs the radiances measured by a radiometer at the TOA. The observed radiances are the combined results from the emission and reflection of the surface and from the absorption, diffusion, and emission of the atmosphere. However, the great variability of the vertical profiles of atmospheric water vapor and temperature makes the atmospheric corrections more difficult. Secondly, the couple of LSEs and

H. Tang and Z.-L. Li, *Quantitative Remote Sensing in Thermal Infrared: Theory and Applications*, Springer Remote Sensing/Photogrammetry,
DOI 10.1007/978-3-642-42027-6_8, © Springer-Verlag Berlin Heidelberg 2014

land surface temperature (LST) in the measured radiances makes it impossible to separate, on a physical basis, the contributions due to LSEs from the contributions due to LST in the observed radiance. For these two reasons, the LSE determination from space requires not only atmospheric corrections but also knowledge of the LST and vice versa. In addition, the interpretation and validation of the LSEs at the pixel scale are still challenging. Considering the spatial resolution of the current onboard systems varying from 10^{-2} to 10 km^2, how to correctly define and interpret LSEs independently of the scale used and how to accurately validate this definition over both homogeneous and heterogeneous surface are critical not only for developing the retrieval methods of LSEs but also for understanding the physical process. These difficulties lead to more studies to be focused on the following subjects in the near future for the accurate estimation of LSEs from space.

8.1.1 Enrichment of the Spectral Emissivity Database

Along with the development of new-generation sensors of high spatial resolution, the possibility of having pure pixels is significantly increased. Because the surface classification can be more accurately performed with the high-spatial-resolution multispectral data, a potential and practical candidate to estimate LSE from space is the CEBM if the spectral emissivity database contains huge amounts of the natural and man-made materials encountered in the world. Although there have been already several spectral libraries, such as the ASTER spectral library and the MODIS UCSB emissivity library, most of the measurements are performed on several specific samples, which implies the representativeness of those spectral libraries is not enough to characterize the land-type variation of LSEs, especially for the urban. Thus, the current available emissivity database is far from meeting the requirements of the CEBM operational use. It is therefore urgent to enrich the spectral emissivity database by both laboratory and field measurements and to document this database well. In addition, an elaborate classification system should be established to make a good balance between the number of classes and the emissivity accuracies.

8.1.2 Modeling of Surface Spectral Emissivity in the Atmospheric Window (3–14 μm)

As stated in Chaps. 4 and 5, there are always $N+1$ unknowns for N spectral measurements even if the atmospheric effects are well corrected for. To overcome this nondeterministic ill-posed problem and to make the solution deterministic, one needs to either decrease the number of unknowns or increase the number of equations. The development of surface spectral emissivity model intends to decrease the number of unknowns and to help estimate surface composition and surface soil moisture from spectral emissivity. This type of model needs several component properties or parameters to simulate the emissivity spectrum in the atmospheric window (3–14 μm). To date, the simulating model in microwave region has been developed and validated, such as the Advanced Integral Equation Method (AIEM) (Chen et al. 2003). The AIEM can simulate the band emissivities of bare soil at different frequencies and polarizations when the volumetric soil moisture and surface roughness parameters are given. Unfortunately, similar models for the TIR region have not been developed yet. The development of a model to simulate the surface spectral emissivity in the atmospheric window (3–14 μm) using the main factors affecting emissivity, such as surface composition, surface soil moisture, structure and roughness, plant species, aerial density, and plant growth state, is therefore urgent.

8.1.3 Development of the Intrinsic Relationship of Emissivities Among Several Channels

As mentioned in Sect. 4.3.2, some studies have already attempted to find extra emissivity constraints to make the LST and LSE retrieval deterministic from the atmospherically corrected TIR data. The temporal and spectral information are often used in LSE retrieval. Compared with temporal information, spectral information will be of great advantage, where accurate geometry registration and similar viewing angles are not required. Furthermore, the cloud-free condition would be more likely to be available in one moment, which shows the superiority of use of spectral information. Consequently, how to make use of the intrinsic relationship of LSEs has attracted more attention. A statistical relationship between multichannel LSEs was found and used as a new equation to separate LSEs and LST from the multispectral TIR data (Liang 2001, 2004). The smoothness characteristic of LSEs was used to separate LSEs and LST from the hyperspectral TIR data (Borel 1997). Recently, the linear change of emissivities with respect to wavelength in a small interval was used to simultaneously retrieve LSEs and LST from the hyperspectral TIR data (Wang et al. 2011). Other reasonable constraints or relationships from the emissivity-selves should be proposed to well separate LSEs and LST in the hyperspectral remote sensing.

8.1.4 Methodology Development for Atmospheric Corrections in Hyperspectral TIR Data

Atmospheric corrections play an important role in the retrieval of LSEs and LST. Accurate atmospheric corrections should be performed before applying most of the

LSEs and LST separation methods. For LSEs retrieved from TIR multispectral data, atmospheric effects are often corrected for by using either radiosounding data or atmospheric profiles retrieved from other sensors or platforms. Along with the development of the new hyperspectral TIR sensor onboard, much more detailed information on the atmosphere and land surface can be acquired. In other words, the narrow bandwidth of the hyperspectral resolution makes atmospheric absorption features prominent in the observed radiance spectrum, which offers an unprecedented opportunity to correct for atmospheric effects using the hyperspectral TIR data itself. Recently, there are some progresses in this issue. For example, an in-scene method (Young et al. 2002; Borel 2008) and an autonomous atmospheric compensation method (Gu et al. 2000) have been proposed to correct for the atmospheric effects. However, both of them assume that the atmospheric state does not change but the LST does over the whole study region. In the near future, research in this topic should therefore be focused on how to perform the atmospheric corrections for the purpose of LST and LSE separation using only the hyperspectral TIR data, without resorting to auxiliary data and assumptions.

8.1.5 Simultaneous Retrieval of LSEs, LST, and Atmospheric Profiles from Hyperspectral TIR Data

The measured radiances at TOA by sensors onboard satellites are dependent on LSEs, LST, and atmospheric states. The coupling of surface-emitted radiance and the atmospheric absorption, diffusion, and emission makes it difficult to retrieve surface parameters (LST and LSEs) and atmospheric profiles separately. The determination of surface parameters from space requires knowledge of the atmospheric profiles and vice versa. It is therefore natural, preferable, and challenging to develop a method to simultaneously retrieve LST, LSEs, and atmospheric profiles without any a priori information about the surface and atmosphere. The developed methods can be regarded as the most promising ways to simultaneously retrieve these parameters in the near future. However, these methods may demand adequate observation channels with narrow bandwidth that can supply enough vertical resolution to extract the atmospheric information and to separate the LST and LSEs. Ma et al. (2000, 2002) made a first attempt to retrieve LSE, LST, and atmospheric profiles without any auxiliary information from multispectral data. However, the broad bandwidth and the rough assumptions make it difficult to improve the retrieval accuracies. The appearance of hyperspectral TIR data with narrow bandwidth to supply enough vertical resolution offers a more attractive opportunity, and efforts in the near future should be therefore made to develop methodologies for simultaneously retrieving LSEs, LST, and atmospheric profiles from hyperspectral TIR data.

8.1.6 Estimation of the Broadband-Hemispherical Emissivity from the Retrieved Narrowband-Directional Emissivities

Knowledge of the surface broadband-hemispherical emissivity is valuable because it is an essential parameter for estimating the surface upward longwave radiation, which is an important component of the surface radiation budget, and also an important parameter for numerical weather predictions and hydrological models (Blondin 1991; Jin and Liang 2006). However, LSEs retrieved from remotely sensed TIR data represent only the narrowband-directional emissivities. They characterize the emission feature of surface in a specific observation direction in a few narrow bands in the atmospheric window (3–14 μm) rather than the broadband-hemispherical emissivity required for the calculation of surface upward longwave radiation (0–∞ μm). The difficulties in estimating the broadband-hemispherical emissivity from remote sensing measurements are that the measurements can only be conducted (1) in a few narrow spectral bands in the atmospheric windows but not over the whole spectrum (0–∞ μm), raising the issue of the spectral integration over the whole electromagnetic spectrum, and (2) in a few number of observation angles, raising the issue of the angular integration over the whole upward hemisphere. The problem to be solved is therefore the determination of the best spectral channels (narrow bands) and observation angles to achieve the most accurate estimation of broadband-hemispherical emissivity. At present, some efforts have only focused on the estimation of broadband-directional emissivity for some specified spectral domains because of the limitation of the measured spectral emissivity in the 3–14-μm region (Ogawa et al. 2003; Wang et al. 2005; Jin and Liang 2006; Tang et al. 2011). To our knowledge, there has been no literature involved in the conversion of the directional emissivity to the hemispherical emissivity.

As demonstrated by Tang et al. (2011), the spectral domains in which we have spectral measurements of the emissivity (3–14 μm) account only for 51.5 % of the emitted radiant flux for a blackbody at 300 K, while the unknown spectrum accounts for the remaining 48.5 % of this flux. The contribution of the spectral interval 3–100 μm to the emitted flux is more than 99 %. Therefore, the broadband-hemispherical emissivity with integrated spectral range reaching 100 μm should be accounted for to minimize the impact of the emissivity signature in the radiant flux determination. Over the last 30 years, a large number of measurements of the spectral directional emissivity of natural media in the atmospheric window (3–14 μm) have been performed in the laboratory and in the field, but no synchronous broadband-hemispherical emissivity is available. Thus, the empirical relationship developed for estimating the broadband-directional emissivity from a few narrow bands in the atmospheric window is limited to 3–14 μm. It is therefore urgent in the near future to (1) develop an instrument that can be used to simultaneously measure the spectral emissivity in the atmospheric window and the broadband-hemispherical emissivity in the whole spectrum (0–∞ μm), (2) solve the problem presented above, and (3) develop some methodologies to estimate the broadband-hemispherical emissivity from the retrieved narrowband-directional emissivities.

8.1.7 Combined Use of Laser CO_2 as an Active Source and TIR Sensor Measurements to Estimate LSEs

Inspired by the principle of the TISI-based method reviewed in Sect. 4.3.1 and ground measurement of LSEs described in Sect. 4.4.2, a possible solution to estimate directional and hemispherical LSE can be obtained based on the simultaneous use of passive and active radiometry. This solution can be realized by viewing the surface with and without the illumination of a CO_2 laser in TIR domain over a very short time interval (Zhang 1988; Nerry et al. 1991). The backscattered coefficient is then extracted by comparing data acquired with and without the CO_2 laser. The surface directional emissivity is therefore obtained if a relationship between backscattered coefficients and directional emissivity is developed. This relationship is a large challenge but very promising way to map LSE. Many efforts should be made in the near future on this topic.

8.1.8 Validation of the LSEs at the Satellite Pixel Scale

Validation, namely, a process of independently assessing the uncertainty of the data products derived from the system outputs, is the most important and urgent issue to be dealt with. Without validation, any methods, models, algorithms, and parameters derived from remotely sensed data cannot be confidently used. The lack of the validation of LSEs at satellite pixel scale limits the improvement and development of the LSE retrieval methods from space. Generally, the direct validation method is only applicable for homogeneous surface, where the "point" scale measurements are the representatives of the whole pixels. For heterogeneous surface, new validation methods should be developed. The indirect method may be a good choice. The nonvalidated LSEs can be indirectly validated with the different satellite-derived LSEs with known accuracies. Up to now, the high-resolution (100 m) North American ASTER Land Surface Emissivity Database (NAALSED) version 2.0 has been released and validated (Hulley and Hook 2009; Hulley et al. 2009). This emissivity database may be used indirectly to validate other LSE products in both homogeneous and heterogeneous surfaces.

8.2 Retrieval of Land Surface Temperature

It is well known that LST is one of the key parameters in the physics of land surface processes at regional and global scales. Consequently, accurately acquiring LSTs at the global scale is crucial to many fields of study including the earth's surface water and energy balances, material and energy exchange in terrestrial ecosystems, and global climate change. As described in Chap. 5, up to now, various methods have been developed to retrieve the LST from multispectral or multiangular TIR data. Due to the fact that the spectral information provided in multispectral data is so limited, all of these methods rely on different approximations to the RTE and on different assumptions and constraints to solve the inherently ill-posed retrieval problem. However, those approximations, assumptions, and constraints might not be true under certain circumstances. Therefore, the optimal approach must be chosen to estimate the LST from space by taking into account the sensor characteristics, the required accuracy, the computational time, the availability of atmospheric temperature and water vapor profiles, and the LSEs. Considering the significant progress made in recent decades in LST estimation from multispectral TIR data, there will be no significant further progress in LST retrieval from multispectral satellite data if there are no innovations in the acquisition of remotely sensed data. To overcome the shortage of multispectral data and to radically improve the accuracy of LST retrieval from space, it is necessary to explore new ideas and break new paths in remote sensing.

The following subjects ought to be focused on to improve LST estimation from space-based measurements in the near future studies.

8.2.1 Refinement of LST Retrieval Algorithms with the Consideration of Aerosol and Cirrus Effects

As discussed in Chap. 5, atmospheric correction is one of the most important issues and usually the first correction step in the LST retrieval algorithms. The errors with respect to atmospheric correction can directly decrease the accuracy of the final derived LST. Because the transmittance of aerosol in the TIR channel (approximately 0.95–0.98 in MODIS TIR channels) (Wan 1999) under normal clear-sky conditions is high and the real-time aerosol estimates (aerosol loading, size distributions, types, and scattering phase functions) are usually lacking, an average aerosol distribution and a constant aerosol loading have been used in the development of all of the LST retrieval algorithms. It is well known that the effect of aerosol on LST retrieval is relatively small compared with the effect of water vapor, but it should not be ignored when highly accurate LSTs are required in some special applications, especially in the presence of heavy aerosol loadings (Jiménez-Muñoz and Sobrino 2006). To further improve the retrieval accuracy of LST, the existing LST retrieval algorithms must be refined, or new algorithms must be developed to correct for the aerosol effect, particularly in the case of heavy aerosol loading.

In addition, the effect of cirrus clouds on the retrieval of LST should also be taken into account. In many LST retrieval algorithms, cirrus clouds are always considered to be cloud contamination and the pixels covered by cirrus clouds are screened out in data preprocessing. Since wavelengths in TIR can penetrate cirrus layers, it is possible to obtain the LST under cirrus cover from TIR data. To this end, new LST retrieval algorithms should be developed to compensate for the effect of the cirrus clouds.

8.2.2 Methodological Development for Simultaneously Retrieving LST and LSE from the New Generation of Geostationary Satellite Data with Multispectral and Multitemporal Data

The new geostationary satellites are prevailing over the polar-orbit satellites in investigating the temporal evolution of land surface and atmospheric information because they provide high-frequency observations at fixed viewing angle for the same location despite their coarser spatial resolutions. Progress can be expected in the development of new methods for extracting the LST from a combination of multispectral and multitemporal TIR data acquired from the multispectral sensors onboard the new generation of geostationary satellites, such as SEVIRI, GOES, and the FY-2 series. Except for the TTM, day/night TISI-based, and physics-based D/N methods in which data measured at two different times (one in daytime and the other in nighttime) are used, all of the methods developed to retrieve the LST from space are based on multispectral data but do not consider temporal information. It is therefore very attractive to develop a new method to simultaneously retrieve the LST and LSE by taking advantage of the multispectral and multitemporal information provided by the geostationary satellites. With the geostationary satellite data, time- and angle-consistent LSTs can be directly produced using these new LST retrieval methods without needing to temporally or angularly normalize the LST.

8.2.3 Methodological Development to Simultaneously Derive LST, LSE, and Atmospheric Profiles (Atmospheric Quantities) from Hyperspectral TIR Data

As stated by Li et al. (2013), the coupling of surface-emitted radiance and the atmospheric absorption, diffusion, and emission makes it difficult to retrieve surface parameters (LST and LSEs) and atmospheric profiles separately. The determination of surface parameters from space requires knowledge of the atmospheric profiles and vice versa. It is therefore natural, preferable, and challenging to develop a method to simultaneously retrieve LST, LSEs, and atmospheric profiles without any knowledge in priori about the surface and atmosphere. Ma et al. (2000, 2002) made a first attempt to retrieve those parameters without any auxiliary information from multispectral TIR measurements. With the appearance of hyperspectral TIR sensors, the narrow bandwidth channels can supply enough vertical resolution to extract the atmospheric profiles (Chahine et al. 2001) or the atmospheric quantities used in atmospheric corrections. The hyperspectral TIR data measured within the atmospheric window can also provide more detailed land surface information, particularly the LSE spectrum rather than the discrete LSEs in multispectral data, as well as more reasonable assumptions or constraints used to radically separate the LST and the LSEs. Although a few studies have been conducted in recent years (Li et al. 2007; Wang et al. 2013), there are still at least two aspects that require increasing attention in the future. First, rapid and accurate RTE models must be developed to meet the requirements of accuracy and speed in the retrieval process. Second, ANNs and physical retrieval methods should also be modified or developed to improve the retrieval accuracies. For example, more details should be considered in the ANNs, including the architectures and learning schemes, the selection of representative training data, and the channels employed. At the same time, additional constraints, such as the linear emissivity constraint proposed by Wang et al. (2011), and mathematical approaches and regularizations should be introduced into physically based retrieval methods to reduce the uncertainties related to the assumptions and stabilize the solutions. Combining ANNs and physics-based methods also represents an option in the near future, because the advantages of these two techniques can complement each other: ANNs can provide initial guesses for the LST, LSEs, and atmospheric profiles (or atmospheric quantities), and then physical retrieval methods can further improve these initial guesses.

Consequently, the exploration of hyperspectral TIR data for LST/LSE separation and the retrieval of atmospheric profiles or atmospheric quantities involved in atmospheric corrections will become one of the hot spots in quantitative remote sensing.

8.2.4 Retrieval of Component Temperatures in Heterogeneous Pixels

For a heterogeneous and nonisothermal pixel, its observed radiance is the ensemble radiance of several components (e.g., soil and vegetation). The pixel-average temperature cannot reflect the real temperature of each component, and the pixel-average emissivity may not be equal to simple averaging of each component's emissivity. On the contrary, if each component is assumed to be isothermal, the component temperature encapsulates more physical meaning than the pixel-average value and provides better parameterizations of the heat fluxes at the land–atmosphere interface. Therefore, component temperatures of a mixed pixel are more important than its corresponding average value. However, the retrieval of component temperatures is not easy because more variables, including component emissivities and atmospheric effects, should be known in advance. Several authors have attempted to retrieve component temperatures from multiple angular data (Li et al. 2001; Menenti et al. 2001; Jia et al. 2003; Timmermans et al. 2009). Although the multiple angular methods are regarded as the most promising way to retrieve component temperatures because the observed radiances of a given pixel under different angles provide more information of the component fractions and temperatures, the corresponding component emissivities must be provided in prior and their angular variations are always ignored. Besides, the retrieval accuracy of this method is also reduced by the spatial misregistration of multiple angular observations. Therefore, the current multiple angular methods are far from satisfying and should be improved in the future. In addition, further investigations should focus on mining the auxiliary information provided by spatial, temporal, and spectral data. Because different VZAs may correspond to different pixel sizes, new algorithms are expected to use hyperspectral TIR data at a given VZA, as the information regarding the component temperatures within a mixed pixel is included in the hyperspectral TIR data.

8.2.5 Methodology for Retrieving LST from Passive Microwave Data and for Combining LSTs Retrieved from TIR and Passive Microwave Data

The TIR data provides the LST with a fine spatial resolution (e.g., several kilometers), but it loses efficiency when the land surface is fully or partly covered by clouds. In contrast, microwaves have the unique advantage of penetrating clouds, allowing for LST retrieval in all weather conditions but with a coarser spatial resolution (up to tens of kilometers) (Aires et al. 2004). TIR and microwave data can thus complement each other, and the combination of the two is a promising line of research for producing long-term LST products in all weather conditions with a spatial resolution as fine as that of TIR data. Future studies are advised to focus on the following subjects.

1. Development of a new physics-based model for retrieving LST values from passive microwave data. Several techniques to retrieve the LST from passive microwave data have been proposed, including (semi) empirical statistical methods, neural networks, and physical models (McFarland et al. 1990; Weng and Grody 1998; Njoku and Li 1999; Aires et al. 2001; Mao et al. 2007; Chen et al. 2011). However, the physical mechanisms underlying those approaches are generally unclear, and their assumptions or simplifications regarding the LSE and atmospheric effects degrade both the feasibility and the accuracy of the derived LST. New physics-based model for LST retrieval from passive microwave data should be developed by focusing on both simplifying the parameterization of the RTM and developing the emissivity relationships between different frequencies and polarizations. A satisfactory model is expected to retrieve the LST from a combination of brightness temperatures measured at different frequencies and polarization modes.
2. Development of a model to derive the skin LST from passive microwave data. As well known, the LST retrieved from microwave data is different from that derived from TIR data. The former reflects an average value of the soil temperature from the land surface to a particular depth (depending on the frequency used to retrieve LST) underneath the surface, whereas the latter is the skin temperature with several microns of depth. To combine these two types of LST and extract the

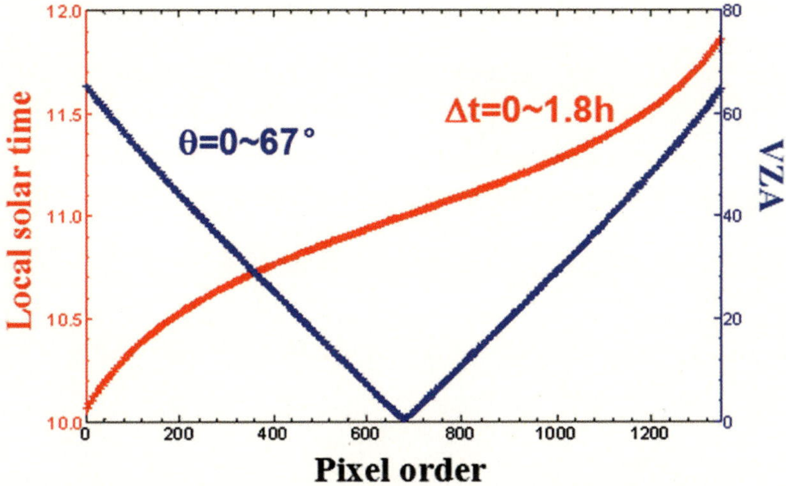

Fig. 8.1 Variations of local solar time and VZA for pixels in one MODIS scan line

skin LST, a model must be developed to extract the skin LST from the LST derived from passive microwave data with the aid of LSTs derived from passive microwave data at different frequencies and of the thermal conductivity equation applied to soil.

3. Development of a microwave–TIR fusion model. An effective model that fuses the LSTs retrieved from TIR and passive microwave data must be developed in the future to produce high-resolution spatial LST data in all weather conditions. The key problem to be resolved is how to recover the LST at the spatial resolution of TIR data when a microwave pixel is fully or partly cloudy.

8.2.6 Methodology for Angular and Temporal Normalization of LST

Due to the intrinsic scanning characteristics of the sensors onboard the polar-orbit satellites, the differences in local solar time and the VZAs for pixels along a given scan line on the same day (see Fig. 8.1) or for the same pixel on different days in one revisit period (see Fig. 8.2) may reach up to 1.5 h and range from nadir to about 67°, respectively. As the LST varies with both time and VZA, there is no comparability among LSTs of one pixel retrieved on different days or LSTs of different pixels on the same day, which significantly limits the applications of the LST products. To address these issues and make the LSTs of different pixels on the same day or of the same pixel on different days comparable, a series of LST models, including angular normalization and temporal normalization, must be developed to produce a long-term, time- and angle-normalized consistent LST product.

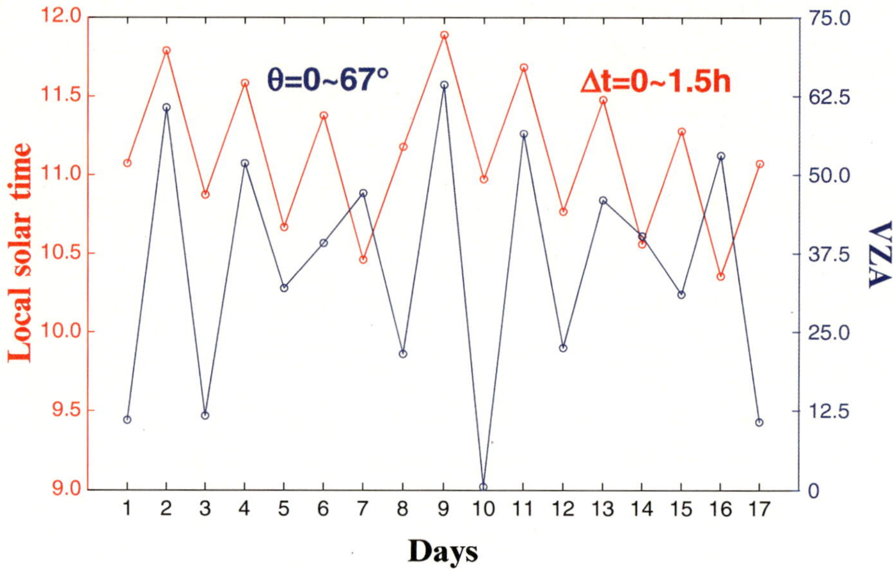

Fig. 8.2 Variations of local solar time and VZA for the same pixel in one MODIS revisit period

8.2.6.1 Temporal Normalization of LST

As illustrated in Fig. 8.1, the difference in the local solar times between the east and the west pixels along the scanning line can be up to 1.5 h, which means that the east pixels are exposed to solar irradiation approximately 1.5 h before the west ones if the sky is clear. One hour of differences in local solar time during the period 10:00–12:00 corresponds to LST differences of approximately 3–5 K. As a result, the LST products derived from the same satellite cannot be compared if the differences in the local solar times of the pixels are significant. This phenomenon also affects LST products acquired by different satellites at different times and significantly limits the applicability of the instantaneous LST products. It is therefore necessary to temporally normalize the satellite-derived LSTs to the same local solar time.

The diurnal temperature cycle (DTC) model shows promising ability to normalize the LST to any time of a cloud-free day. However, only DTC models with six parameters have been developed to describe the diurnal variation of the LST on cloud-free days (Göttsche and Olesen 2001; Schädlich et al. 2001; Jiang et al. 2006). Because polar-orbit satellites generally pass a given location only once or twice per day (four times total for MODIS Terra and Aqua), either a new DTC model with a minimum number of unknown parameters (<4) or a combination of polar-orbit and geostationary satellites must be developed in the future to temporally normalize the polar-orbit satellite-derived LSTs. Considering that geostationary satellites observe the same location with high temporal frequency, LSTs derived from geostationary satellite data can be used to determine a typical DTC model. Assuming that the LSTs derived from polar-orbit satellite data exhibit the same diurnal pattern as those derived from geostationary data, the once- or twice-daily LSTs derived from polar-orbit satellite data can be interpolated to any time of a cloud-free day utilizing the DTC model developed using geostationary data. However, on a partly cloudy day (no cloud contamination when the polar-orbit satellite overpasses), these DTC models will not be applicable, and a local LST variation model should therefore be developed in the future with the aid of temporal information provided by geostationary satellites to normalize the instantaneous polar-orbit satellite-derived LSTs. This type of local LST variation model only requires that there be no cloud contamination when the polar-orbit satellite overpasses, rather than nearly clear-sky conditions throughout the day.

8.2.6.2 Angular Normalization of LST

The LST varies with VZA and the difference in the LSTs measured in nadir and off-nadir observations can be as large as 5 K for bare soils and even 10 K for urban areas (Lagouarde et al. 1995, 2004; Chehbouni et al. 2001; Li et al. 2004; Lagouarde and Irvine 2008). Because most polar-orbit satellites scan the land surface in the cross-track direction with different VZAs varying from nadir to about 67° as illustrated in Figs. 8.1 and 8.2, angle-dependent variations in the retrieved LST are inevitable, making the LSTs of different pixels in the same orbit incomparable and causing erroneous results in application. This effect must also be considered for LSTs obtained from different sensors or at different times. Therefore, it is very crucial to normalize the satellite-derived instantaneous LSTs at various VZAs to a reference VZA (e.g., nadir).

Li et al. (1999) proposed to perform angular normalization on satellite-derived LSTs by simply attributing the angular variation of the measured effective temperature derived from area-weighted emitted radiances to the directional behavior of the pixel emissivity. The directional emissivity, however, defined in this manner is usually not measurable from space, and the assumption that there is no downward environmental thermal radiance may cause some unexpected errors in the normalized result. Another technique for normalizing the satellite-derived LST relies on a simplified directional thermal RTM that considers the component temperatures and fractions within the pixel. New methods can begin by parameterizing the directional thermal RTM with the minimum number of unknowns based on the directional fraction of vegetation cover and the component temperatures of pixels. Then, the method should establish relationships between the directional radiative temperatures observed from different directions. The off-nadir LST can be normalized to a reference direction (e.g., at nadir) by determining the fractions of various components and the corresponding component temperatures or their ratios from multiangle or multichannel observations. The fraction of components under a specific viewing angle can be calculated using the bidirectional reflectance distribution function (BRDF) model in the visible and near infrared spectral regions. Although angular variations in the LST have been demonstrated or simulated at the pixel scale in the literature (Pinheiro et al. 2004, 2006; Rasmussen et al. 2010, 2011), there is no any practical way to perform angular normalization of satellite-derived LSTs due to the complexity of this normalization. This issue therefore requires further investigation in the future. To validate the normalization model, angular measurements of the thermal radiation at ground level must also be conducted.

8.2.7 Physical Meaning of Satellite-Derived LST and Its Applications

The LST must be physically defined with absolute certainty. However, no agreement has been reached on the definition of the LST because of the unclear physical meaning of the satellite-derived temperature, especially over heterogeneous and nonisothermal surfaces. The definition of the LST also depends on that of the LSE because the LST and LSE are coupled in the total radiance. There are currently several definitions of the LSE, such as the *r-emissivity* (Becker and Li 1995), the *e-emissivity* (Norman and Becker 1995), and the *apparent emissivity* (Li et al. 1999). These definitions are the same for homogeneous surfaces at thermal equilibrium, but because natural surfaces observed from space are usually heterogeneous, the assumptions of homogeneity and thermal equilibrium are often violated in reality, especially in measurements with low spatial resolution. Therefore, the differences between these definitions are evident in many cases. The *r-emissivity* definition is recommended for LST and LSE retrieval from space-based measurements because the *r-emissivity* is measurable from space.

Whatever definition of the LST is used, the satellite-derived LST, also known as the radiometric temperature or the skin temperature, can only capture the thermal radiation information from a very thin depth underneath the surface and therefore cannot be directly substituted for the thermodynamic or aerodynamic temperatures in estimating surface fluxes or other relevant applications. Instead, a conversion must be made between these different temperatures. However, current studies seldom consider these differences and treat the skin temperature as the thermodynamic or aerodynamic temperature without any conversion. This simplification causes unexpected uncertainties in their results. Therefore, further attention should be paid to this problem in the future by considering the physical definition of different temperatures and the accuracy requirements of relevant applications.

8.2.8 Validation of Satellite-Derived LST

Although many studies for the validation of satellite-derived LST have been reported in the literature, it should be still a continuing topic. The most important problem in LST validation might be the accuracy and representativeness of the ground-truth LST at the satellite pixel scale. Although ground-based validation is considered to be the most reliable validation technique, measurements of the ground-truth LST are limited by the difficulty of finding a homogeneous region as large as the satellite pixel size and by the difficulty and associated costs of sampling over heterogeneous landscapes. The first difficulty might be solved by improving the spatial resolution of TIR data, while the second will require the development of a new sampling scheme for ground-level LST measurements, such as a wireless-net observation system or multiple-scale observation methods with corresponding new instruments. As for cross-validation techniques, it is necessary to develop an appropriate scaling procedure to remove the effects of spatial, temporal, and angular effects on the LST. The indirect validation using predicted LST from the land data assimilation systems or climate models to evaluate retrieved LST from space must pay more attention to improving the accuracy of the output of these models and dealing with the scale mismatch issue in both space and time between the satellite measurements and the assimilated or predicted LST. Therefore, LST validation is still an ongoing subject of research.

8.3 Estimation of Regional Evapotranspiration

Radiances measured by satellite sensors at the top of the atmosphere (TOA) are in general representative of areal quantities over a grid size of 10^1–10^3 m. In this case, several questions may be posed: Can the macroscopic surface parameters (e.g., T_S, albedo) over such a large scale be retrieved from the measured radiance? Do these parameters exist? How to define them? Is the physical interaction process relevant to these parameters at the local scale applicable to that over the larger scale? Answering these questions may fundamentally require a definition of the macroscopic parameters and a study of the spatial-scaling effect. To retrieve the macroscopic parameters from the TOA radiances measured by the satellite sensors, one needs to correct for the atmospheric effects and build a connection through physical models.

From the overview given in Chap. 6, one can see that the main restricting factors in the estimates of actual evapotranspiration (ET) over regional scale from remote sensing techniques are actually the retrieval accuracies of surface variables retrieved from satellite data, the acquisition of near-surface meteorological data over regional scales, the temporal and spatial data/model scaling among different scales, and the validation of ET obtained from models at satellite pixel scale. Future studies should be conducted to solve these problems.

8.3.1 Modeling of Land Surface Processes at the Soil–Biosphere–Atmosphere Interface at Regional Scale

To answer the aforementioned questions, one needs to model the exchange processes relevant to the macroscopic parameters at the soil–biosphere–atmosphere interface. This modeling should be able to specify the physical meaning of the satellite measurements and the passage of local to regional/global scales and concerns a semi-phenomenological analysis that could lead to a new method of effectively assimilating satellite data over land surfaces.

8.3.1.1 Dialectical Approaches to Model the Spatial–Temporal Variations of Land Surface Processes at Various Scales

Two modeling methods can be developed to investigate what occurs at regional and global scales.

Integrating Method

The integrating method is to model the process for each of the elements that compose a pixel and extrapolates these models by a process of "surface integration" to deduce what occurs to large scales. It is about a type of upscaling. Because of the nonlinearity of the processes, this integration is complex with assumptions that are always difficult to control. This method is very useful to understand what occurs and can direct the research of the "integrating" variables (parameters) directly at the scale considered. It is however difficult with this method (1) to benefit from "simplifications" which must appear at large scale, due to the fact that one cannot measure all the characteristics of the elements composed the pixel; and (2) to highlight "the good" variables representative of the system at large scales.

This method can lead to models having a very great number of parameters and variables. However, it is impossible to determine these parameters and variables at large scale without arbitrary because of their extreme local variability.

"Autonomous" Method at Large Scale

The integrating method, though rich and useful, must be supplemented by a method that is founded on the principle of "scale autonomy" and can be applied to analyze and model the observations from the satellite sensors. In the autonomous method, the process at a given scale is described and understood in an autonomous way without making reference to the phenomena and processes intervening at a lower scale, even if they are the consequence. The passage from one scale to others permits to describe the parameters and variables defined in a given scale in function of the variables and processes of under systems intervening on a lower scale.

Questions arise on the existence of the large-scale autonomous description and the availability of the necessary and sufficient measurements. Because only satellite measurements are available, the question becomes whether one can define the necessary and sufficient parameters using these satellite measurements to describe the land surface state and processes at the satellite pixel scale. The answer is not obvious (Morel 1985) and even not really known. However, experiments showed that it is possible to highlight spatial indicators which are sensitive to the variations of the state of the biosphere (Rasool 1987; Roerink et al. 2000; Roerink and Menenti 2000; Moody and Johnson 2001). It is not possible to currently give an exhaustive list of these indicators. One can nevertheless quote a certain number of it: surface temperature (T_S), temporal sums of temperature, albedo, complex inertia, vegetation index (VI), microwave polarization difference indices (MPDI), precipitation indices, moisture indices, roughness indices, resistance indices, and temporal sums of some of these indices.

These indices are not independent and laws between their variations are possible to be established. This method ("autonomous" method) is found to constitute an original approach of describing the processes at regional/global scale without resorting to the details of the local scale. For example, the NDVI/T_S relation can be an indicator of the evapotranspiration resistance and surface moisture (Whitehead et al. 1986; Nemani and Running 1989; Nemani et al. 1993; Carlson et al. 1994), and the albedo/T_S relation provides a means to monitor the surface desertification (Becker and Séguin 1985; Séguin et al. 1987). Moreover, the interannual variations of the soil surface resulting from the hydrous deficit can be characterized by the correlations between visible reflectances and MPDI (Choudhury 1990, 1991).

The autonomous method is now feasible, though still very little developed. The earth observation satellites are now providing huge space measurements to the scientists, and an ensemble of important surface parameters and spatial indicators can be derived.

8.3.1.2 Reformulation of the Energy Balance at Large Scale

Efforts will be made by introducing "integrating parameters" and a parameterization of the diurnal variation of surface temperature with a minimum number of parameters into the reformulation of the energy balance at large scales (Göttsche and Olesen 2001) in order to use the temporal information provided by satellite data. To simulate complicated phenomena, one can try to introduce measurable parameters from space, such as a complex inertia or complex coefficients of transfer (Abdellaoui et al. 1986; Raffy and Becker 1986).

8.3.1.3 Phenomenological Analysis of the Spatial–Temporal Variations of the Spatial Indicators at Satellite Pixel Scale

The phenomenological analysis is able to describe the phenomenological relations between surface parameters and spatial indicators, to reveal the new parameters relevant to the land surface states and processes, to highlight the thresholds of the release of certain phenomena, to establish the laws and properties that account for the variations of the spatial indicators, and to study the laws and the stability of the processes as a function of the parameters that control these laws.

The signal processing methods can be used in the phenomenological analysis to solve the nonlinear problems. One will focus on how the variations of the spatial indicators change with the parameters controlling the equations that are to be established.

8.3.1.4 Modeling and Assimilation of the Data

It is essential to associate the measurements from satellite sensors with land surface models to build a connection between them. The existing land surface models can be modified with efforts on the assimilation of satellite data and the new parameterization of ET and evaporative fraction (EF) based on the parameters retrieved from space. It is undoubted that certain equations need to be reformulated to introduce parameters that are directly accessible from space measurements.

8.3.2 Further Improvement of the Accuracy of Land Surface Parameters Retrieved from Remotely Sensed Data

Land surface temperature directly indicates the magnitude of the energy and water that are available over the land surface. Along with the remotely sensed surface parameters of surface albedo, emissivity, NDVI, soil moisture, fractional vegetation cover, and LAI, it has a significant impact on the accurate estimation of the surface energy components and especially the retrieved ET. Though great progress has been made, the retrieval of some of the surface parameters from remote sensing data still needs to be further improved, and more attention should be paid to the physical meaning of these surface parameters.

8.3.3 Research in Depth on the Impact of the Advection

Horizontal advection is one of the uncertainty sources in the partition of surface available energy into sensible heat flux and latent heat flux and often occurs in the urban and desert areas. It is another source of energy to evaporate the water from surface, which causes the surface energy imbalance especially over small spatial scales. It is still uncertain about the scale at which advection has to be accounted for and the mechanism of the energy exchange between neighboring pixels.

8.3.4 Calibration of Land Surface Process Models with the Remote Sensing ET

Remote sensing ET models are able to provide the spatial distributions of instantaneous ET under clear-sky conditions, whereas the long-term evolution of the ET and soil water content can be simulated by the land surface process models. It seems that the limitation of the remote sensing ET models can be overcome by the land surface process models. However, because of the large grid size and the uncertainties in the model inputs, it is difficult to accurately estimate the ET at large scale from the land surface process models without introducing satellite data. Acquisition of remote sensing retrieved surface parameters over a spatial scale of several meters to several kilometers is an effective way to reduce the uncertainties of the current land surface process models. Methodologies should therefore be developed to calibrate the ET simulated by the land surface models with the remote sensing ET and to incorporate the remote sensing retrieved parameters as many as possible. Data assimilation is therefore a promising technique to integrate the models, the data, and the optimization methods to estimate the ET continuously.

8.3.5 Validation of ET and Land Surface Parameters at Satellite Pixel Scale

Validation is an indispensable step to use the methods, models, algorithms, and parameters derived from remotely sensed data in confident. It is by definition a process of assessing the system outputs with independent means. Both the sensor measurements and the derived geophysical parameters must be properly validated. The validations of the remote sensing data, the retrieved surface parameters, and the remote sensing ET estimates can help to optimize models and improve the accuracy of the outputs.

Because of the inconsistent source areas and the advection effects, validation of the remote sensing ET estimates is one of the most troublesome problems. It may be questionable to validate the ET estimates at the coarse satellite pixel scale with the "point" scale measurements from the Bowen ratio system, lysimeter, and eddy correlation system over nonuniform and heterogeneous surfaces. As pointed out in Chap. 6, the newly developed LAS is a promising tool to provide the ground truth of ET at much larger scales.

References

Abdellaoui, A., Becker, F., & Olory-Hechinger, E. (1986). Use of Meteosat for mapping thermal inertia and evapotranspiration over a limited region of Mali. *Journal of Climate and Applied Meteorology, 25*(11), 1489–1506.

Aires, F., Prigent, C., Rossow, W. B., & Rothstein, M. (2001). A new neural network approach including first guess for retrieval of atmospheric water vapor, cloud liquid water path, surface temperature, and emissivities over land from satellite microwave observations. *Journal of Geophysical Research, 106*(14), 14887–14907.

Aires, F., Prigent, C., & Rossow, W. B. (2004). Temporal interpolation of global surface skin temperature diurnal cycle over land under clear and cloudy conditions. *Journal of Geophysical Research, 109*, D04313.

Becker, F., & Li, Z.-L. (1995). Surface temperature and emissivity at various scales: Definition, measurement and related problems. *Remote Sensing Reviews, 12*, 225–253.

Becker, F., & Séguin, B. (1985). Determination of surface parameters and fluxes for climate studies from space observation. Methods, results and problems. *Advances in Space Research, 5*(6), 299–317.

Blondin, C. (1991). Parameterization of land-surface processes in numerical weather prediction. In T. J. Schmugge & J. Andre (Eds.), *Land surface evaporation: Measurement and parameterization* (pp. 31–54). New York: Springer.

Borel, C. (1997, May 6–8). Iterative retrieval of surface emissivity and temperature for a hyperspectral sensor. In *First JPL workshop on Remote Sensing of Land Surface Emissivity* (pp. 1–5), Pasadena, CA (United States). Los Alamos National Lab., NM (United States).

Borel, C. (2008). Error analysis for a temperature and emissivity retrieval algorithm for hyperspectral imaging data. *International Journal of Remote Sensing, 29*, 5029–5045.

Carlson, T. N., Gillies, R. R., & Perry, E. M. (1994). A method to make use of thermal infrared temperature and NDVI measurements to infer surface soil water content and fractional vegetation cover. *Remote Sensing Reviews, 9*(1–2), 161–173.

Chahine, M. T., Aumann, H., Goldberg, M., McMillin, L., Rosenkranz, P., Staelin, D., Strow, L., Susskind, J., & Gunson, M. (2001). AIRS-team retrieval for core products and geophysical parameters. In *Algorithm theoretical basis document*. JPL D-17006: AIRS Level 2 ATBD Version 2.2.

Chehbouni, A., Nouvellon, Y., Kerr, Y. H., Moran, M. S., Watts, C., Prevot, L., Goodrich, D. C., & Rambal, S. (2001). Directional effect on radiative surface temperature measurements over a semiarid grassland site. *Remote Sensing of Environment, 76*, 360–372.

Chen, K. S., Wu, T. D., Tsang, L., Li, Q., Shi, J. C., & Fung, A. K. (2003). Emission of rough surfaces calculated by the integral equation method with comparison to three-dimensional moment method Simulations. *IEEE Transactions on Geoscience and Remote Sensing, 41*, 90–101.

Chen, S., Chen, X., Chen, W., Su, Y., & Li, D. (2011). A simple retrieval method of land surface temperature from AMSR-E passive microwave data—A case study over Southern China during the strong snow disaster of 2008. *International Journal of Applied Earth Observation and Geoinformation, 13*, 140–151.

Choudhury, B. J. (1990). A comparative analysis of satellite-observed visible reflectance and 37 GHz polarization difference to assess land surface change over the Sahel zone, 1982–1986. *Climatic Change, 17*(2–3), 193–208.

Choudhury, B. J. (1991). Synergistic use of visible, infrared and microwave observations for monitoring land surface change. In *5ème colloque International Mesures Physiques et Signatures en Télédétection* (pp. 469–473). Paris, France: Agence spatiale européenne.

Göttsche, F. M., & Olesen, F. S. (2001). Modelling of diurnal cycles of brightness temperature extracted from METEOSAT data. *Remote Sensing of Environment, 76*(3), 337–348.

Gu, D., Gillespie, A. R., Kahle, A. B., & Palluconi, F. D. (2000). Autonomous atmospheric compensation (AAC) of high resolution hyperspectral thermal infrared remote-sensing imagery. *IEEE Transactions on Geoscience and Remote Sensing, 38*, 2557–2570.

Hulley, G. C., & Hook, S. J. (2009). Intercomparison of versions 4, 4.1 and 5 of the MODIS land surface temperature and emissivity products and validation with laboratory measurements of sand samples from the Namib desert, Namibia. *Remote Sensing of Environment, 113*, 1313–1318.

Hulley, G. C., Hook, S. J., & Baldridge, A. M. (2009). Validation of the North American ASTER Land Surface Emissivity Database (NAALSED) version 2.0 using pseudo-invariant sand dune sites. *Remote Sensing of Environment, 113*, 2224–2233.

Jia, L., Li, Z.-L., Menenti, M., Su, Z., Verhoef, W., & Wan, Z. (2003). A practical algorithm to infer soil and foliage component temperatures from bi-angular ATSR-2 data. *International Journal of Remote Sensing, 24*, 4739–4760.

Jiang, G. M., Li, Z.-L., & Nerry, F. (2006). Land surface emissivity retrieval from combined mid-infrared and thermal infrared data of MSG-SEVIRI. *Remote Sensing of Environment, 105*, 326–340.

Jiménez-Muñoz, J. C., & Sobrino, J. A. (2006). Error sources on the land surface temperature retrieved from thermal infrared single channel remote sensing data. *International Journal of Remote Sensing, 27*, 999–1014.

Jin, M., & Liang, S. (2006). Improved emissivity parameterization for land surface modeling using global remote sensing observations. *Journal of Climate, 19*, 2867–2881.

Lagouarde, J. P., & Irvine, M. (2008). Directional anisotropy in thermal infrared measurements over Toulouse city centre during the CAPITOUL measurement campaigns: First results. *Meteorology and Atmospheric Physics, 102*, 173–185.

Lagouarde, J. P., Kerr, Y. H., & Brunet, Y. (1995). An experimental study of angular effects on surface temperature for various plant canopies and bare soils. *Agricultural and Forest Meteorology, 77*, 167–190.

Lagouarde, J. P., Moreau, P., Irvine, M., Bonnefond, J. M., Voogt, J. A., & Solliec, F. (2004). Airborne experimental measurements of the angular variations in surface temperature over urban areas: Case study of Marseille (France). *Remote Sensing of Environment, 93*, 443–462.

Li, X. W., Strahler, A. H., & Friedl, M. A. (1999). A conceptual model for effective directional emissivity from nonisothermal surfaces. *IEEE Transactions on Geoscience and Remote Sensing, 37*, 2508–2517.

Li, Z.-L., Stoll, M. P., Zhang, R. H., Jia, L., & Su, Z. B. (2001). On the separate retrieval of soil and vegetation temperatures from ATSR data. *Science in China Series D: Earth Sciences, 44*, 97–111.

Li, Z.-L., Zhang, R., Sun, X., Su, H., Tang, X., Zhu, Z., & Sobrino, J. A. (2004). Experimental system for the study of the directional thermal emission of natural surfaces. *International Journal of Remote Sensing, 25*, 195–204.

Li, J., Weisz, E., & Zhou, D. K. (2007). Physical retrieval of surface emissivity spectrum from hyperspectral infrared radiances. *Geophysical Research Letters, 34*, L16812.

Li, Z.-L., Wu, H., Wang, N., Qiu, S., Sobrino, J. A., Wan, Z., Tang, B. H., & Yan, G. (2013). Land surface emissivity retrieval from satellite data. *International Journal of Remote Sensing, 34*(9–10), 3084–3127.

Liang, S. (2001). An optimization algorithm for separating land surface temperature and emissivity from multispectral thermal infrared imagery. *IEEE Transactions on Geoscience and Remote Sensing, 39*, 264–274.

Liang, S. (2004). *Quantitative remote sensing of land surfaces* (pp. 377–387). Hoboken: Wiley.

Ma, X. L., Wan, Z. M., Moeller, C. C., Menzel, W. P., Gumley, L. E., & Zhang, Y. L. (2000). Retrieval of geophysical parameters from Moderate Resolution Imaging Spectroradiometer thermal infrared data: Evaluation of a two-step physical algorithm. *Applied Optics, 39*, 3537–3550.

Ma, X. L., Wan, Z. M., Moeller, C. C., Menzel, W. P., & Gumley, L. E. (2002). Simultaneous retrieval of atmospheric profiles, land-surface temperature, and surface emissivity from Moderate-Resolution Imaging Spectroradiometer thermal infrared data: Extension of a two-step physical algorithm. *Applied Optics, 41*, 909–924.

Mao, K., Shi, J., Tang, H., Guo, Y., Qiu, Y., & Li, L. (2007). A neural-network technique for retrieving land surface temperature from AMSR-E passive microwave data. In *IEEE International Geoscience and Remote Sensing Symposium* (pp. 4422–4425). Barcelona: IEEE.

McFarland, M. J., Miller, R. L., & Neale, C. M. U. (1990). Land surface temperature derived from the SSM/I passive microwave brightness temperatures. *IEEE Transactions on Geoscience and Remote Sensing, 28*, 839–845.

Menenti, M., Jia, L., Li, Z.-L., Djepa, V., Wang, J., Stoll, M. P., Su, Z., & Rast, M. (2001). Estimation of soil and vegetation temperatures with multiangular thermal infrared observations: IMGRASS, HEIFE, and SGP 1997 experiments. *Journal of Geophysical Research, 106*, 11997–12010.

Moody, A., & Johnson, D. M. (2001). Land-surface phenologies from AVHRR using the discrete Fourier transform. *Remote Sensing of Environment, 75*(3), 305–323.

Morel, P. (1985). Perspectives spatiales dans le domaine des recherches sur le climat. *CNES-Séminaire de Prospective* (pp. 25–30). Deauville, France.

Nemani, R. R., & Running, S. W. (1989). Estimation of regional surface resistance to evapotranspiration from NDVI and thermal-IR AVHRR data. *Journal of Applied Meteorology, 28*(4), 276–284.

Nemani, R., Pierce, L., Running, S., & Goward, S. (1993). Developing satellite-derived estimates of surface moisture status. *Journal of Applied Meteorology, 32*(3), 548–557.

Nerry, F., Stoll, M. P., & Kologo, N. (1991). Scattering of a CO_2 laser beam at 10.6 μm by bare soils: experimental study of the polarized bidirectional scattering coefficient; model and comparison with directional emissivity measurements. *Applied Optics, 30*(27), 3984–3995.

Njoku, E. G., & Li, L. (1999). Retrieval of land surface parameters using passive microwave measurements at 6–18 GHz. *IEEE Transactions on Geoscience and Remote Sensing, 37*, 79–93.

Norman, J. M., & Becker, F. (1995). Terminology in thermal infrared remote sensing of natural surfaces. *Agricultural and Forest Meteorology, 77*, 153–166.

Ogawa, K., Schmugge, T., Jacob, F., & French, A. (2003). Estimation of land surface window (8–12 μm) emissivity from multispectral thermal infrared remote sensing: A case study in a part of Sahara Desert. *Geophysical Research Letters, 30*, 1067.

Pinheiro, A. C. T., Privette, J. L., Mahoney, R., & Tucker, C. J. (2004). Directional effects in a daily AVHRR land surface temperature dataset over Africa. *IEEE Transactions on Geoscience and Remote Sensing, 42*, 1941–1954.

Pinheiro, A. C. T., Privette, J. L., & Guillevic, P. (2006). Modeling the observed angular anisotropy of land surface temperature in a Savanna. *IEEE Transactions on Geoscience and Remote Sensing, 44*, 1036–1047.

Raffy, M., & Becker, F. (1986). A stable iterative procedure to obtain soil surface parameters and fluxes from satellite data. *IEEE Transactions on Geoscience and Remote Sensing, 3*, 327–333.

Rasmussen, M. O., Pinheiro, A. C., Proud, S. R., & Sandholt, I. (2010). Modeling angular dependences in land surface temperatures from the SEVIRI instrument onboard the geostationary Meteosat Second Generation satellites. *IEEE Transactions on Geoscience and Remote Sensing, 48*, 3123–3133.

Rasmussen, M. O., Göttsche, F. M., Olesen, F. S., & Sandholt, I. (2011). Directional effects on land surface temperature estimation from Meteosat Second Generation for Savanna landscapes. *IEEE Transactions on Geoscience and Remote Sensing, 49*, 4458–4468.

Rasool, S. I. (1987). Potential of remote sensing for the study of global change- COSPAR report to the International Council of Scientific Unions (ICSU). Oxford/Elmsford: Pergamon Press. *Advances in Space Research, 7*(1), 103.

Roerink, G. J., & Menenti, M. (2000). *Time series of satellite data: Development of new products.* Delft: Netherlands Remote Sensing Board (BCRS), Programme Bureau, Rijkswaterstaat Survey Department.

Roerink, G. J., Menenti, M., & Verhoef, W. (2000). Reconstructing cloudfree NDVI composites using Fourier analysis of time series. *International Journal of Remote Sensing, 21*(9), 1911–1917.

Schädlich, S., Göttsche, F. M., & Olesen, F. S. (2001). Influence of land surface parameters and atmosphere on METEOSAT brightness temperatures and generation of land surface temperature maps by temporally and spatially interpolating atmospheric correction. *Remote Sensing of Environment, 75*, 39–46.

Séguin, B., Assad, E., Freteaud, J. P., Imbernon, J., Kerr, Y., & Lagouarde, J. P. (1987). *Suivi du bilan hydrique à l'aide de la télédétection par satellite. Application au Sénégal*. Report to the EECDG8.

Tang, B. H., Wu, H., Li, C. R., & Li, Z.-L. (2011). Estimation of broadband surface emissivity from narrowband emissivities. *Optics Express, 19*, 185–192.

Timmermans, W. J., Su, Z., & Olioso, A. (2009). Footprint issues in scintillometry over heterogeneous landscapes. *Hydrology Earth System Sciences Discussion, 6*, 2099–2127.

Wan, Z. (1999). *MODIS land-surface temperature algorithm theoretical basis document*. Greenbelt, MD: NASA/GSFC.

Wang, K. C., Wan, Z. M., Wang, P. C., Sparrow, M., Liu, J. M., Zhou, X. J., & Haginoya, S. (2005). Estimation of surface long wave radiation and broadband emissivity using Moderate Resolution Imaging Spectroradiometer (MODIS) land surface temperature/emissivity products. *Journal of Geophysical Research, 110*, D11109.

Wang, N., Wu, H., Nerry, F., Li, C. R., & Li, Z.-L. (2011). Temperature and emissivity retrievals from hyperspectral thermal infrared data using linear spectral emissivity constraint. *IEEE Transactions on Geoscience and Remote Sensing, 49*, 1291–1303.

Wang, N., Li, Z.-L., Tang, B. H., Zeng, F. N., & Li, C. R. (2013). Retrieval of atmospheric and land surface parameters from satellite-based thermal infrared hyperspectral data using a neural network technique. *International Journal of Remote Sensing, 34*(9–10), 3485–3502.

Weng, F., & Grody, N. C. (1998). Physical retrieval of land surface temperature using the special sensor microwave imager. *Journal of Geophysical Research, 103*, 8839–8848.

Whitehead, V. S., Johnson, W. R., & Boatright, J. A. (1986). Vegetation assessment using a combination of visible, near-IR, and thermal-Jr AVHRR data. *IEEE Transactions on Geoscience and Remote Sensing, 1*, 107–112.

Young, S. J., Johnson, B. R., & Hackwell, J. A. (2002). An in-scene method for atmospheric compensation of thermal hyperspectral data. *Journal of Geophysical Research, 107*, 4774.

Zhang, R. H. (1988). A proposed approach to determine the infrared emissivities of terrestrial surfaces from airborne or spaceborne platforms. *International Journal of Remote Sensing, 9*(3), 591–595.

Index

H. Tang and Z.-L. Li, *Quantitative Remote Sensing in Thermal Infrared: Theory and Applications*, Springer Remote Sensing/Photogrammetry,
DOI 10.1007/978-3-642-42027-6,

Printed by Printforce, the Netherlands